SCIENCE

IN 100 KEY BREAKTHROUGHS

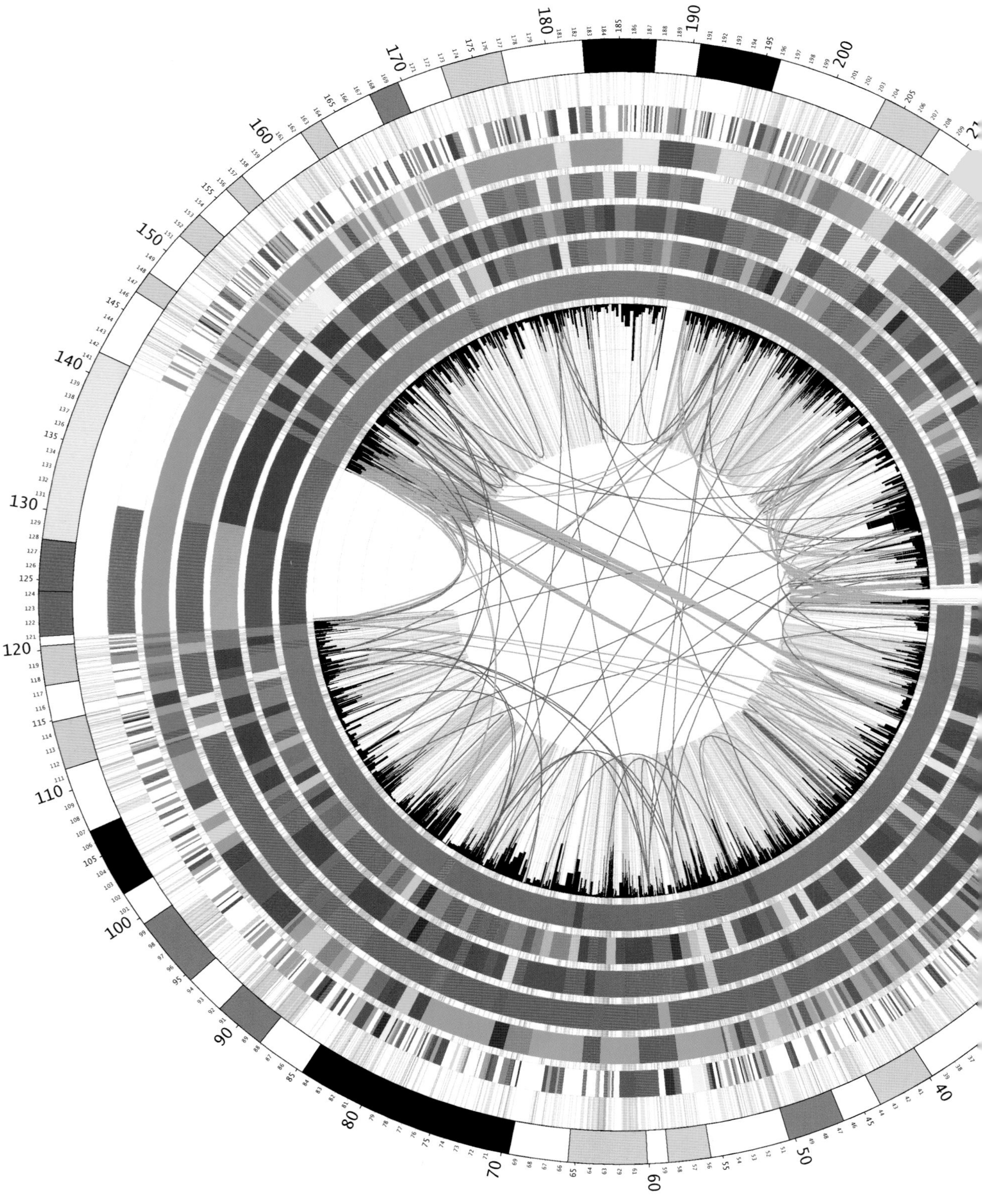

40
45
50
55
60
65
70
75
80
85
90
95
100
105
110
115
120
125
130
135
140
145
150
155
160
165
170
175
180
185
190
195
200
205

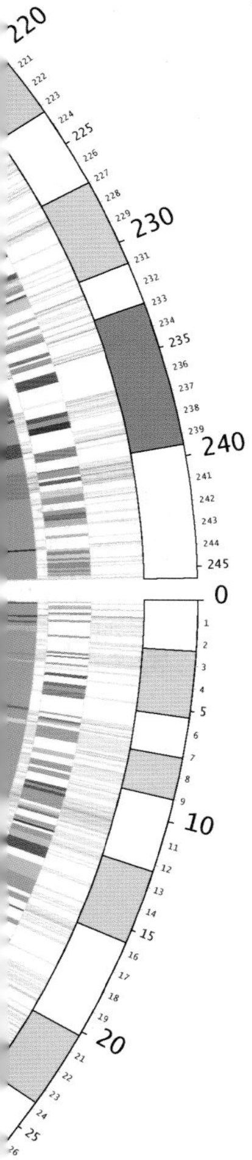

SCIENCE

IN 100 KEY BREAKTHROUGHS

PAUL PARSONS

FIREFLY BOOKS

Contents

Introduction

Every now and again scientists make a discovery that changes the world forever. No one wants to go back to the days of surgery without anaesthesia, nights without electricity or life without a computer.

Scientific discoveries open our eyes to the dangers facing our world, such as asteroids and climate change, and supply us with the means to prevent them from causing harm. They also answer incredible questions, such as the origin of space and time, how life arose on Earth and how interactions between particles govern the behaviour of matter. All of these breakthroughs hail from the genius and hard work of scientists. Some are the product of years of research. Others, such as the discovery of antibiotics or the detection of evidence for the Big Bang, happened by chance. These 'serendipitous discoveries' act as proof of the value of blue-sky research (projects that have no pre-established goal).

This book takes a look at the 100 most important scientific discoveries that have shaped our world and given us perspective on the universe at large. The discoveries are arranged chronologically, starting with the absolute basics – counting – and finishing with the creation of synthetic life in 2010.

Some of the breakthroughs you will have heard of, such as lasers, antiseptics and the discovery of ice ages. Others might be new – for example, the Navier-Stokes equations, which lie at the heart of the theory of fluid dynamics, or fullerenes, which are new kinds of carbon that have remarkable engineering properties. The 'greatest hits' are all here, too – seminal contributions to knowledge such as relativity, quantum theory, computers and evolution.

Perhaps the greatest breakthrough of all might be the scientific method itself – the idea of formulating theories and designing experiments to test them. It is from this simple premise, which was first set out by Iraqi scientist Ibn al-Haytham during the 11th century, that everything else followed.

What of the future?

Science is evolving and there will be many more discoveries to come. For example, astronomers are looking for life on extrasolar planets beyond our solar system. Knowing that the Earth is not the only place in the universe to harbour living creatures alters our perspective on life, religion and how we conduct ourselves as a society. The pace of research is such that the discovery of alien life – if it exists – may take place within the next few decades. Other researchers are building computers that use the laws of quantum theory to carry out tasks that are impossible on the 'classical' computers that sit on our desks today. A quantum computer could crack otherwise impenetrable codes, make lightning-fast searches and even change the way banks do business by altering the way information is handled. Theorists believe that these computers derive their power by harnessing copies of themselves running in parallel universes. Prototype quantum computers have been built in labs; working desktop machines are predicted to be only 20 years away.

Meanwhile, physicists are working to make what could be the most important breakthrough of all: the quest for a quantum theory of gravity. Einstein's general theory of relativity – our best theory of gravity – is incompatible with modern versions of quantum theory, the physics of tiny subatomic particles. But a quantum treatment of gravity must exist in order to describe the Big Bang in which the universe was born. Deducing the correct theory of quantum gravity will be a giant step towards wrapping up gravity with the other fundamental forces, such as – electromagnetism, and the strong and weak nuclear forces – to arrive at a 'theory of everything'.

Crystal balls

Of course, the most exciting future discoveries will be the ones that we cannot predict. For example, in 2010, in a Californian lake, bacteria were found that thrive on arsenic – a chemical that is toxic to every other known form of life. Until now, every lifeform has been based on six chemicals – hydrogen, carbon, oxygen, nitrogen, sulphur and phosphorous. But in these bacteria, arsenic has taken over all the functions normally performed by phosphorous. If life on Earth can display such variety, then there is no reason why life on other worlds should resemble it in any way at all.

Some developments are predicted even if their consequences are not. Futurologist Ray Kurzweil believes this is true of artificial intelligence (computers that mimic the human capacity for decision-making and thought). He argues that once a computer can outpace human intellect it will redesign itself until the rate at which it improves tends towards the infinite. Kurzweil calls this state the 'singularity' – after a term coined by science fiction author Vernor Vinge. What life will be like on the other side of the singularity – if it exists – is anyone's guess. Nevertheless, Kurzweil believes it will happen before the century is out.

That's the future, though. Although there are many breakthroughs yet to come, I think you'll agree that science today is already based on some incredible accomplishments. I hope you enjoy reading about them.

PAUL PARSONS

1 Counting

DEFINITION THE DEVELOPMENT OF NUMBERS, AND THE ABILITY TO ASSIGN THEM TO PHYSICAL QUANTITIES AND RECORD THEIR VALUE

DISCOVERY EARLIEST EVIDENCE IS THE LEBOMBO BONE, MARKED WITH TALLY NOTCHES AND DATING FROM AROUND 35,000 BC

KEY BREAKTHROUGH STONE-AGE HUMANS FIGURED OUT THAT THEY COULD TRANSFER BASIC FINGER COUNTING TO INANIMATE OBJECTS

IMPORTANCE ESSENTIAL FOR TIMEKEEPING, FINANCE, EARLY MATHEMATICS – AND ULTIMATELY THE WHOLE FUTURE OF SCIENCE

From adding up the number of loaves you have sold at market to computing the distribution of weight on the structure of a skyscraper – learning to count was an absolutely crucial development in the history of humanity, without which modern civilization simply could not exist. This advancement in human thinking took place at least 37,000 years ago.

Once upon a time, the full extent of human counting ability could be encapsulated in two words: 'one' and 'many'. At some point during the course of the Stone Age, that changed and early human beings gained the ability to accurately gauge and record large numbers.

Scientists know this due to a single artefact recovered from a cave in Africa in the 1970s. It is called the Lebombo bone, after the Lebombo Mountains between South Africa and Swaziland where it was found. The bone is a fibula – a lower leg bone – from a baboon. That in itself was not particularly remarkable. What sparked the archaeologists' interest were the 29 notches cut into the bone. These are almost certainly 'tally marks' – the owner of the bone was keeping a count of something, though quite what isn't clear.

Archaeologist Peter Beaumont of the McGregor Museum, Kimberley, South Africa, commented that the bone is reminiscent of the 'calendar sticks' still used by modern-day tribes in Namibia to track the passing of days. Indeed, 29 is remarkably close to the number of days in the lunar month. Whatever its specific use, many historians believe that the Lebombo bone – which was subsequently dated to around 35,000 BC – is the world's oldest known mathematical artefact.

LEFT Trading tokens, Iran, 3500–3100 BC. These pictorial 'coins' were one of the earliest forms of currency. As the amount of commerce increased, trading cultures developed standardized written symbols to represent larger numbers.

Other such 'tally sticks', as they are called, have been discovered elsewhere around the world. In 1937, an archaeological dig in Moravia, Czechoslovakia, turned up a wolf bone marked with 55 notches, arranged in groups of five. This is thought to date from around 30,000 years ago.

Beyond simple tallies

Of particular interest is one tally stick called the Ishango bone, which was recovered from a site near Ishango in the Democratic Republic of the Congo. Like the Lebombo bone, it was a baboon's fibula with sequences of notches carved into its length. But the pattern of the notches on the Ishango bone is much more complicated and goes beyond simple counting.

The Ishango bone notches are arranged into three columns and each one consists of several groups. The notches in the centre column seem to demonstrate an understanding of multiplication and division – with a group of three followed by a group of six, then a group of four followed by a group of eight (the first and third groups are both multiplied by a factor of two) and then a group of ten followed by a group of five (the second group has been divided by two).

'A wily finger-counter who opts for binary rather than tallying ones will be able to count up to an impressive 1023.'

The notches in the left and right columns are each divided into four groups. These form odd numbers, the total of which on both sides, adds up to 60. All the numbers in the left-hand column (11, 13, 17 and 19) are primes – that is, numbers that can only be divided exactly by themselves and one. Meanwhile, the numbers in the right-hand column (11, 21, 19 and 9) are all derived from the formulae 10 ± 1 and 20 ± 1. The Ishango bone has been dated at around 20,000 years old.

Base-10

Of course, simple tally counts are not the best way to record large numbers – as anyone who has tried to count higher than ten on their fingers will know. Modern systems of counting get round this by specifying numbers using a combination of high-value and low-value increments. For example, our own system uses units, tens, hundreds and so on – with one digit (ranging from 0–9) to represent the number of each. We call this system base-10, because there are ten basic digits. This system lets us write big numbers concisely, such as '126', rather than a very long tally of 1s (or even a shorter, but still cumbersome, tally of nines – 14 sets of them in the case of the number 126).

Other bases exist as well. The simplest is base-2, or binary. This uses only two digits, 0 and 1, but is nevertheless far more efficient than a simple tally. Indeed, a wily finger-counter who opts for binary rather than tallying will be able to count up to an impressive 1023. Invented in India between the fourth and second centuries BC, binary is the counting system used today by all computers – and pocket calculators (because 0 and 1 are easy to represent in the states of an electronic switch).

Sixty seconds

The forerunner of the modern pocket calculator was the abacus – a calculating device invented by the Ancient Sumerians (who lived in the region known today as Iraq) in or around 2700 BC. They, along with the Ancient Babylonians, pioneered the world's first advanced counting systems.

The Babylonians embraced a sexagesimal number system – that is, base-60. If that sounds obscure, think again – it is from where we inherited our hours-minutes-seconds system of timekeeping. Sexagesimal numbering is also used in geometry, where the interior angle of a circle is divided into 360°, each of which splits into 60 arcminutes and each of those into 60 arcseconds.

This system was useful to the Babylonians, who studied astronomy and needed a reliable way to keep track of angles on the celestial sphere. For the rest of the world, it was more workaday concerns that drove the early development of counting – timekeeping, navigation and, probably more so than anything else, trading and the beginnings of formal commerce.

ABOVE The Ancient Mesopotamian empires each had their own refined counting systems. This limestone inventory tablet excavated in Iraq dates from the fourth century BC. Each section shows separate commodities with the broad, triangular indents representing the number of items.

2 Geometry

DEFINITION GEOMETRY DETERMINES THE MATHEMATICAL RELATIONSHIPS BETWEEN THE SHAPES AND SIZES OF OBJECTS

DISCOVERY THE PRINCIPLES OF GEOMETRY WERE FIRST SET OUT BY THE GREEK PHILOSOPHER EUCLID OF ALEXANDRIA, *C.* 300 BC

KEY BREAKTHROUGH EUCLID BUILT GEOMETRY AS A FORMAL SCIENCE AND ESTABLISHED ITS FUNDAMENTAL AXIOMS

IMPORTANCE GEOMETRY HAS A MULTITUDE OF APPLICATIONS, AND IS THE BASIS FOR EINSTEIN'S THEORY OF GENERAL RELATIVITY

Geometry (translated as 'Earth measuring') is the study of the shapes of figures in two, three and sometimes higher dimensions. It makes use of the mathematics of angles and line lengths, and the delicate interplay between them. It was first studied by the surveyors of land and buildings, but is now also used by engineers, designers of computer graphics, nanotechnologists, molecular chemists and theoretical physicists – to name just a few.

Geometry as a science was born over 2000 years ago, with the work of the Greek philosopher Euclid of Alexandria. Very little is known about Euclid himself – which is surprising given how influential his work has become to mathematicians and scientists. But what is known is that he lived, roughly speaking, between 300–260 BC and, at some point during his life, wrote a book called *Elements* in which he set out the principles that would dominate geometry until the 19th century.

Elements was published in no less than 13 volumes. Many of the ideas it included were already known, but Euclid added his own proofs to them. His great achievement was in pulling together many disparate strands and from them weaving a unified and coherent picture of what geometry is and how it works, and presenting a framework for how it can be applied. Prior to the publication of *Elements*, geometry was a random collection of ideas and results. Euclid assembled them in a logical and systematic way to transform geometry into a formal scientific discipline.

The strong mathematical flavour of *Elements* would later influence some of the greatest philosophers and scientists through time, including Newton,

LEFT Curved grid representing the geometric curvature of an 'open' or 'saddle' universe. The universe's curvature is determined by the density of matter and energy. In an open universe, the amount of matter is insufficient to prevent the universe expanding forever. The curvature is said to be negative and the volume infinite.

Copernicus, Galileo and Kepler. Einstein, too, is said to have kept a copy, which he affectionately called the 'holy little geometry book'. At the heart of *Elements* were Euclid's five 'axioms' – postulates setting out the basic geometrical rules. The first four axioms are fairly self-evident (making it all the more amazing what Euclid managed to derive from them). For example, the first says that you can draw a straight line between any two points. The second states that it is possible to extend this straight line and that when you do this the line remains straight. The third axiom asserts that any point with a line connected to it defines a circle, with the point as its centre and the line as its radius. And the fourth says that where two perpendicular lines meet, the angle they make is a right angle – and that all right angles are equal.

Parallel postulate

The fifth axiom is the only one of Euclid's geometric rules that is not quite so obvious. Also known as the 'parallel postulate', it states that given a line and a point that is not on the line, then there is only one straight line that you can draw passing through the point that does not intersect the first line anywhere along its length. It means parallel lines are always parallel – they do not diverge and they never meet.

From these five principles, Euclid was able to prove powerful geometrical theorems – for example, that the internal angles of a triangle add up to 180°, that a cone has exactly one-third the volume of a cylinder of the same base area and height, that a triangle formed by the diameter of a circle and a point on the circumference of the circle must always contain a right angle (known as Thales's theorem), and that the sum of the squares of the shortest two sides of a right-angled triangle add up to the square of the longest side (Pythagoras's theorem).

Fifth element

Some scientists have argued that even Euclid himself was not entirely convinced by the parallel postulate – his fifth axiom. Indeed, in many of the proofs presented in *Elements* he seems to avoid using it. This led some geometers to question whether it was really necessary.

In 1830, Russian mathematician Nikolai Lobachevsky decided to find out what would happen if he deliberately violated the fifth axiom. As he followed the steps in his analysis, he was expecting to reach a logical contradiction that would thus prove to him that the fifth axiom was necessary. Instead, he succeeded in creating a perfectly valid new kind of geometry in which – given a line and a point – you can draw more than one line (many, in fact) through the original point and never intersect the first line. Lobachevsky's creation became known as 'hyperbolic geometry'. Euclid's geometry wasn't wrong, but it only applies in what we would now call 'flat space' – for example, on a

'Geometry was a random collection of ideas and results. Euclid assembled them in a logical and systematic way to turn geometry into a scientific discipline.'

ABOVE Geometry is not all about the drawing of straight lines and circles – some impressive works of modern architecture rely upon it. A prime example is the Grande Pyramid at the Louvre Museum, Paris, which stands 20 metres (22 yards) tall and constructed from 603 interlocking, diamond-shaped components.

flat table top. Hyperbolic geometry is what you get when the table top is deformed – bent into a saddle shape so that two lines that start out parallel gradually curve away from each other and steadily diverge.

Curved Earth

Hyperbolic geometry was not the only example of a non-Euclidean geometry. There also exist so-called 'elliptic geometries', where the opposite happens. Here, given a point and a line, there are no new lines through the point that never intersect the first line – and so parallel lines must eventually meet. The surface of the Earth, or any sphere, has elliptic geometry. For example, two lines that are parallel at the equator (but themselves perpendicular to the equator) will cross one another at the north and south poles.

In the 1850s, German mathematician Bernhard Riemann formulated the study of different geometries (both Euclidean and non-Euclidean), known as 'differential geometry'. It replaced Euclid's diagrammatic approach with one based on algebra and calculus (see page 25) and was necessary because some convoluted geometries are difficult to picture – and for some of the higher dimensional spaces (differential geometry can be generalized to multiple dimensions), visualization is utterly impossible.

Sixty years later, a German physicist named Albert Einstein was looking for a mathematical framework upon which to build a new theory he was toying with. His theory described gravity as geometry on the curved surface of space and time by using Riemann's differential geometry. He created his general theory of relativity – a cornerstone of 20th-century physics.

3 Curvature of the Earth

DEFINITION THE EARTH IS NOT A FLAT DISK, AS WAS ONCE BELIEVED, BUT A CURVED SPHERE

DISCOVERY ERATOSTHENES OF CYRENE WAS THE FIRST TO MEASURE THE CIRCUMFERENCE OF THE EARTH, IN THE YEAR 240 BC

KEY BREAKTHROUGH HE DEDUCED THAT CHANGES IN SHADOWS ACROSS THE PLANET'S SURFACE REVEAL HOW IT CURVES

IMPORTANCE MARKED THE BEGINNING OF GEOGRAPHY AS A SCIENTIFIC DISCIPLINE

Earth is our home in the universe. In the third century BC, Greek philosopher Eratosthenes figured out what our home world actually looks like, finding that it is a curved sphere and making a reliable measurement of its circumference. It looked like the end of the road for Flat Earthers.

Many years ago, human beings – even those who were well-travelled and literate – believed that the Earth was flat. Sail too far in the wrong direction and, they feared, you are likely to literally fall off the edge of the world.

The first scholars to realize that the Earth might not be flat lived in Ancient Greece. They included the sixth-century-BC philosopher Pythagoras, followed by Parmenides in the fifth century BC. Aristotle, in the fourth century BC, had his own suspicions but lacked conclusive proof. That proof would be provided by Eratosthenes, who took some of the greatest strides forward in our understanding of the spherical Earth.

Eratosthenes was born in 276 BC in Cyrene (modern-day Libya). He moved to Alexandria to study, and in 236 BC became librarian of the city's Great Library – at the time, the greatest repository of knowledge in the world. Colleagues apparently nicknamed him 'Pentathlos' because he was renowned for being mildly competent in many disciplines – maths, astronomy, athletics and poetry – but, they joked, master of none. He would prove them wrong.

LEFT Greek scholar Eratosthenes proved over 2000 years ago that the Earth is spheroidal rather than flat. His findings are reinforced today by stunning images taken from space illustrating the Earth's curvature.

Casting shadows

In 240 BC, Eratosthenes used a cunning method to prove that the surface of the Earth is curved. He discovered that on the summer solstice – the day of the year with the most hours of daylight (usually 20–21 June) – in the

Egyptian city of Syene (now known as Aswan), the Sun at midday would be directly overhead. He knew this because tall structures, such as obelisks, would cast no shadows at this exact time.

However, Eratosthenes also discovered that in his home town of Alexandria, the same was not true. Here, the Sun at midday on the solstice was not directly overhead – Egyptian obelisks and other tall objects threw distinct shadows on the ground. By measuring the size of a shadow and comparing it to the size of the object casting it, he was able to work out the angle between the Sun's position on the sky and the point directly overhead, known as the 'zenith'. He found this angle to be about 7°.

'Eratosthenes's figure for the Earth's circumference is 46,620km (28,968 miles). The true value is just over 40,000km (24,855 miles) – so it is not a bad estimate for calculations made in the third century BC.'

Calculating curves

Eratosthenes speculated this difference between the two cities was due to the Earth's curvature, and that this tilted vertical structures in Alexandria with respect to those in Syene. He realized that if that was the case – and, indeed, it was – then he could use the information to deduce the size of the entire Earth. To do this, he drew a diagram showing a cross-section through the curved Earth onto which he marked the positions of Syene and Alexandria (see diagram opposite). At Syene, the Sun is directly overhead and so an incoming ray of sunlight makes an angle of 90° with the ground. At Alexandria, however, the Sun is 7° below vertical – and so a ray of sunlight makes an angle of 83° with the ground.

Because the Sun is so distant, its rays appear roughly parallel with each other. Eratosthenes used this fact to work out, from geometrical principles, that lines drawn from Alexandria and Syene to the centre of the Earth's globe must subtend an angle equal to the displacement of the Sun from the zenith at Alexandria – which he had already established was an angle of 7°. There are a total of 360° in a full circle, therefore the distance from Alexandria to Syene represents a fraction equal to 7/360 of the Earth's total circumference. All Eratosthenes had to do was measure this distance.

From a camel journey he had taken, Eratosthenes was able to estimate the separation of the two cities at approximately 5000 stadia – a 'stadion' is a Greek unit of distance equal to the length of a stadium. Dividing by 7/360 then implied that the circumference of the Earth was about 257,000 stadia.

Geography starts here

How accurate was the original estimate made by Eratosthenes? The answer depends how long a stadion actually was, for which historical records are not absolutely clear. The commonly used 'Attic stadion' measured exactly 185 metres (202 yards). Using this value, Eratosthenes's figure for the Earth's circumference comes to 46,620km (28,968 miles). The true value is just over 40,000km (24,855 miles) – so it is not a bad estimate for calculations made in the third century BC. But if we venture that Eratosthenes was using the alternative 'Egyptian stadion', equal to 157.5 metres (172 yards), then his result becomes 39,690km (24,662 miles) – remarkably close to the real value of the Earth's circumference.

Eratosthenes did not stop there with his research. He went on to devise a refined system of coordinates for charting navigational positions on the surface of the terrestrial sphere. These worked using two angles subtended at the Earth's centre from fixed reference points on the planet's surface. We still use these coordinates today – they are called latitude and longitude. Using this system, Eratosthenes also drew up a new map of the world based on the best available knowledge. For these reasons, Eratosthenes is often regarded as the father of geography. He even came up with a rudimentary date calendar, which included the correct placing of leap years.

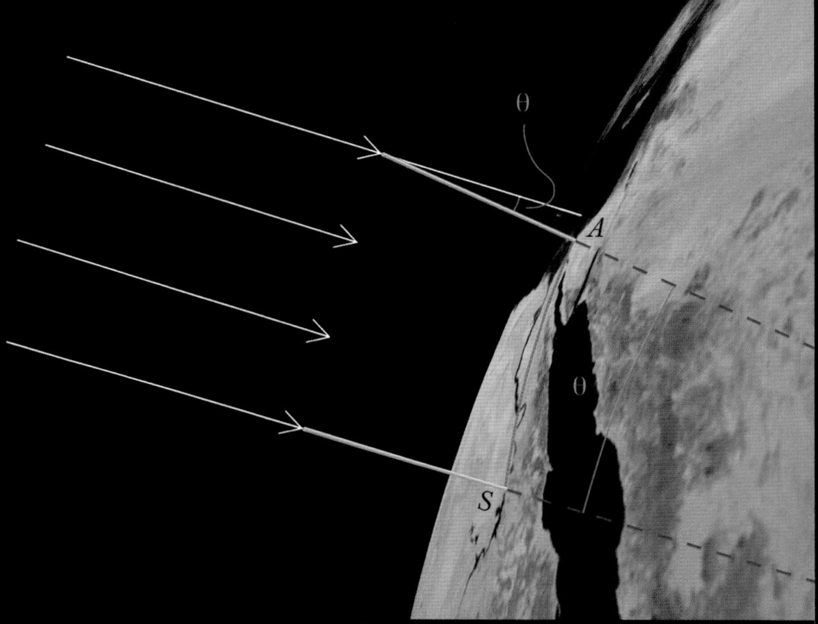

Flat rejection

Using accurate measurements taken from spacecraft, we now know that the Earth is not perfectly spherical. The planet's rotation causes its midriff to bulge out slightly in the centre, giving the Earth a slightly squashed shape. This means that someone standing at sea level at the North or South Pole is actually about 21km (13 miles) closer to the planet's centre than someone standing at sea level on the equator.

Incredibly, despite these fundamental revelations about the nature of our world, the Flat Earth Society – an organization for those who claim they believe the Earth to be disk-shaped – remains in existence today. They argue that the North Pole is, in fact, the icy centre of the disk, while Antarctica and the southern polar cap form a solid wall of ice around its outer edge that prevents us from falling off the planet.

As Albert Einstein, who proved it is not just the Earth that's curved but space too, once put it: 'Two things are infinite: the universe and human stupidity; and I'm not sure about the universe.'

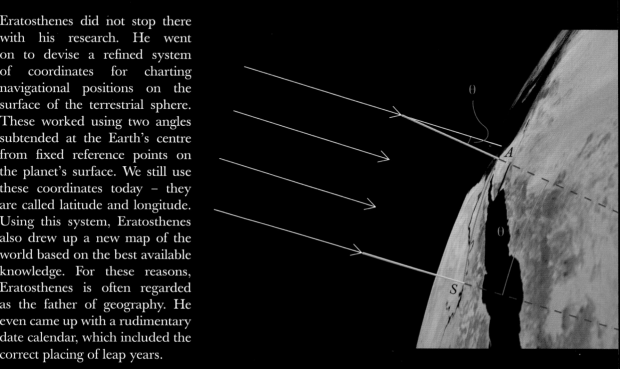

ABOVE Diagrammatic representation of Eratosthenes's experiment to measure the curvature of the Earth. Two towers (S and A) are separated by distance (D). Earth's natural curvature means that while the Sun's light is parallel to one tower and casts no shadow, the light makes an angle (θ) with the other tower and leaves a shadow on the ground. Geometry implies that the total circumference of the Earth is θ × D / 360°.

4 Astrometry

DEFINITION THE FIELD OF SCIENCE CONCERNED WITH MEASURING THE POSITIONS OF CELESTIAL OBJECTS

DISCOVERY HIPPARCHUS OF RHODES PUBLISHED THE FIRST STAR CATALOGUE IN THE SECOND CENTURY BC

KEY BREAKTHROUGH HIPPARCHUS RELIED ON A STAR-MAPPING DEVICE CALLED AN 'ARMILLARY SPHERE'

IMPORTANCE ASTROMETRY WAS CRUCIAL IN ESTABLISHING THE COSMIC DISTANCE SCALE, AND FOR TIMEKEEPING AND NAVIGATION

Astrometry concerns the positions of astronomical objects on the sky and was pioneered by the Greek mathematician Hipparchus. In the 19th century, improved instruments and observational techniques enabled astronomers to measure distances to the stars while today, astrometry is essential for timekeeping, astronomy research and even guarding against killer asteroids.

The Greek polymath Hipparchus lived on the island of Rhodes in the second century BC. He carried out the first systematic studies of the motion and size of the Sun and Moon, and was responsible for developing many of the techniques in trigonometry – relating the angles and side lengths of triangles – that are crucial to much of mathematics today.

In addition to all this, Hipparchus compiled the first-ever star catalogue, around the year 135 BC. It included the positions and brightnesses of approximately 850 stars – and much of the star catalogue data printed in Ptolemy's *Almagest* book of AD 150 is believed to have derived from Hipparchus's original observations.

LEFT Astrometry is the science of measuring the positions of stars and other celestial objects on the night sky. Pictured is a field of over 100,000 stars in the core of the Omega Centauri star cluster taken by the Hubble Space Telescope. Throughout history, measuring star positions was essential for the development of navigation and timekeeping.

Armillary sphere

This work predated the invention of the telescope by more than 1700 years. Instead, Hipparchus used an 'armillary sphere', consisting of several large metal circles that could be aligned with the circles on the sky that make up the celestial sphere. Sighting devices located on the armillary sphere then enabled the user to align it with a specific point on the sky and read off its coordinates. It was a slow and painstaking procedure, confounded by the difficulties of trying to make such delicate observations through the turbulent and hazy atmosphere of the Earth. This made Hipparchus's results accurate only to about a third of a degree – the full Moon, by comparison, spans around half a degree.

The results obtained using the armillary sphere were accurate enough to also detect an effect known as 'precession' – a kind of wobbling of the Earth's rotation axis rather like the way a spinning top wobbles, sweeping out a cone shape as it turns. The effect is tiny – it takes 26,000 years for the Earth's axis to make one complete sweep. But Hipparchus obtained the remarkably accurate figure of 28,000 years. Egyptian philosopher Ptolemy later had to adjust his geocentric view of the universe, introducing a tiny degree of rotation to the sphere of fixed stars, to account for this effect (see page 37).

Parallax

Hipparchus also developed the technique of parallax, by which some astronomical distance observations are now made. Parallax is where the apparent position of an object against a fixed background appears to change when viewed from different directions – the shift in position becomes greater the closer the object is.

You can demonstrate the parallax effect for yourself. Hold a finger up about 30cm (12in) in front of you. Now close one eye and line up your finger with a reference point on a nearby wall – say, a light switch. Move your head from side to side and your finger appears to move relative to the reference point. Now move your finger away from you so that it is at arm's reach, line it up again with the reference point and repeat – the amount by which your finger moves relative to the switch will now be much less. Gauging this degree of movement is thus a measure of how far away your finger is.

Parallax had already been noticed by another Greek astronomer, Aristarchus of Samos. Hipparchus applied the technique to calculate the distance to the Moon using observations – made at two different points on the Earth – of the amount of the Sun's surface covered by the Moon during a solar eclipse. He obtained a result of between 59 and 67 Earth radii. This was very accurate given the primitive nature of his observations – the true distance from the Earth to the Moon is now known to be 60 Earth radii.

Hipparchus's catalogue was destroyed in antiquity – possibly in the fire at the Great Library of Alexandria.

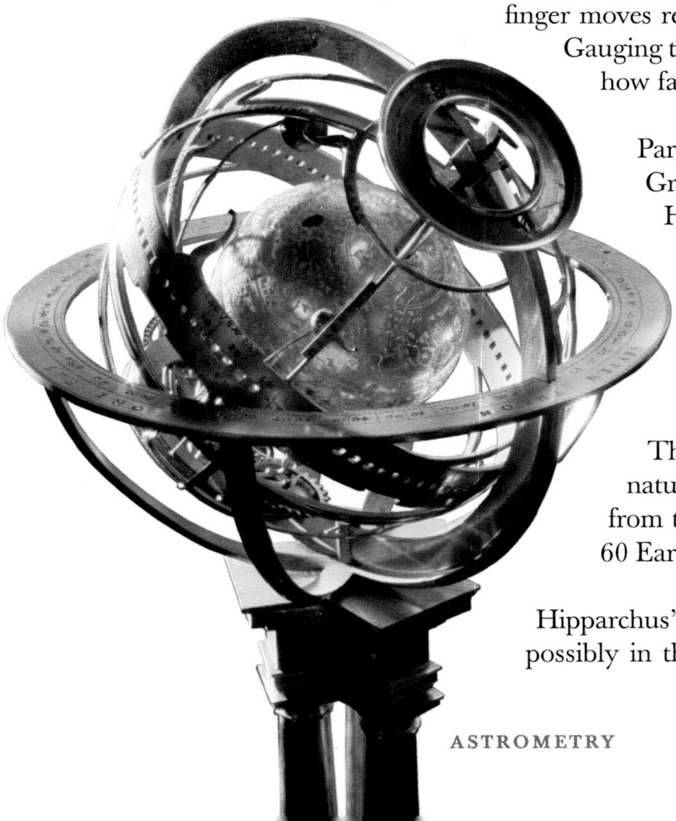

BELOW One of the first instruments for measuring the positions of objects on the night sky was the armillary sphere, invented by the Ancient Greeks.

During the Middle Ages, Persian astronomers conducted their own surveys of the heavens, and these were still being used by astronomers and mathematicians many centuries later. Indeed, a great number of stars today are still known by derivatives of their Arabic names, such as Aldebaran (Al-Dabaran – 'the follower') and Algol (Al-Ghul – 'the ghoul').

'Hipparchus compiled the first-ever star catalogue, around the year 135 BC. It included the positions and brightnesses of approximately 850 stars.'

In the late 15th century, Danish astronomer Tycho Brahe gathered together what were the most accurate astronomical observations ever made. He used his 'mural quadrant' – probably the largest astronomical device of its day, consisting of a 3-metre-high (3.3-yard) quarter-circle arc, marked with fine graduations so that celestial objects could be sighted and their positions accurately read. The data gathered with the mural quadrant were crucial in formulating Kepler's laws of planetary motion (see page 45).

Distance ladder

Up until this time, all astronomical observations were plagued by the distortion of light as it passed through the Earth's atmosphere. In 1810, German mathematician Friedrich Bessel was able to quantitatively explain this effect and so create a way to correct it. Soon after, Bessel exploited the improved accuracy this brought. By using parallax – the method Hipparchus had first employed to gauge the distance to the Moon – he made the first distance measurement to a nearby star. Bessel calculated the star 61 Cygni to be 9.8 lightyears away (its true distance is now known to be 11.36ly).

This led to the creation of a new astronomical distance unit – the 'parallax second', or parsec. A star located 1 parsec away appears to change its position on the sky through parallax by 1 second of arc (1/3600 of a degree) as the Earth moves from one side of its orbit to the other. A parsec is equal to 3.26 lightyears. Bessel's calculations showed 61 Cygni to be 3 parsecs away. Bessel's accurate measurements also enabled him to detect minuscule wobbles of the stars Sirius and Procyon, which he correctly ascribed to the gravitational influence of unseen companion stars.

Cosmic time

A similar technique is used today to detect extrasolar planets (see page 381). Other uses for astrometry include tracking near-Earth asteroids and even gauging when clock time has begun to drift away from our natural day–night cycle – requiring the addition of a leap second to international timekeeping. Astrometry was taken to a new level in 1989 when the European Space Agency (ESA) launched its Hipparcos satellite – the first space mission dedicated solely to astrometry. It recorded the positions of 120,000 stars – each accurate to less than a millionth of a degree. And in 2012, ESA launches the GAIA spacecraft – measuring more star positions to an unprecedented accuracy of billionths of a degree.

5 Algebra

DEFINITION SYSTEMATIC USE OF SYMBOLS TO REPRESENT THE RELATIONSHIPS BETWEEN NUMBERS AND MATHEMATICAL OBJECTS

DISCOVERY THE PUBLICATION IN AD 820 OF THE BOOK *AL-JABR WA'L MUQABALAH* BY MUHAMMAD IBN MUSA AL-KHWARIZMI

KEY BREAKTHROUGH THE BOOK MARKED THE BEGINNING OF ALGEBRA AS A SYSTEMATIC DISCIPLINE IN ITS OWN RIGHT

IMPORTANCE IT IS A FUNDAMENTAL STRAND OF PURE MATHS; HAS LED TO GROUP THEORY, ESSENTIAL IN THEORETICAL PHYSICS

Algebra is one of the most fundamental disciplines in the whole of mathematics. It was invented by a Persian scientist working in Iraq in the ninth century AD – while the western world languished in the Dark Ages. Today, an advanced algebraic technique called group theory has become a powerful tool in theoretical physics.

Algebra is one of the central branches of pure mathematics. Whereas arithmetic is concerned solely with numerical operations, such as addition and multiplication, algebra replaces numbers with symbols – also known as 'variables'. This allows mathematical relationships between variables to be expressed in the form of an equation. Algebra then posits rules by which these symbols can be manipulated – and this allows the equation to be rearranged, or 'solved'.

An example is the equation $5 = 3x - 1$. What is x? The laws of algebra tell us that we can alter the form of this equation so long as we perform the same steps on both sides. So we can add 1 to both sides to give $6 = 3x$, and then divide both sides by 3, to yield $x = 2$. We are then said to have solved the equation for x.

In antiquity, Greek scholars developed a system of solving equations in this way. Following the work of Euclid and others, their methods were characteristically geometric. This involved framing equations as diagrams – with the sizes of quantities indicated by the lengths of lines – from which the solution could be read off. For instance, two numbers multiplied together is given by the area of the rectangle formed by two perpendicular lines.

LEFT Algebra was the beginning of modern mathematics – underpinning esoteric disciplines, such as calculus, complex numbers and chaos theory. Chaos theory led to the idea of fractals – geometric representations of chaotic systems. Pictured is part of a fractal called the Mandelbrot set.

In the third century AD, the Greek mathematician Diophantus of Alexandria developed more sophisticated techniques – reminiscent of the algebra used today. But his approach was very utilitarian – focused on solving practical problems at hand, rather than developing the mathematical techniques needed to solve these equations as a scientific pursuit in its own right.

Persian progress

That approach to solving solutions had to wait a further five centuries, when the Persian scholar Muhammad ibn Musa al-Khwarizmi published *Al-jabr wa'l muqabalah* (The Compendious Book on Calculation by Completion and Balancing) in the year AD 820. The book presented systematic techniques for solving algebraic equations, using two main approaches. Balancing ('al-muqabalah') involved subtracting the same quantity from both sides of an equation, for example, subtracting 1 from $x + 1 = 3$ to reveal $x = 2$. Completion ('al-jabr') meant adding the same quantity to both sides of an equation, for instance, adding 5 to $x - 5 = 10$ to arrive at the solution $x = 15$.

'*Al-jabr*, the Arabic phrase for "completion", is the root of our modern word "algebra"'.

Al-jabr, the Arabic phrase for 'completion', is the root of our modern word 'algebra'. It is thought to have been first introduced in the title of the Latin translation of Al-Khwarizmi's book, *Liber algebrae et almucabola*, in the 12th century. Al-Khwarizmi's name itself has similarly survived, westernized to 'algorithm' – a word used today by mathematicians and computer scientists to describe the fundamental sequence of steps required to perform a calculation.

Al-jabr wa'l muqabalah also included solutions for quadratic equations – those with terms involving x^2 as well as x. Given $x^2 - 6x + 8 = 0$, for example, what is x? The book presents a number of methods for solving quadratic equations – in this case, deducing that x can be either 2 or 4.

As Euclid had done for geometry over 1000 years before (see page 13), Al-Khwarizmi began with a handful of basic axioms and from these developed a rigorous mathematical framework for solving algebraic problems. The book included discussion of some applications of his methods – including commerce and the calculation of inheritance. Crucially, however, these were not his primary motivations. Al-Khwarizmi studied the equations in their own right, rather than as a means to solving other problems – and in so doing established algebra as a major pillar of pure mathematics. Other Persian mathematicians, such as Omar Khayyám (AD 1050–1123), extended Al-Khwarizmi's analysis further to cubic equations (those with terms involving x^3) and other more complex mathematical relationships.

Numerical rhetoric

Al-Khwarizmi presented his equations as what is called 'rhetorical algebra' – writing out in words 'three lots of "thing" plus 1' rather than in symbolic form, '$3x + 1$'. A more symbolic treatment was developed by the Arab scientists

Ibn al-Banna and Al-Qalasadi between the 13th and 15th centuries. Meanwhile, the translation of Al-Khwarizmi's book into Latin in the 12th and 13th centuries had stimulated the study of algebra in Europe. It took hold first in Italy but spread across the continent as the Renaissance brought a new wave of scientific enquiry to follow the stagnation during the Dark Ages.

After the Renaissance, powerful new algebraic methods were developed in Europe, such as the theory of matrices – a way to represent systems of linked equations using a two-dimensional array of numbers, which could then be solved using a logical, systematic procedure.

Group theory

Algebra thus far was simply a set of rules for manipulating symbols that represent numbers. In the 19th century, French mathematician Évariste Galois recognized that there was no reason why numbers should be the only mathematical objects that algebra could be applied to. All you need to do, he reasoned, is calculate the rules governing the particular thing in which you are interested and it should be possible to construct a detailed algebra describing it.

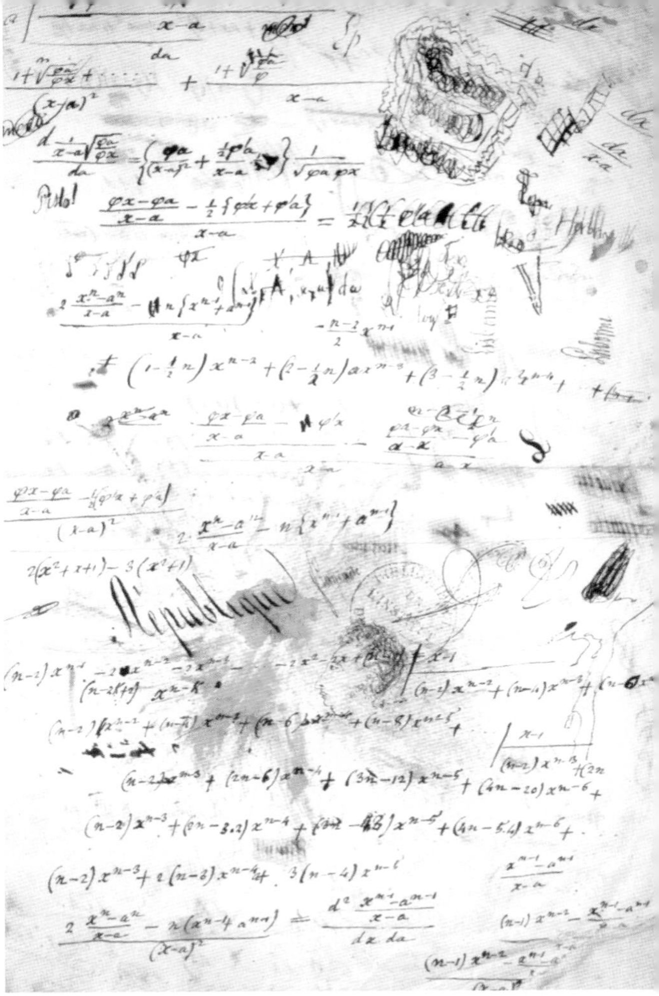

ABOVE Mathematician Évariste Galois realized that algebra need not apply exclusively to numbers, but to anything obeying mathematical rules. His work was the foundation of a branch of mathematics called group theory. Pictured is an original sheet of his handwritten manuscript.

This has led to a branch of maths known as 'group theory', where different sets of algebraic rules define 'groups' of mathematical objects. These objects can be anything within the realm of mathematics. The set of all possible twists of a Rubik's cube has been shown to form a group in this sense, with the algebra describing it being used to investigate solutions to the cube. Group theory has found other applications in code-breaking and even analyzing the mathematical structure of music.

Group theory has also been proven to be an indispensable tool in theoretical physics, where particular groups are found to encapsulate the different 'symmetries' of physical theories – shifts in the parameters that specify each theory which leave its fundamental properties unchanged.

These incredible applications of algebra are all thanks to Al-Khwarizmi and, paradoxically, his decision to forget about applications altogether – and to study a seemingly obscure branch of pure mathematics just to see where the equations might take him.

6 Aviation

DEFINITION CONSTRUCTION OF DEVICES ENABLING HUMANS TO TRAVEL THROUGH THE AIR SUPPORTED PURELY BY AERODYNAMIC FORCES

DISCOVERY FIRST SUCCESSFUL HEAVIER-THAN-AIR GLIDE CARRIED OUT BY ABBAS IBN FIRNAS IN AD 875

KEY BREAKTHROUGH ALTHOUGH INJURING HIMSELF ON LANDING, IBN FIRNAS PROVED THAT HEAVIER-THAN-AIR HUMAN FLIGHT IS POSSIBLE

IMPORTANCE HE SET THE STAGE FOR AVIATION PIONEERS, SUCH AS LEONARDO DA VINCI AND, ULTIMATELY, THE WRIGHT BROTHERS

It is a dream that has captured the imagination ever since human beings first glanced skywards – to fly through the air like a bird. The quest to be airborne has cost many aviators their lives. But in the ninth century AD, the intrepid scholar Abbas ibn Firnas succeeded where all before had failed.

Mention the discovery of manned flying machines and most people think of the Wright brothers and their famous flights at Kitty Hawk, North Carolina, in 1903. But, in fact, the first heavier-than-air flight is thought to have taken place over 1000 years earlier in the northeastern corner of present-day Spain.

This breakthrough was achieved by Islamic scholar Abbas ibn Firnas. In the ninth century AD, modern-day Algeria, Gibraltar, Morocco, Portugal and Spain formed a combined state: the Caliphate of Córdoba. Living in the Caliphate's Spanish region, Ibn Firnas developed ideas such as a musical metronome, a way of making clear glass (from which he fashioned reading lenses), and other innovations in astronomy and medicine.

In AD 875, Ibn Firnas attempted his boldest project yet. He built a glider. Some accounts say that he constructed it from wood and fabric; others, probably exercising a little too much creative licence, report that he strapped the wings of two giant birds to his arms. Whatever the nature of his design, all accounts agree that his flight was largely successful. Ibn Firnas leapt from a tower and soared through the air for several hundred feet before landing.

The 17th-century Moroccan historian al-Maqqari recounted the event: 'He flew a considerable distance, as if he had been a bird but, in alighting again on the place whence he had started, his back was very much hurt.' Indeed, it

LEFT Wind tunnel testing is crucial in modern aviation. Here, a low-speed airfoil (wing) is shown in a smoke-flow-visualization tunnel. When the airfoil is at a large angle-of-attack to the oncoming air, it stalls and the wing experiences separated airflow and turbulence on its upper surface. If this happens to a real aircraft, the loss of lift causes a sudden drop in altitude.

is believed that Ibn Firnas's glider had no tail stabilizer. Birds use their tails to stall their flight just before landing, allowing them to drop down softly onto the ground. Ibn Firnas had neglected this necessity in his calculations, leading to a landing that was somewhat harder than anticipated.

Historians today believe Ibn Firnas's glider was the first successful attempt at heavier-than-air flight – building a machine that can fly despite being heavier than the equivalent volume of air. In contrast, lighter-than-air flight – using devices such as balloons and airships that are lighter than the same volume of air, and thus rise up by simple buoyancy – had already been achieved in China during the third century BC.

Ibn Firnas's flight inspired other glider attempts in both the Islamic world and Europe. Most of these were amateur creations – and several are believed to have resulted in the deaths of their pilots. The first serious theoretical designs for flying machines were produced by Italian polymath Leonardo da Vinci in the 16th century. He produced plans for gliders and even helicopters. Da Vinci never lived to see any of his machines fly, but in 2003 a British TV documentary team was able to build a working glider from his plans.

Aerodynamic lift

In the 18th century, the first scientific theories of flight began to emerge. It is now known that planes fly because of a principle in physics known as the Bernoulli effect, named after Swiss scientist Daniel Bernoulli who first published it in his book *Hydrodynamica* in 1738. In essence, Bernoulli states that a flowing fluid or gas will speed up when it meets a constriction, while at the same time the pressure in the fluid decreases. For example, a fluid flowing through a pipe will speed up if the pipe gets narrower, causing its pressure to drop.

'Ibn Firnas leapt from the top of a high tower, and soared through the air for several hundred feet before landing.'

This discovery led engineers to the design of aerofoils for aircraft wings – essentially a humped profile resembling a constriction in a pipe but with the upper half of the pipe removed. Air passing over the aerofoil accelerates, and this – according to Bernoulli's principle – causes its pressure to drop. This, in turn, sets up a pressure difference between the top and bottom of the wing, and creates a force pushing from the area of high pressure (beneath the wing) to the area of low pressure (above the wing) – which is what creates the lift that keeps the aeroplane airborne.

Pressure falling

You can demonstrate the Bernoulli effect for yourself by simply holding a sheet of paper to your lips and blowing across the top of it. The fast-moving air that you exhale creates low pressure above the paper, causing it to rise up. The low pressure above aircraft wings sometimes causes droplets of moisture to condense out from the air, producing dramatic vapour clouds

around the aircraft. The appearance of this vapour is particularly impressive around aircraft moving at, or close to, sound-speed – when it is known as a 'Prandtl-Glauert singularity'.

ABOVE An F-22 Raptor fighter aircraft performs a banking manoeuvre at near sound-speed. Clouds of water vapour are visible, forming in areas of low pressure around the plane – a phenomenon known as a Prandtl-Glauert singularity.

First flights

And so the stage was set for Wilbur and Orville Wright to make history at Kitty Hawk on 17 December 1903. With Orville at the controls, the *Wright Flyer I* – which was a powered aircraft, not just a glider – made its first flight, skimming above the ground for 12 seconds. Wilbur made the final of four flights that day, which lasted almost a minute and covered a total distance of 260 metres (284 yards).

The progress of aviation from that moment was incredible. Frenchman Louis Blériot made the first oceanic crossing – flying over the English Channel in 1909. During the First World War, aircraft were being used for reconnaissance, bombing and air-to-air combat. American Charles Lindbergh made the first solo crossing of the Atlantic in 1927.

In 1969 – just 66 years after the first powered flight – human beings flew to the Moon. How appropriate, then, that Abbas ibn Firnas's seminal contribution to our conquest of the skies has been honoured in the naming of a lunar impact crater after him. Ibn Firnas Crater spans 89km (55 miles) and lies in the northeastern quadrant of the Moon's far side.

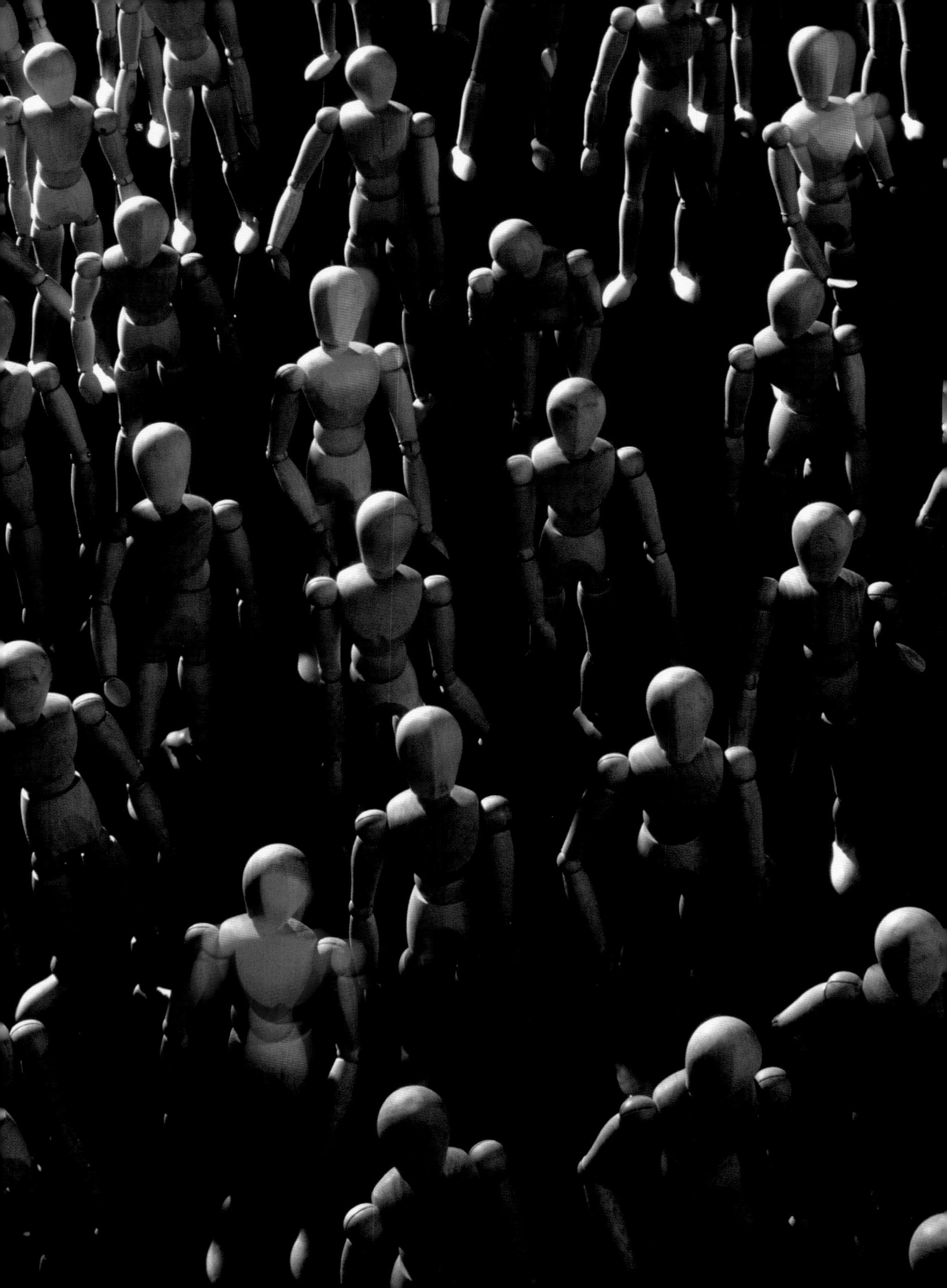

7 Scientific method

DEFINITION THE SCIENTIFIC METHOD IS THE DOCTRINE OF FORMULATING THEORIES AND TESTING THEM BY EXPERIMENTS

DISCOVERY THE METHOD WAS FIRST EXPOUNDED BY IRAQI SCIENTIST ABU ALI AL-HASAN IBN AL-HAYTHAM IN AD 1021

KEY BREAKTHROUGH THE REALIZATION THAT OBJECTIVE TRUTHS CAN ONLY BE DEDUCED FROM HARD EVIDENCE

IMPORTANCE IT WAS THE FRAMEWORK BY WHICH THE REST OF SCIENCE WOULD BE DISCOVERED

Today, the interplay between abstract theory and practical experiment in science is so fundamental that it is hard to imagine one without the other. And yet it has not always been this way. In the 11th century AD, a Persian philosopher quite literally wrote the book on the modern scientific method. His teachings have been guiding scientists ever since.

How do you decide what you believe? Is it enough to have heard something from a friend? Maybe you need to have seen it reported in the media? Or even to have read a technical article about it in a specialist journal? Or are you such a sceptic that you need to carry out your own investigation?

These are the questions that scientists are presented with on a daily basis. Their job is to make sense of the torrent of information pouring in about the world around us. Sorting truths from untruths, deciding which scientific theories are valid and which are not, and which experimental results to trust and which are flawed is no easy task. But it is a lot easier than it could be, due to the work of Abu Ali al-Hasan ibn al-Haytham.

Ibn al-Haytham, also known as Alhazen, was born in Basra (modern-day Iraq), in AD 965. He was a true polymath, making major contributions to the fields of physics, psychology, ophthalmology, medicine and astronomy. His life, however, was nearly cut short by a failure of his engineering skills when commissioned to design a dam. He had agreed to help the ruler of the Islamic 'Fatimid Caliphate' – a region encompassing Egypt and north Africa – to design and build a dam to prevent the River Nile from flooding. But he grossly underestimated the difficulty of the task and the project failed.

LEFT Uniformity within a group of individuals is an illusion carefully maintained in placebo drug trials. Participants are given either the test drug or an inactive drug, which has no effect. Crucially, neither the participants nor the clinicians know who receives the active treatment and who receives the placebo. This removes the possibility of bias in the test results.

Rather than risk being accused of incompetence by the ruling Caliph, and suffer whatever dreadful penalty that might bring, Ibn al-Haytham feigned madness. As a result he was placed under house arrest in Cairo for ten years – but, nevertheless, his life was spared.

Book of Optics

Ibn al-Haytham spent his time in custody productively, writing a seven-volume book called *Kitab al-Manazir* (Book of Optics), which was published in 1021. In the book he presented many principles regarding the properties of light and its interaction with matter, as well as speculating on the nature of human visual perception.

But it was not what Ibn al-Haytham said that was significant, so much as how he said it. Ibn al-Haytham is known to many as the world's first true scientist because he was the first to have adopted the methods and principles by which modern scientists work. Foremost among these was the idea of formulating hypotheses and then testing them against experimental evidence. He demonstrated this guiding mantra in *Kitab al-Manazir* – with every technical statement supported by either experimental evidence or mathematical proof.

'Ibn al-Haytham is known to many as the world's first true scientist because he was the first to have adopted the methods and principles by which modern scientists work.'

Ibn al-Haytham's highly rigorous scientific approach to research had been born, rather incongruously, out of his staunch religious views. As a devout Muslim he believed that humans were flawed (and that only God could achieve perfection), and he therefore sought to remove any element of human fallibility from his work – relying only on hard and fast physical evidence rather than the subjective opinions and interpretations of other individuals. Over the centuries that followed, *Kitab al-Manazir* was translated into Latin and then circulated across Europe – where it was well received by the scholars of the Renaissance. These scholars included giants such as René Descartes and Francis Bacon, who refined the ideas of Ibn al-Haytham into the scientific method employed by researchers today.

Theory and experiment

Ibn al-Haytham's approach breaks down broadly into four main steps. The first step is to understand the nature of a scientific problem. Given a particular phenomenon to explain, the scientist must learn as much as they can about it by preliminary research, reading the work of other scientists and making accurate measurements. Second, the scientist must formulate a clear hypothesis – that is, a conjectured theory that they believe could possibly explain the phenomenon. With a hypothesis in hand, the third step is to investigate the hypothesis and figure out what observable predictions it makes. And finally, the last step is to design and carry out experiments to put these predictions to the test.

Step two – the formulation of a hypothesis – is where the scientist's own imagination and experience must guide them. Constructing the hypothesis can be a daunting task. But in the 14th century, a Franciscan friar in England called William of Ockham advocated a principle to make this part of the process easier. The principle, which has since become known as 'Ockham's razor', offers a methodical approach to the formulation of hypotheses. In essence, it says that you should try the simplest possible hypothesis first – only introducing extra levels of complexity and additional assumptions once experiments have proven your first theory wrong. And you keep going like this until you eventually arrive at the correct theory. Or rather, until you have a theory that you cannot prove to be false.

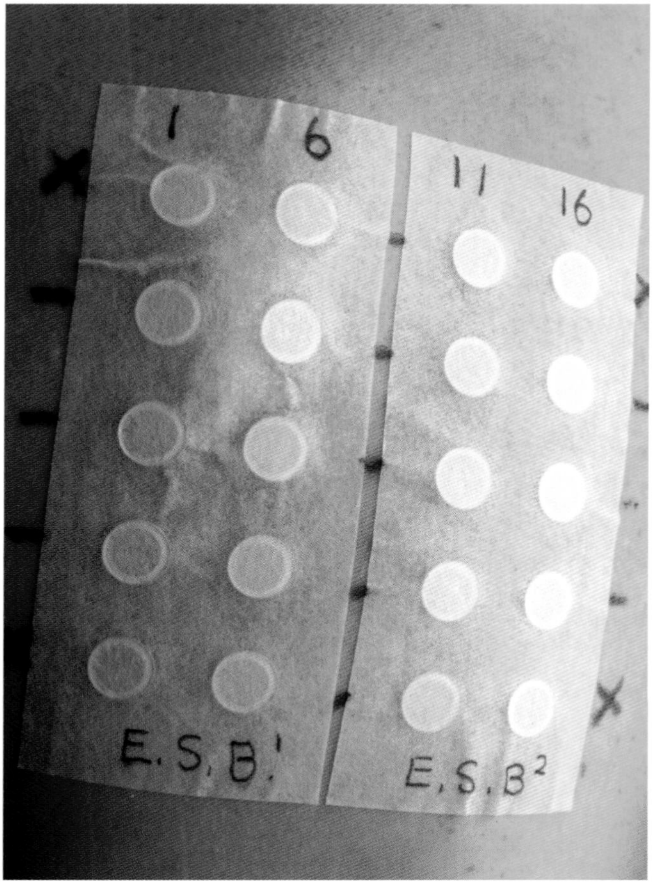

ABOVE The scientific method links effects to their causes. In medicine, this led to treatments based on clinical testing, rather than folklore. With an allergy test, for example, patches are stuck on a patient's skin (as shown), each containing a different allergen, such as pollen or dust. Subsequent skin reactions denote to which allergen the patient is sensitive.

Critical rationalism

It is a seemingly unusual feature of theories in science – a discipline that prides itself on precision – that they can never be proven 100 percent correct. All scientists can do is design better and better tests that place tighter and tighter limits on a theory – but it is impossible to prove a theory is right. The great German physicist Albert Einstein knew this much while formulating his theory of relativity, when he said: 'No amount of experimentation can ever prove me right; a single experiment can prove me wrong'.

The 20th-century science philosopher Karl Popper later championed this view, advocating what he called 'critical rationalism' – the idea that scientific theories should be tested to destruction.

Ibn al-Haytham's healthy scepticism has now become the *modus operandi* for scientists across the globe. But there is no reason that it should be solely the preserve of scientific endeavour. Indeed, in an age when it seems everyone has something to tell us – both in conversation and through the media – we might do well to remember his fundamental lesson. Namely, that we should believe nothing and question everything.

8 Heliocentric solar system

DEFINITION THE NOTION THAT THE SOLAR SYSTEM REVOLVES AROUND THE SUN, AND NOT AROUND THE EARTH

DISCOVERY PROPOSED BY POLISH ASTRONOMER NICOLAUS COPERNICUS IN 1543

KEY BREAKTHROUGH HE REALIZED THE HELIOCENTRIC VIEW OFFERED A MORE NATURAL WAY TO EXPLAIN THE MOTION OF THE PLANETS

IMPORTANCE THE THEORY MARKED THE BEGINNING OF THE MODERN AGE OF ASTRONOMY

The heliocentric view of the solar system put forward by Nicolaus Copernicus ushered in a new chapter in astronomy. It led to the view that there is nothing special about the Earth's location in space – a principle that is the keystone of modern cosmology and our knowledge of the universe.

It is hard to believe that the world's most learned scholars once believed the Earth lay at the centre of the universe – and that the Sun, the stars and all the other planets revolved around us. But they did, and it was a view that persisted for thousands of years.

The theory was first put forward by the Ancient Greeks. In the fourth century BC, Eudoxus of Cnidus put forward a model in which the planets were set into a series of concentric spheres which turned around the Earth. The Moon occupied the nearest sphere, followed by Mercury, Venus, the Sun, Mars, Jupiter and then Saturn. Outside this was an eighth sphere into which were embedded the distant stars. But there was a problem with this theory: astronomers had noticed that sometimes the planets appear to change direction on the sky. The word 'planet' comes from the Greek word for 'wanderer' because the planets are seen to move across the starry background from night to night. But occasionally the direction of this motion would, for a time, reverse and the planet would be seen to sweep across the sky in the opposite direction, before ultimately resuming along its original path. This is known as 'retrograde motion'.

In the year AD 150, the Egyptian philosopher Ptolemy proposed an ingenious solution to this problem in his book *Almagest* (The Great Compilation). He imagined that in addition to its motion around the Earth, each planet

LEFT The Sun during a total solar eclipse, in which the Moon passes between the Sun and the Earth. Copernicus realized that the Sun lies at the centre of our solar system, counter to the accepted view of the day.

underwent smaller circular motions, called 'epicycles'. When the rotation of the sphere and the epicyclic motion are going in the same direction then the planet exhibits normal, or 'prograde', motion. But when the epicyclic motion is counter to the motion of the sphere then the movement of the planet appears to slow down – and in extreme cases can reverse.

Ptolemy had offered an appealing and elegant theory, which was why astronomers and philosophers were more than happy to accept it for nearly 1400 years. But in the late 15th century, a young astronomer in Poland had ideas of his own that rejected this earlier notion in its entirety.

Nicolaus Copernicus was born in 1473, in the Polish town of Torún. In the early 1490s, he studied at the University of Kraków where he learnt about the astronomy of the Ancient Greeks and of Ptolemy's epicycles as a means for explaining retrograde motion. But Copernicus saw these models as clumsy and inefficient – with logical holes in their construction. He set about developing his own model.

Over the years leading up to 1514, Copernicus recast the orbits of the planets so that they revolved not around the Earth, but around the Sun. Copernicus found that a natural order appeared with Mercury orbiting closest to the Sun, followed by Venus, then the Earth, Mars, Jupiter and Saturn (outer worlds Uranus and Neptune had not yet been discovered). This explained why Mercury and Venus never seemed to stray far from the Sun.

Copernicus also noticed that much of the reliance on Ptolemaic epicycles disappeared. In his model the Earth was moving, allowing retrograde motion to emerge naturally as it periodically laps the other worlds of the solar system in their orbits. It is like overtaking a car on the road – relative to distant objects, the car appears to move backwards as it is overtaken. Similarly, a planet that the Earth overtakes appears to move backwards relative to the distant stars.

'Copernicus found that a natural order appeared with Mercury orbiting closest to the Sun, followed by Venus, then the Earth, Mars, Jupiter and Saturn.'

Endless epicycles

Copernicus's model was revolutionary and very elegant, but it did not dispense with epicycles altogether. He still needed to include some in order to make the theory square with observations. That was because Copernicus was still working with the idea of planets moving on circular orbits. The need for epicycles would only disappear entirely with the laws of planetary motion that would be developed by Johannes Kepler (see page 45) – and which replaced perfect circular motion with orbits that are elliptical.

Copernicus published the heliocentric theory in his *magnum opus* work *De revolutionibus orbium coelestium* (On the Revolutions of the Heavenly Spheres) in 1543 – the year of his death. The idea of a Sun-centred planetary system

LEFT The position of Mars (1), Venus (2) and Saturn (3) during a planetary conjunction. The Sun is positioned just below the horizon. Copernicus's theory neatly explained why the positions of Venus and Mercury never stray far from the Sun – because they orbit closer to the Sun than the Earth does.

– what we now call the solar system – was not immediately accepted by other astronomers. This was due in part to the Catholic Church, which was opposed to the theory. But it was also down to the fact that the theory was no more easy to use than Ptolemy's and – in its early incarnations – was no more accurate either.

The theoretical work of Johannes Kepler and Isaac Newton, along with the telescopic observations of Galileo, certainly helped to change this view during the 17th century. But it was not until highly detailed astronomical observations became possible in the 18th and 19th centuries that the heliocentric solar system became an undeniable truth.

Cosmic backwater

Copernicus clung to a certain degree of grandeur in his original heliocentric model – displacing the Earth from the centre of the universe, only to put the Sun there instead. Today, Copernicus's ideas have enabled us to appreciate the Sun and solar system for what they really are – an average star, with an average bunch of planets circling around it, in an average corner of an average galaxy.

This realization that there is nothing special about our world, or its place in the universe – now known as the 'Copernican principle' – is what allows us to practise cosmology (the study of the large-scale universe). For if there was anything special about our view of the heavens then it would be impossible to apply astronomical observations to make any sort of wider inferences about the universe in general. And in that case, the entire history of modern cosmology might never have happened.

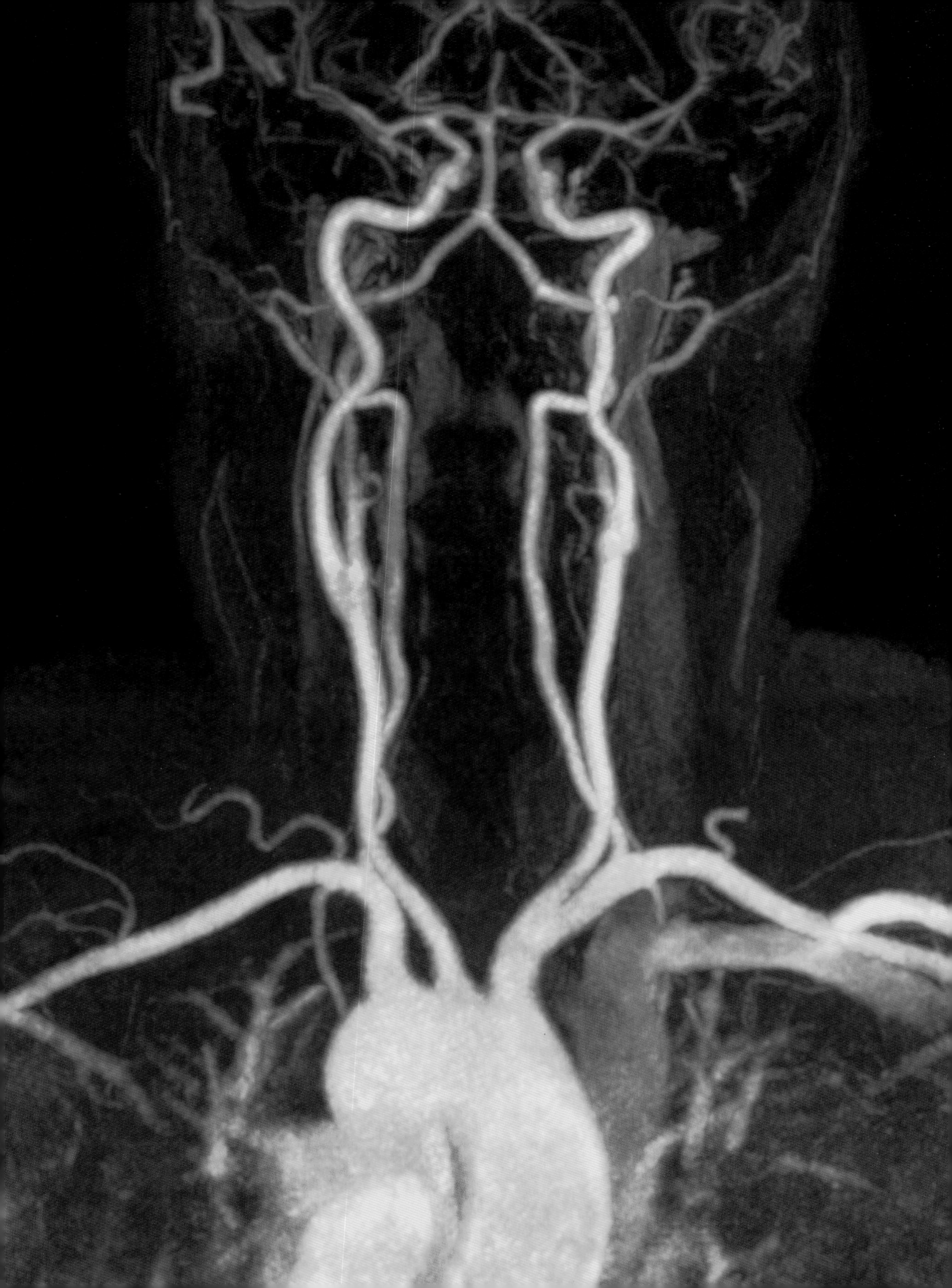

9 Human anatomy

DEFINITION THE PHYSICAL LAYOUT OF THE ORGANS AND OTHER COMPONENTS MAKING UP THE HUMAN BODY

DISCOVERY THE FIRST ACCURATE DESCRIPTION OF HUMAN ANATOMY WAS ANDREAS VESALIUS'S *DE HUMANI CORPORIS FABRICA* IN 1543

KEY BREAKTHROUGH VESALIUS APPLIED SCIENTIFIC RATIONALE TO THE DISSECTION AND STUDY OF HUMAN CORPSES

IMPORTANCE THE BOOK'S PRECISE DESCRIPTIONS AND LAVISH ARTWORK BROUGHT THE SCIENCE OF HUMAN ANATOMY TO LIFE

Anyone who has been to see Professor Gunther von Hagens's *Body Worlds* – a travelling exhibition of preserved human corpses – may well have experienced the same reactions as the first readers of Andreas Vesalius's book *De Humani corporis fabrica.* Both works combined dissected cadavers with stunning artistry and were seminal achievements in communicating the science of human anatomy to a wider audience. Only Vesalius got there first by nearly 500 years.

Anatomy is the roadmap of the human body and thus an essential field of study for all medical students. One of the first anatomists was the Roman physician Galen, who lived in the second century AD. He was, however, constrained by the Roman law that made it illegal for him to dissect human cadavers – an admittedly grisly, yet essential, business for anyone hoping to understand exactly how the human body works. Instead, Galen resorted to dissecting pigs and apes, inferring by analogy the anatomy of humans from his animal findings.

Between the 11th and 13th centuries, during the Golden Age of Islamic science, Muslim physicians were able to carry out human dissections. They did make some discoveries that were at odds with Galen's writings – for example, revealing that the lower human jaw was made of one bone and not two (as Galen had inferred from his animal studies). But by and large, the work of the Arab scientists served to reinforce the teachings of Galen.

LEFT Magnetic resonance imaging shows the arteries of the neck and head carrying oxygenated blood from the heart. Before the 20th century, the only way to study anatomy was by dissection.

But Galen's influence began to wane in the 16th century, with the rise to prominence of a master anatomist by the name of Andreas Vesalius. He was born in 1514, in Brussels. In 1528, aged 14, he entered the University of

Louvain, where he studied medicine. After a spell at the University of Paris, he moved to the University of Padua to study for a doctorate – which he was granted just a year later in 1537. Soon after, he was offered a teaching position in Padua, specializing in surgery – and anatomy, which had become a particular interest for him.

Top of Vesalius's teaching syllabus were the writings of Galen, which had by now governed anatomical thinking for the best part of 1400 years. In Padua, it was perfectly legal to cut down the corpses of hanged criminals from the gallows and dissect them for study purposes – so that is exactly what Vesalius did. He used the steady supply of bodies both to illustrate his anatomy lectures and to conduct his own research. As a result, it soon became clear to him that Galen's animal studies had led him way off the mark about a number of things regarding the anatomy of humans.

'In Padua, it was perfectly legal to cut down the corpses of hanged criminals from the gallows and dissect them – so that is exactly what Vesalius did.'

For example, Vesalius overturned Galen's established idea that blood is pumped around the body by the liver. He correctly stated that the kidneys do not filter urine, but rather that they filter the blood and then channel the filtered material into the urine for excretion. Vesalius also dispelled common anatomical myths, such as men having one less rib than women.

Parallel dissection

Many of these new revelations were achieved through what were called 'parallel dissections', in which Vesalius would simultaneously dissect the body of a human and an animal side-by-side to highlight the differences in their anatomical make-up, and show where Galen had therefore gone wrong in his original assumptions.

It seemed that many anatomists of Vesalius's time, and before, had been so awed by the reputation of Galen that they had not dared challenge his claims publicly. Vesalius felt no such deferential obligations and examined everything Galen had written with a fresh, critical eye. It was a progressive attitude he also tried to cultivate in his students, encouraging them to continually question their own findings – and, indeed, to question even the claims made by Vesalius himself. He urged them to always seek evidence to support their beliefs – and, as such, he could well be regarded as the world's first true medical scientist.

Vesalius also made pioneering new discoveries that Galen had not even touched upon. He made the first comprehensive map of the human muscular system. He explained how the intricate structure of the stomach functions, and how it works in concert with the colon and other internal organs. And, although the details remained a mystery to him, he established the basic purpose of the nervous system (see page 177).

Fabric of the flesh

Vesalius turned the notes of his lectures on human anatomy at the University of Padua into a book, *De Humani Corporis Fabrica* (On the Fabric of the Human Body), published in 1543. It was an instant success. In addition to Vesalius's lucid text, in which he described the results of his research, the book also stood out for its striking illustrations. These were not simple sketches by Vesalius himself, but detailed works of fine art, depicting semi-dissected corpses in almost rapturous poses.

The illustrations are thought to have been produced by students of the Italian painter Titian. From the level of detail, the artists must have been present at Vesalius's dissections (rather than working from his notes). In fact, these artworks turned out to be so central to the book's success that Vesalius later published a textually abridged version with greater emphasis on these stunning graphical elements.

Using the new communication tools of the Renaissance, such as printing and techniques for reproducing images, *Fabrica* was without doubt the greatest anatomy textbook that had ever been written. And so it would remain for hundreds of years, until 1858 when English surgeon Henry Gray published the first edition of *Gray's Anatomy* – the standard student's reference today.

BELOW A 'plastinated' cadaver created by Gunther von Hagens. It is made by replacing water and fat in a corpse with a polymer compound that preserves the tissue. Muscle tissue is pink, while the ligaments and tendons are white. The blend of anatomy and art is reminiscent of Vesalius.

10 Kepler's laws

DEFINITION THREE MATHEMATICAL LAWS THAT DICTATE THE MOTION OF THE PLANETS AROUND THE SUN

DISCOVERY THE FIRST TWO LAWS WERE PUT FORWARD IN 1609, AND THE THIRD IN 1619 – ALL BY JOHANNES KEPLER

KEY BREAKTHROUGH KEPLER GAINED ACCESS TO ACCURATE DATA ON PLANETARY MOTION GATHERED BY TYCHO BRAHE

IMPORTANCE REPRESENTED THE BEGINNING OF ASTROPHYSICS – THE APPLICATION OF THE LAWS OF PHYSICS TO CELESTIAL OBJECTS

Kepler's laws of planetary motion are three rules governing the behaviour of planets, and other bodies, in orbit around the Sun. Kepler formulated his laws entirely from observational data 80 years in advance of Newton's law of universal gravitation, placing the heliocentric view of the solar system on a sound footing and transforming astronomy into an exact science.

Johannes Kepler was born in the German town of Weil der Stadt in 1571. He was a gifted child who showed strong mathematical aptitude from an early age. In 1589, he went to the University of Tübingen and it was there, in his mathematics classes, that he learned about the two opposing views of our solar system – the ancient Earth-centred Ptolemaic system and the radical new theory put forward by Copernicus (see page 37), in which the planets all orbit around the Sun.

Kepler was instantly drawn to the Copernican heliocentric view, and began applying his mathematical knowledge to come up with a rational theory describing it. His first attempt was to use geometry alone. He tried to explain the orbital distances of the planets in terms of regular polyhedral solids – cubes, tetrahedrons and so on – nestled inside one another.

In 1600, Kepler made the acquaintance of the Danish astronomer Tycho Brahe, now living in Prague, who had made probably the most accurate planetary observations of the time – logging in great detail how the planets changed their position from night to night. Brahe's observations of the heavens were made using a device of his own construction, called the 'mural quadrant' – a giant 90-degree arc several metres in radius and graduated along its edge in

LEFT Kepler's laws describe the motion of the planets around the Sun and explain why the solar system's outer planets, such as Saturn (pictured), take longer to complete each orbit than inner worlds, such as Earth. Saturn orbits only once every 29.5 years.

sixtieths of a degree. By sighting the quadrant on the night sky, Brahe was able to make unprecedentedly accurate astronomical measurements. Brahe was impressed by Kepler's mathematical ability and was so intrigued by his theoretical ideas about the workings of the solar system – in particular, the heliocentric theory – that he subsequently employed Kepler to analyze the data he had so painstakingly gathered. When Brahe died suddenly in 1601, Kepler was made his successor as mathematician to Emperor Rudolph II. Kepler then used his unrestrained access to Brahe's data to refine his own theory of the Copernican solar system. By a laborious process of trial and error, he was able to formulate three key mathematical laws describing the motion of the planets.

Kepler's first law is a simple one. It states that the planets of the solar system all move on elliptical orbits with the Sun at one of the two 'foci'. An ellipse is an elongated oval shape, formed by taking a circle and stretching it in one direction. A circle is the special case of an ellipse where both the foci coincide. You can draw a circle yourself by hammering a nail into a board. Slip a loop of string around both the nail and a pencil – move the pencil while keeping the string taut and the result is a circle. Now hammer two nails into a board and repeat. The result is an ellipse with the two nails as the foci.

The second law states that a line joining the Sun to an orbiting planet sweeps out equal areas in equal times as the planet moves. The upshot of this law is that planets move faster as they approach the Sun – as expected when they fall inwards – and slower when they are further away.

Kepler spreads the news

Both Kepler's first and second laws were publicized through his book *Astronomia nova* (A New Astronomy), published in 1609 – the same year Galileo invented the astronomical telescope. Kepler's third and final law was not published until ten years later, in his 1619 book *Harmonices Mundi* (The Harmony of the World). This third law states that the square of the orbital period of a planet – the time it takes to make one complete revolution around the Sun – is proportional to the cube of the orbit's 'semi-major axis'. This is equal to half of the length of a line drawn from one side of the ellipse to the other and that passes through both of its foci.

'Amazingly, Kepler formulated all three of his laws of planetary motion well in advance of any fundamental theory for how the solar system works.'

As an example, the third Kepler law implies that if planet A lies five times further away from the Sun than planet B, then planet A takes 125 times as long to complete a single orbit. Although Kepler's interpretation of planetary orbits in terms of polyhedral solids is now known to be incorrect, his three laws accurately describe the workings of the solar system to a very good approximation. Amazingly, Kepler formulated all three of his laws of planetary motion well in advance of any fundamental theory for how the solar system works. That would follow in 1687 with Newton's law of universal gravitation, and any skilled mathematician can use Isaac Newton's law of gravity to derive Kepler's laws. The laws also apply to other two-body gravitational systems when one of the bodies is very much more massive than the other.

ABOVE Kepler tried to explain the orbiting planets as spheres spaced according to five platonic solids – octahedron, icosahedron, dodecahedron, tetrahedron and cube – nestled within each other. Each solid was inscribed within a sphere, and within that was the next sphere and the next solid. The theory is now known to be incorrect.

Newton's laws, however, also allow for the existence of additional trajectories – not encompassed within Kepler's original theory – for celestial objects moving within a gravitational field. For example, some comets falling in from the outer solar system can travel along an open trajectory, making one close pass by the Sun before heading back out into space, never to return again. In contrast, Kepler's laws (with the exception of the second one) deal exclusively with closed, periodic orbits.

Whether a celestial orbit is open or closed is determined by a parameter called the 'eccentricity', denoted by the letter e. If e lies between 0 and 1 then the orbit will be an ellipse, as described by Kepler's laws. If $e = 0$ then the orbit forms a perfect circle. But if e grows to be equal to or greater than 1 then the orbit is becomes open.

The modern view

Today, we also know that Kepler's laws break down in the limit of strong gravitational fields. This is where Newton's theory of gravity is also invalid, and Einstein's theory of general relativity must be used instead. General relativity makes its presence felt in our solar system – for example, influencing the orbit of the inner planet Mercury. Despite these drawbacks, Kepler's laws are perfectly adequate for describing the motion of planets, asteroids and spacecraft plying the outer solar system, where the Sun's gravity is significantly weaker. Kepler's laws remain required reading for physics students and are also used by astronomers to calculate the orbits of extrasolar planets around their host stars (see page 381). But back in the 17th century, all three laws were a genuine revolution that helped cement Copernicus's Sun-centred view of the solar system – from which all of modern astronomy and astrophysics was born.

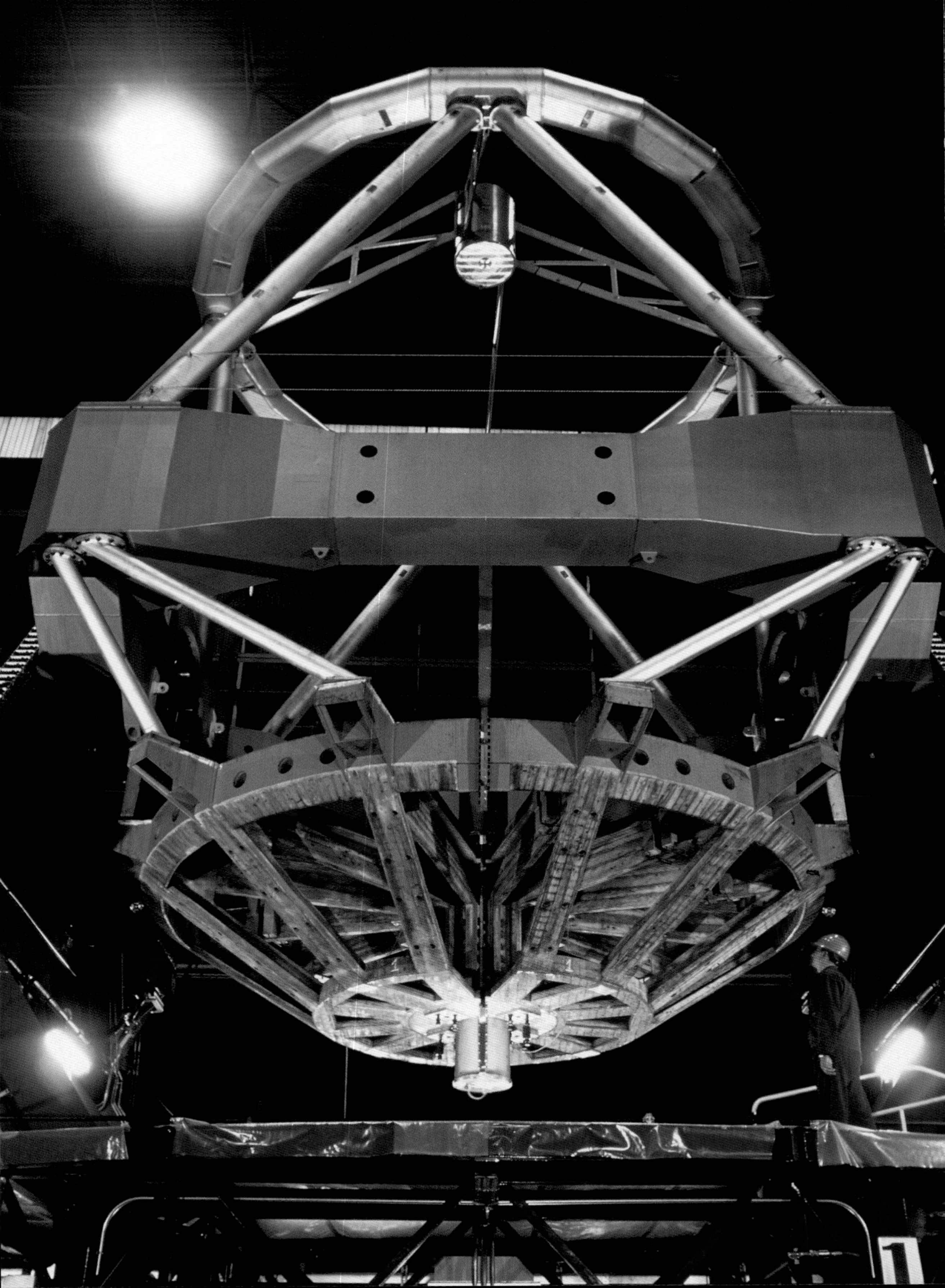

11 Astronomical telescope

DEFINITION A CONFIGURATION OF LENSES OR MIRRORS USED BY ASTRONOMERS TO MAGNIFY IMAGES OF CELESTIAL OBJECTS

DISCOVERY THE FIRST ASTRONOMICAL TELESCOPE WAS BUILT BY GALILEO GALILEI IN PADUA, ITALY, IN 1609

KEY BREAKTHROUGH GALILEO LEARNED ABOUT THE INVENTION OF THE SIMPLE LENS TELESCOPE AND ADAPTED IT FOR ASTRONOMY

IMPORTANCE BROUGHT A WEALTH OF NEW OBSERVATIONS OF SPACE AND PROVIDED A TEST-BED FOR THE FIELD OF ASTROPHYSICS

For millennia, astronomers studied the heavens using nothing more than the naked eye. Then in 1609, Italian scientist Galileo took a new invention called the telescope and turned it skywards. Suddenly, the universe snapped into focus, bringing about a revolution in our understanding of it. Now astronomers are building the most powerful telescopes ever – instruments that promise revelations even Galileo could not have dreamed of.

The telescope was invented by three Dutchmen – lens-makers Hans Lippershey and Sacharias Jansen, and optician Jacob Metius – in 1608, all working independently. It is unclear who was first, though Lippershey seems to be the favourite. He fitted a convex lens to the front end of a tube and a concave lens at the back end. The resulting device enabled the user to see images of distant objects at 3× magnification.

Word of this amazing new invention spread across Europe fast. In 1609, the Italian polymath Galileo Galilei learned of its existence and instantly saw that it could be used for astronomy. Up until that point all astronomical study was conducted with the naked eye. The only instruments available were sextant-like devices, known as 'astrolabes', to chart the positions of stars and other objects on the sky.

By all accounts, Galileo had not even seen a design for the Dutch 'perspective glass', as it was known. Armed just with the knowledge that it consisted of two lenses at opposite ends of a tube, he worked out the rest for himself – pretty much overnight. He improved on the design, taking its magnifying power up from 3× to over 30×. In 1610, Galileo turned his new device on

LEFT Construction of the Very Large Telescope (VLT) at the European Southern Observatory, Chile. It comprises four separate telescopes linked together to work as one unit. Each telescope focuses light using a giant mirror.

the night sky. He made the first sketches of mountains on the Moon and saw the moon-like phases of the planet Venus. He also made some of the first observations of cool patches on the surface of the Sun, known as sunspots (though do not ever look at the Sun through a telescope yourself – you could be blinded forever). And he observed the Milky Way, finding it composed of hundreds of stars rather than gas clouds, as had been previously supposed.

One of Galileo's first targets to view was the giant planet Jupiter. As well as being struck by the detail visible on the planet's surface, Galileo observed 'three fixed stars, totally invisible by their smallness' positioned in a line either side of Jupiter. A few days later, he discovered a fourth star. Observations over subsequent nights showed these tiny dots to be moving relative to the planet, and he soon realized that they were orbiting around it. Galileo had discovered Jupiter's four largest satellites – Io, Europa, Ganymede and Callisto – which are still known today as the 'Galilean moons'.

'In 1610, Galileo turned his new device on the night sky. He made the first sketches of mountains on the Moon and saw the moon-like phases of the planet Venus.'

These four moons were the first astronomical objects to be discovered that were orbiting a body other than the Earth. This, together with the changing phases of Venus that he had seen, led Galileo to publicly support the Sun-centred view of the solar system that had been put forward by Polish astronomer Nicolaus Copernicus. Sensibly, Copernicus had arranged for his theory to be published posthumously. Galileo was not as cautious, however, and his outspokenness led to him being tried for heresy by the Catholic Church in 1633. He spent the remainder of his life under house arrest.

Sold to sailors

Galileo summed up his early telescopic observations in a book, published in March 1610, called *Siderius Nuncius* (Starry Messenger). He also made a tidy profit selling his telescopes to mariners. The success of Galileo's instrument way outstripped that of its Dutch predecessors – so much so, and perhaps unfairly, that the design is known today as the Galilean telescope.

Astronomer Johannes Kepler was the first to replace the concave eyepiece lens with a convex lens. This widened the telescope's field of view, but also inverted the image (switching the left–right and up–down sense). For astronomy, image-inversion is not a big problem. The designs of astronomical telescopes used by Galileo and Kepler are known as 'refracting', because they use lenses that refract (bend) the path of light rays passing through them.

Big is best

There is a limit to the size of telescope that can be built in this way – large-diameter telescopes are preferable for astronomy because they gather more light than smaller instruments, enabling the user to see fainter objects. In 1668, Isaac Newton came up with a different design that used curved mirrors instead of lenses to focus the light. Newtonian reflecting telescopes were cheaper to make, could be larger in diameter and did not suffer from 'chromatic aberration' – a problem with refractors, causing them to focus light of different colours at different points. The largest refracting telescope ever used is the 101cm (40in) Yerkes telescope, in Wisconsin, USA. It weighs 20 tons, a large proportion of which is accounted for by its front lens – a lens made any bigger would sag under its own weight and be useless.

By comparison, the largest reflecting telescope today is the Gran Telescopio Canarias, on the island of La Palma. Its main mirror is 10.4 metres (11.4 yards) across – made of 12 interlocking, computer-controlled segments. In Chile, the Very Large Telescope (VLT) is made with a network of four, 8.2-metre (9-yard) telescopes with the combined observing power of a 16-metre (17.5-yard) telescope. The European Southern Observatory (ESO), which operates the VLT, is building the European Extremely Large Telescope (E-ELT), with a mirror spanning a colossal 42 metres (46 yards) and made of 984 interlocking hexagonal components. It will be operational by 2018.

Space telescopes

The main problem with telescopes on Earth is that they suffer from an effect called 'atmospheric seeing'. This is caused by thermal currents in the atmosphere that distort the light passing through it. 'Seeing' greatly hampers the performance of telescopes because it is rather like looking at fish in a pond through the undulating surface of the water.

The solution is to situate telescopes in space. The first of these was the 2.4-metre (2.6-yard) Hubble Space Telescope, which has sent back stunning views of the cosmos. Hubble has been joined by other observatories and, by 2015, NASA hopes to launch its successor – the 6.5-metre (7.1-yard) James Webb Space Telescope. These developments were all started by Galileo's desire to train a simple set of lenses upon the heavens. And anyone can join in – you can buy a refracting telescope for the cost of a pair of shoes.

BELOW The operational Very Large Telescope, Chile, incorporates 'adaptive optics' that correct thermal fluctuations in the atmosphere, yielding crisper images that rival those from the Hubble Space Telescope.

12 Calculus

DEFINITION A POWERFUL TECHNIQUE FOR HANDLING THE RATES OF CHANGE OF MATHEMATICAL FUNCTIONS

DISCOVERY DISCOVERED INDEPENDENTLY BY ENGLISH PHYSICIST ISAAC NEWTON AND GERMAN MATHEMATICIAN GOTTFRIED LEIBNIZ

KEY BREAKTHROUGH NEWTON AND LEIBNIZ DEVELOPED TECHNIQUES BASED ON THE SMALL VARIATIONS OF A FUNCTION AT A POINT

IMPORTANCE CALCULUS IS THE MOST SIGNIFICANT BRANCH OF MATHEMATICS, WITH APPLICATIONS IN ALL THE SCIENCES

Calculus is a method for working out how mathematical quantities change with respect to one another. In its most basic form it is relatively simple to implement. But more advanced treatments, used for complex real-world problems – such as those encountered in quantum theory and mathematical finance – are fiendishly difficult. Calculus was pioneered in the 17th century and has been a central strand of modern mathematics ever since.

Of all the branches of mathematics, calculus is arguably the most powerful. It is used for analyzing the properties of mathematical functions, such as how a function changes in response to changes in its variables. It has applications in mathematical analysis – charting the behaviour of algebraic functions – but its biggest payoffs are in applied science.

All of modern science relies on calculus, where it is used for working out how physical systems change with time – and for working backwards to calculate the underlying physical quantities responsible for observed changes in a system. In chemistry, calculus is used to define the time evolution of chemical reactions. In biology, it can explain the relative population sizes of species competing for shared resources. Even in economics – disrespectfully referred to as the 'dismal science' – it is used for computing how the prices of stocks and shares vary in response to changing market forces.

Calculus deals with two principal operations: 'differentiation' and 'integration'. The simplest of the two operations is differentiation, and is concerned with determining the gradient or slope of a mathematical function at a particular point. This is useful, for example, in physics – given a mathematical equation

LEFT The marine mollusc *Nautilus pompilius* takes a structure known as a logarithmic spiral, in which the rate of change of the radius is directly proportional to the radius itself.

for the position of an object with time, differentiation tells you the object's speed as a function of time. It works by taking a function plotted as a curve on a graph and chopping it up into a number of intervals. A straight line drawn from one end of an interval to the other gives an approximation to the curve's gradient at the midpoint of that interval. And as the interval tends towards zero, this approximation becomes increasingly accurate.

This is how mathematicians define differentiation. Given a variable x, which depends on another variable t, written $x(t)$, then differentiating x with respect to t means dividing the change in x by the change in t, as the size of the change in t tends to zero. For simple power laws, such as $x = t^2$, the differential just boils down to subtracting one from the power and multiplying by the old power. For example, if $x = t^2$ then the differential is just $2t$. In other words, at any point on the curve t^2, a straight line touching the curve will have a slope equal to $2t$. Such a straight line is sometimes called a 'tangent'. The differential of the variable x, with respect to another variable t, is usually denoted dx/dt. If this were a physics problem, then x might represent the position of an object in space with time, t. Working out dx/dt then tells you the speed of the object at any given time.

'All of modern physics relies on calculus.'

Infinitesimal acorns

Integration is the opposite of differentiation – a principle called the fundamental theorem of calculus. Given that $dx/dt = 2t$, integration then tells you that $x = t^2$. In general, given some function, integration provides the area under the curve you get when you plot this function in the form of a graph. Like differentiation, it can be thought of in terms of chopping up the curve into intervals, but this time – rather than the slope of each interval – you are working out the entire area of the strip under each part, and then taking the limit as the interval tends to zero. Adding up the area of each infinitesimal strip then gives the total area under the whole curve.

Integration has multiple applications in the mathematics of geometry, where it is possible to work out not just areas but the volumes of complex shapes. For example, the volume of a sphere can be calculated by first working out the volume of a hollow shell – the surface area of a sphere multiplied by a tiny interval in radius – and then integrating (equivalent to adding up) the volumes of all these infinitesimally thin shells between the sphere's centre and its surface.

Whereas differentiation is quite straightforward – involving a few methodical rules – integration can be extremely hard to carry out on paper. Many contemporary problems requiring integration in real-world science have to be solved numerically on a computer – where the computer literally calculates the contribution from each infinitesimal slice and then adds them all up to provide an answer.

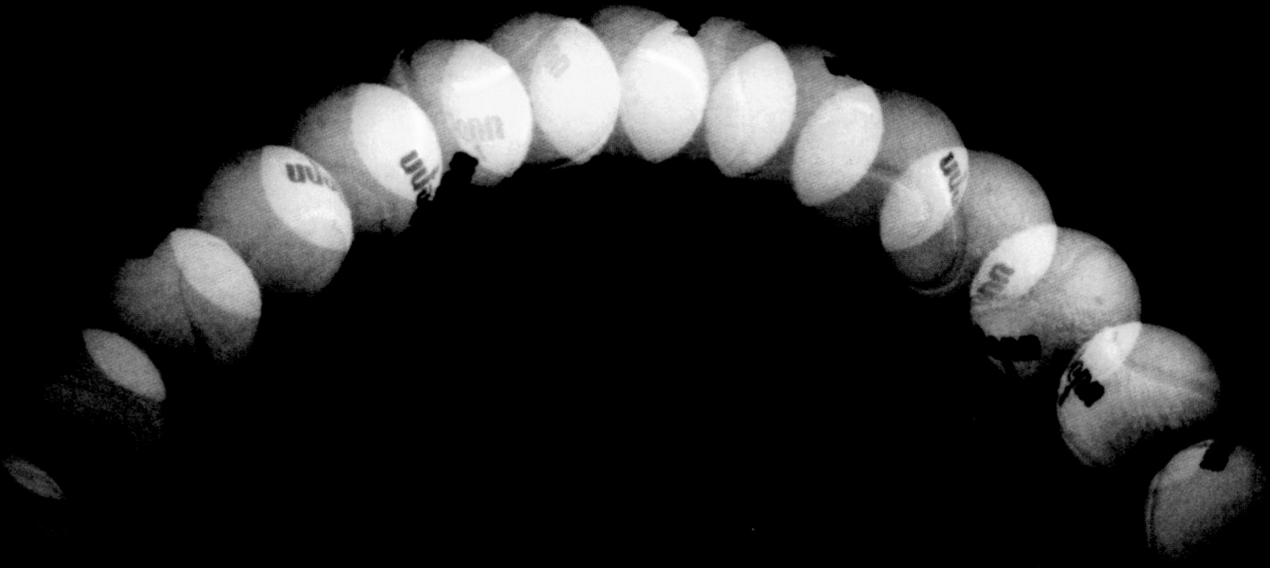

Priority dispute

Calculus was independently invented by the English physicist Isaac Newton and the German mathematician Gottfried Leibniz. Newton is thought to have developed his calculus – which he called the 'method of fluxions' – between 1665 and 1667, while staying at the family home in Lincolnshire to avoid the Great Plague. However, he did not publish his findings for many years. Leibniz, on the other hand, developed his own approach in the early 1670s and published it relatively promptly, in 1684.

When Newton learned of Leibniz's work, he was furious – convinced that the German had plagiarized his own findings (although Newton had not formally published the method of fluxions, he had circulated copies of his notes to colleagues). As with Newton's dispute over the law of gravity with Robert Hooke, a deep rift resulted between the two men that lasted until Leibniz's death in 1716.

Of the two formalisms, Leibniz's was more useful. Newton was first and foremost a physicist. For him, calculus was a means to an end – a way to figure out how nature works. And he thus developed it only as much as was needed to serve his purposes. In contrast, Leibniz was a mathematician, who took interest in mathematical techniques for their own sake. He developed his method carefully and systematically – as a field of science in its own right.

Scientists and mathematicians today prefer to use Leibniz's version of the technique and also the universal notation that he developed. Even our use of the name 'calculus' is down to him.

ABOVE The motion of a ball thrown into the air can be derived using calculus, by setting the rate of change of its velocity equal to the acceleration due to gravity at the Earth's surface, then solving the resulting equation.

13 Newton's *Principia*

DEFINITION A SET OF BOOKS SETTING OUT NEWTON'S LAWS OF MOTION AND THE THEORY OF UNIVERSAL GRAVITATION

DISCOVERY PUBLISHED IN 1687 BY ENGLISH PHYSICIST SIR ISAAC NEWTON, REPRESENTING THE SUM OF MANY YEARS' WORK

KEY BREAKTHROUGH NEWTON SOUGHT PRECISION AND MINIMALISM IN HIS THEORIES – WHICH HE FOUND IN RIGOROUS MATHEMATICS

IMPORTANCE THE PUBLICATION OF NEWTON'S *PRINCIPIA* WAS THE BIRTH OF MODERN THEORETICAL AND MATHEMATICAL PHYSICS

In his *Principia*, Isaac Newton brought about a complete revolution in science. In a single publication, he set out the laws governing the behaviour of moving bodies and detailed his theory of gravitation – which applied to everything from the planet Jupiter to falling apples. A mathematician as well as a physicist, Newton framed all his theories in rigorous equations.

Isaac Newton was born in January 1643 in Lincolnshire. In 1661, he went up to Trinity College, Cambridge, to study mathematics and natural philosophy. By 1669, he was the university's Lucasian Professor of Mathematics – a position that would much later be held by Stephen Hawking.

Newton made seminal contributions to science – unpicking the nature of light (revealing, for example, that white light is made up of a rainbow of different colours), building one of the world's earliest reflecting telescopes and pioneering the development of calculus.

But his greatest achievements were set out in a three-volume set of books called *Philosophiae Naturalis Principia Mathematica* (The Mathematical Principles of Natural Philosophy) – more commonly referred to as Newton's *Principia* – published in 1687. In these books, he established a systematic way for scientists to analyze the world using the mechanical tools of mathematics.

Newton took the verbose and ambiguous blusterings of 'natural philosophers', as they liked to be called, and replaced them with rigorous, succinct equations. 'I contrive no hypotheses,' he wrote. Indeed, all his claims were proven to the letter in watertight mathematical calculations.

LEFT The forward motion of a bicycle and its rider offers a perfect illustration of Newton's laws of physics – bringing into play the powerful forces of acceleration, action and reaction, momentum and inertia.

The book offered the first exposition of Newton's laws of motion – three fundamental principles that dictate the behaviour of moving objects. The first law says that every object remains at rest or in a state of uniform motion unless acted on by a force. In other words, stationary objects remain stationary and moving objects continue moving in a straight line – unless a force acts on them. That force might be a collision with another object, or it might be due to the influence of a force of nature – such as gravity making a ball roll downhill.

The second law says that when a force acts on an object then the object experiences an acceleration, so as to satisfy the equation: force = the acceleration multiplied by the object's mass. Newton's second law is really saying that the heavier an object is the harder it is to move – apply the same force to two objects, one weighing ten times as much as the first, and the lighter object will accelerate ten times faster. Newton called this concept 'inertia'.

'Newton took the verbose and ambiguous blusterings of "natural philosophers", as they liked to be called, and replaced them with rigorous, succinct equations.'

Newton's third law of motion asserts that for every force, there is an equal force pushing back in the opposite direction. Also known as the law of action and reaction, it is the reason why a rifle kicks back against your shoulder when you fire it. Burning the powder in the bullet casing generates gases that produce a force as they expand. This force propels the bullet forwards, while an equal but opposite force propels the rifle back towards you. Newton's second law then makes the bullet – which is lighter than the rifle – accelerate to a much higher speed than the body of the rifle.

Newton arrived at his laws of motion after experiments and theoretical trial and error. They apply equally to billiard balls, cars negotiating a bend or rockets accelerating through space. In the mid-1700s, Newton's laws of motion were extended to describe the rotation of rigid bodies by the Swiss mathematician Leonhard Euler – describing everything from spinning gyroscopes to the rotation of the Earth.

Matters of gravity

The other big thesis expounded in the *Principia* was Newton's law of universal gravitation. This is a law of physics that gives the size of the gravitational force acting between two bodies. It is equal to their masses multiplied together divided by their separation squared, multiplied by a constant of nature ('Newton's constant') equal to 6.67×10^{-11} (or, 6.67 divided by 100 billion – very small).

Newton's law of gravitation, together with Newton's three laws of motion, provides an accurate description of the motion of planets (as well as asteroids, spacecraft and pretty much anything else) through our solar system. Indeed, *Principia* includes a full derivation of Kepler's laws from these basic principles. It also explains terrestrial phenomena, such as balls rolling downhill and the behaviour patterns of the ocean tides.

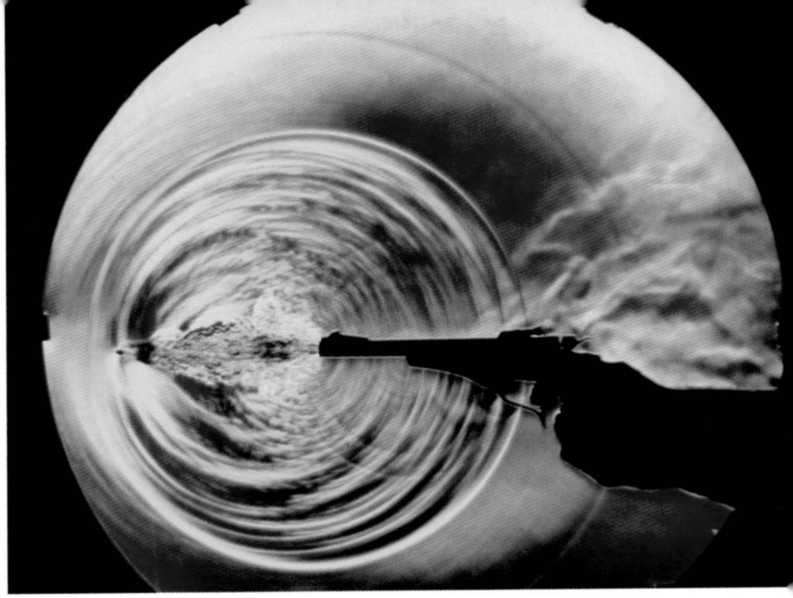

Halley's comments

Newton was inspired to write the *Principia* following conversations with English astronomer Edmond Halley. Halley, in turn, had been in discussions with physicist Robert Hooke about gravity, in which Hooke claimed that he had calculated gravity to be a force that diminishes with the distance squared between the two gravitating objects (exactly the form of Newton's law). When Halley put this to Newton, Newton answered that he already knew this – having carried out the same calculations himself.

ABOVE When a bullet is fired, the gun itself kicks back. This is an example of Newton's third law of motion – the law of action and reaction. The law also explains why rockets can fly – high-speed exhaust gas travels downwards, causing the rocket to lurch up.

The meeting with Halley stimulated Newton to revisit his work on gravity and, as a result, he produced a nine-page manuscript on orbital motion for Halley to present to London's Royal Society in 1684. Halley was so impressed with Newton's paper that he asked for more. In response, Newton set to work on the *Principia*, which is thought to have turned into an obsession that consumed him until spring 1686. On its completion, Halley was amazed at what Newton had produced – to the point that when the Royal Society was unable to pay the publication costs, Halley stumped up the money from his own pocket. *Principia* was published on 5 July 1687.

Hooke was understandably aggrieved that Newton had scooped the glory for the law of gravity, although to this day it remains unclear quite which of the two men had arrived at the theory first. Some historians have suggested that the idea of a square law for gravity had been well-known for some time. Indeed, Newton even stated in a letter to Hooke that he had arrived at his conclusions 'by standing on the shoulders of giants'. Others have wondered whether this comment was, in fact, a veiled swipe at Hooke's diminutive stature. Either way, they remained enemies for the rest of their lives.

Further editions of the *Principia* were published in 1713 and 1726 – as well as two translations into English (from the original Latin) several years later. Today, many original copies of these editions survive in libraries around the world – while modern reprints are still available to buy. However, even Newton's ideas were not destined to last forever. In the early 20th century, they were superseded by Albert Einstein's theory of relativity – which rewrote both the laws of motion and those of gravity.

14 Linnaean taxonomy

DEFINITION A SYSTEMATIC WAY OF CLASSIFYING SPECIES IN THE NATURAL WORLD

DISCOVERY FIRST SUGGESTED BY SWEDISH BOTANIST CARL LINNAEUS IN HIS 1735 BOOK *SYSTEMA NATURAE*

KEY BREAKTHROUGH LINNAEUS PROPOSED A SCIENTIFIC WAY TO CLASSIFY ORGANISMS BASED ON THEIR PHYSICAL APPEARANCE

IMPORTANCE LINNAEUS'S WORK SET THE BAR FOR THE MODERN METHOD OF CLASSIFYING SPECIES BY THEIR GENETIC ANCESTRY

Is it animal, vegetable or mineral? This traditional children's game stems from the work of genius Swedish naturalist Carl Linnaeus, who was the first to try to classify objects in the natural world according to their appearance, by dividing it into three kingdoms. He was the first scientist to make the order of nature understandable to human beings.

Carl Linnaeus was born in Sweden on 23 May 1707. He developed a fascination with plants from an early age and, at the request of his father, was tutored in botany by a local doctor to develop this interest. Aged 21, he enrolled at Lund University, in Scania, but later moved to Uppsala to pursue dual interests in botany and medicine.

It was during his time at Uppsala that his talent as a botanist began to show. A thesis he had written on plant sexuality so impressed one of the university's professors that he requested Linnaeus to begin giving lectures – even though he was still only a student.

In 1735, Linnaeus travelled to the University of Harderwijk in the Netherlands to take his doctor's medical degree. Later that year, he would publish the first edition of a book that was ultimately to revolutionize the way scientists view the natural world.

The book was called *Systema Naturae* (System of Nature), and set out Linnaeus's theory of plant classification based on sexual characteristics. Different types of plants were referred to as taxa (singular 'taxon') – and their study became known as Linnaean taxonomy. In this system, flowering plant

LEFT In this tulip, the male reproductive organs (stamens) are the central black protrusions, while the female ovaries are the white stigma positioned in the centre. Fertilization may occur when pollen lands on the stigma and moves towards the ovaries.

species were divided into separate classes according to the exact number of male reproductive organs in the flower (the stamens) and according to the number of female organs (the pistils).

The book was not just an attempt to classify the botanical world. Linnaeus had made a pact with his friend, the naturalist Peter Artedi, that if either of them died, the other would complete their unfinished work. In 1735, Artedi drowned and so Linnaeus completed his work on the classification of fish – which also formed part of *Systema.*

In 1741, Linnaeus became Professor of Medicine at the University of Uppsala. But since he was so committed to the cause of studying naturalism, he swapped jobs in 1742 with another professor at the university – Nils Rosen – to become Professor of Botany, Dietetics and Material Medica.

The first edition of *Systema* was a meagre 11 pages long, but Linnaeus kept amending and expanding it every few years, supplementing his data with material collected while travelling in Sweden and Europe. By the tenth edition, published in 1758, Linnaeus's book was over 1000 pages in length and it classified 7700 species of plants and 4400 species of animals.

'By the tenth edition, published in 1758, Linnaeus's book was over 1000 pages in length and classified 7700 species of plants and 4400 species of animals.'

Rank and file

Linnaeus divided up the natural world in a hierarchical fashion. The highest rank was 'kingdom', and he posited that the world was divided into three kingdoms – literally, animal, mineral and vegetable. Next down was 'class'. So, for example, in the tenth edition of the book the animal kingdom was divided up into six classes – mammals, birds, amphibians, fish, insects and 'vermes' (a class that included worms, slugs and leeches).

Next down was 'order'. For instance, the mammals class contains the orders primates, rodents and cats – to name but a few. Each order was subdivided into a number of 'genus' categories, and each genus finally divided up into different 'species'. Human beings, for example, are members of the animal kingdom, mammal class, order primates, genus *Homo* and species *sapiens.*

Binomial classification

Using full classifications to describe every biological organism could prove cumbersome, so Linnaeus pioneered a simplified way of referring to them all – which has become known as 'binomial classification'. Effectively, it boils down to using just the genus and species to identify each organism. So, human beings are then known simply as *Homo sapiens* (the format is that the genus takes a capital letter while the species is lower case). The system is still in use today. Linnaeus's plant classification scheme, however, was not so successful, and has long been superseded. But, his method of dividing

the animal world was closer to the mark. It was extended further at the turn of the 18th century by the French zoologist and naturalist Georges Cuvier into a system that more or less describes the animal kingdom as we know it today. Linnaeus formulated his classification out of a passionate desire to make the natural world comprehensible to humans. He never regarded his classification of species to be any sort of reflection on how those species had arisen or how they were related to each other. In 1859, however, Charles Darwin published his book *The Origin of Species* – and overnight it became clear that species were related.

ABOVE Linnaean classification is based on physical appearance. The natural world is capable of throwing up some very odd appearances indeed, as demonstrated by these booted racket-tail hummingbirds. Linnaeus's book of species grew to over 1000 pages in length.

Cladistics

In 1958, nearly 100 years after the birth of Darwin's theory, evolutionary biologist Julian Huxley put forward the notion of 'cladistics' – a method of classifying the natural world according to ancestry rather than physical appearance. Cladistics deals with 'clades', branches of the tree of life that are drawn up according to genetics and which classify species on the basis of how they have evolved from other species.

Although Linnaeus was not directly involved in Huxley's creation, his work on *Systema Naturae* – as well as his many other books, including *Philosophia Botanica* (published in 1751) and *Species Plantarum* (1753) – was instrumental in establishing the classification of the natural world as a scientific discipline. For his efforts, Linnaeus was knighted in 1761. He retired from teaching duties in 1774 following a stroke and died in 1778, aged 70.

15 Kinetic theory

DEFINITION A THEORY IN PHYSICS FOR CALCULATING PROPERTIES OF A GAS FROM THE MOTION OF ITS CONSTITUENT PARTICLES

DISCOVERY THE FOUNDATIONS OF KINETIC THEORY WERE LAID DOWN BY SWISS MATHEMATICIAN DANIEL BERNOULLI IN 1738

KEY BREAKTHROUGH BERNOULLI EQUATED THE KINETIC ENERGY OF ALL THE PARTICLES IN A GAS TO ITS TOTAL THERMAL ENERGY

IMPORTANCE KINETIC THEORY HAS GIVEN PHYSICISTS RADICAL NEW INSIGHTS INTO THE FUNDAMENTAL PROPERTIES OF MATTER

It is obvious when you think about it – the large-scale behaviour of a gas must ultimately derive from adding up the small-scale behaviour of its composite particles. The first physicist to turn this into a workable idea was Daniel Bernoulli: his kinetic theory formed the basis for statistical mechanics and bridged the esoteric laws of quantum theory and everyday physics.

Kinetic theory is based on the idea that the temperature and pressure of a gas can be derived from the motions of its constituent atoms and molecules.

The theory follows from setting the kinetic energy – the energy of motion of all the particles in a gas – equal to the gas's thermal energy, which depends solely on its temperature. In this way, the temperature of a gas is just a manifestation of how energetically all its particles are moving about. The theory explains the existence of the absolute zero of temperature, which is −273.15°C. This is the temperature at which all the particles in a gas are stationary, and have zero kinetic energy. Since kinetic energy cannot be negative it is impossible to have a temperature colder than this. The theory also explains the pressure a gas exerts on the walls of its container as being caused by the force of collisions of these particles with the container walls.

LEFT Manganese atoms jostling around illustrate the fact that the temperature and pressure of gases are the result of collisions between the atoms and molecules from which they are formed. The harder the collisions, the hotter the temperature and the higher the pressure.

The basics of kinetic theory were put forward by the Swiss mathematician Daniel Bernoulli in his 1738 book *Hydrodynamica*. Despite its scientific significance, the theory failed to catch on immediately. The German physicist Rudolf Clausius developed the kinetic theory further in the 1850s. In its simplest incarnation, as put forward by Bernoulli, molecules were treated like spheres that bounced off one another like billiard balls. Clausius realized that molecules generally look more complicated than balls – having an irregular shape formed from atoms bonded together (think of those stick-and-ball molecular models from school). This meant that as well as simple energy of motion, a molecule could possess internal kinetic energy caused by vibrations of these atomic bonds. This internal vibration takes energy, and Clausius was able to modify the equations of kinetic theory to account for it.

During the 1860s and 1870s, Scottish physicist James Clerk Maxwell and Austrian physicist Ludwig Boltzmann took kinetic theory to a new level. Working independently, they deduced that particles in a gas do not all move at the same speed – some move fast and some move slower, with the average of all their speeds defining the gas's temperature. Maxwell and Boltzmann were able to derive a mathematical function giving the proportion of particles moving within any given range of speeds. This 'distribution function' depends on the mass of the particles and the gas's average temperature.

In making these discoveries, Maxwell and Boltzmann had pioneered a new field in physics – called statistical mechanics, a kind of kinetic theory on steroids. It works by applying sophisticated mathematics from the fields of probability and statistics to the 'microphysics' of particles in a gas to then make powerful inferences about the gas's 'macrophysics' – its bulk properties.

Mistaken identity

The Maxwell-Boltzmann approach to statistical mechanics works as long as the gas particles are big enough not to be affected by the laws of quantum physics – the science of subatomic particles that emerged in the 20th century. But when the particles are small, quantum effects must be accounted for.

Quantum uncertainty makes it impossible to tell apart two or more particles in the gas. This inability to differentiate has implications in analyzing the particles using probability theory and statistics. The situation is like tossing two coins. What is the probability that you will get a head and a tail? For ordinary coins there are four outcomes: HH, HT, TH and TT. Two of the four possible outcomes give a head and a tail, so the probability is a half. But if these were 'quantum coins' – say with H and T replaced by the quantum states of subatomic particles – then HT and TH would be indistinguishable because of uncertainty. They now count as just one outcome, from a possible three. One in three translates to a probability of a third, which is quite different to the nonquantum answer of a half.

'The temperature of a gas is just a manifestation of how energetically all its particles are moving about.'

Quantum spin

Physicists soon discovered there was another complication with the Maxwell-Boltzmann theory. The behaviour of quantum particles is influenced by a property known as their 'quantum spin'. Quantum spin is analogous to spin in the everyday world – in that it is to do with a particle's rotation – but

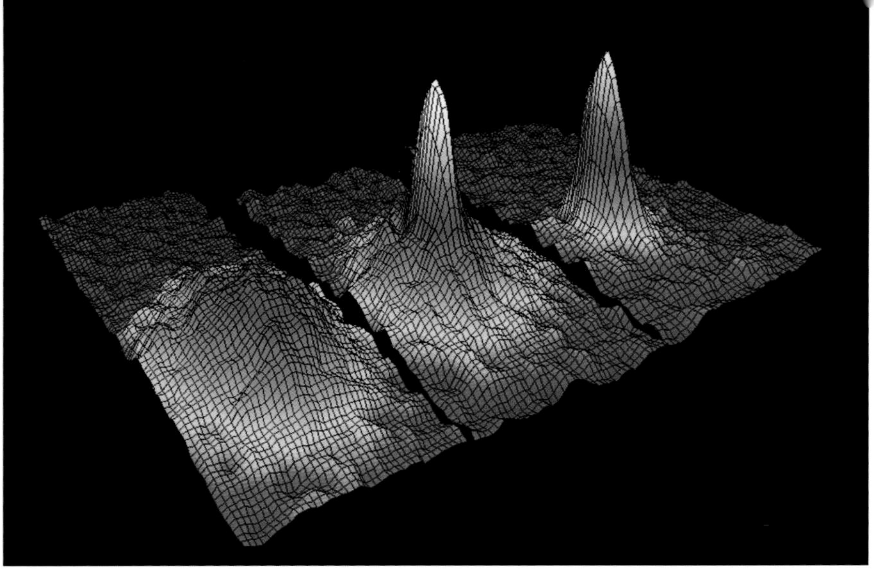

LEFT Representation of the Bose-Einstein condensate. The two peaks correspond to lower temperatures, when the atoms form a single 'superatom'. This unique formation of matter shows how certain atoms have peculiar quantum mechanical properties.

differs slightly because the spin of quantum particles is fixed, depending only on the particular particle species with which you are dealing. Electrons, for example, have spin 1/2. Photons, particles of light, have spin 1.

Fermi-Dirac and Bose-Enstein

Particles like electrons with spin that is an odd whole-number multiple of 1/2 – such as 1/2, 3/2, 5/2 – are called fermions. These obey a quantum law called the 'exclusion principle', which prevents two or more fermions coexisting in the same quantum state. This leads to an extra force that pushes fermions apart, and which is accounted for in the statistical law describing them, deduced by Enrico Fermi and Paul Dirac in 1926. Fermi-Dirac statistics are used to describe matter inside neutron stars (compressed neutron particles) and exotic stars known as white dwarfs (made of compressed electrons).

On the other hand, particles with whole-number spin – such as 0, 1, 2 or 3 – are known as 'bosons'. Boson gases obey a law known as Bose-Einstein statistics, after the scientists – Satyendra Nath Bose and Albert Einstein – who worked out its correct form. Bosons do not feel the effect of the exclusion principle, and so there is no limit to the number of particles that can stack up in the same quantum state. If a gas of bosons is cooled close to absolute zero, all the particles can collapse into the lowest-energy quantum state – the 'ground state' of the system – to form a 'Bose-Einstein condensate'.

A Bose-Einstein condensate is a macroscopic blob of matter that behaves according to quantum laws. It is a 'superatom' – many atoms joined together and behaving as one. Bose and Einstein predicted this state of matter in the 1920s, and the first one was created experimentally in 1995 at the University of Colorado. Bose-Einstein condensates remain an active area of research, and have the potential for some astonishing applications – including 'atom lasers', that can literally beam matter across space.

16 Principle of least action

DEFINITION PHYSICAL SYSTEMS FOLLOW PATHS THAT MINIMIZE THE 'ACTION' – A MEASURE OF THE TOTAL ENERGY IN THE SYSTEM

DISCOVERY FIRST PUT FORWARD IN 1744 BY THE FRENCH MATHEMATICIAN PIERRE-LOUIS MAUPERTUIS

KEY BREAKTHROUGH MAUPERTUIS WAS INSPIRED BY THE IDEA THAT LIGHT TRAVELS THE QUICKEST PATH BETWEEN TWO POINTS

IMPORTANCE THE PRINCIPLE OF LEAST ACTION IS NOW A FUNDAMENTAL TENET AT THE HEART OF THEORETICAL PHYSICS

Marbles roll down hills. And they do so via the shortest possible route. They do not roll back up again, they do not roll sideways and then downwards: a marble released from rest rolls straight down the steepest slope available. That is a statement of the principle of least action. Put simply, it says that nature is lazy. This single rule is probably the most powerful principle in mathematical dynamics.

'Nature is thrifty in all its actions.' So declared the French mathematician Pierre-Louis Maupertuis. In 1744, he put forward a principle that is absolutely central to mathematical physics today. It is called the principle of least action and, as Maupertuis had intimated, the general gist of it is that nature likes to do as little work as possible.

Maupertuis had been inspired by an earlier piece of work in the field of optics, by his fellow countryman Pierre de Fermat. Fermat's principle states that a light ray follows a path between two points that can be traversed in the least possible time. Fermat considered two pieces of glass butted up against one another. The speed of light through the two pieces of glass is different – they are said to have different 'optical densities'. So what is the quickest route a light ray can take to get from one piece of glass to the other? It is not the simple straight line that you might expect – unless the light ray is coming in exactly perpendicular to the surface of the first piece of glass.

LEFT The natural flow of water is an example of the principle of least action. The principle is used in all areas of physics, from the study of gravity to quantum theory.

To see why light does not always travel in simple straight lines, imagine a lifeguard on a beach trying to decide the quickest path to reach a swimmer in distress in the sea. As with the light rays, if a straight line drawn from the lifeguard to the swimmer is perpendicular to the shoreline then that represents the shortest route. But if the swimmer is in the water some way along from where he is keeping watch, then the lifeguard must move on a diagonal path – first across the sand towards the water and then through the water to where the swimmer is located.

The lifeguard knows that he can run along the beach faster than he can swim through the water – so the quickest route is a careful balance of running and swimming. If the lifeguard tries to move in a straight line directly to the swimmer he is going to spend too much time moving slowly through the water; if he runs along the beach until he is as close as possible to the swimmer before he jumps in the water then he will waste time travelling further than he needs to. The lifeguard's optimal path to the swimmer in distress is somewhere between the two, and requires running along the beach at a diagonal first, then making a sharp change of course towards the swimmer as soon as he reaches the water and starts swimming.

BELOW A ruler appears to bend as it enters water – an optically dense substance. Pierre de Fermat used the principle of least time to work out how paths of light rays bend as they slow down upon travelling into water or glass, resulting in refraction.

Refraction and reaction

Fermat realized that the exact same analysis may be applied to his light rays. Through a series of experiments, he found that requiring a light ray to travel along the quickest route through two pieces of glass reproduced the well-known mathematical principle governing the phenomenon of refraction. This is where rays of light really do bend as they pass between media that have different optical densities, and in which the speed of light thus varies.

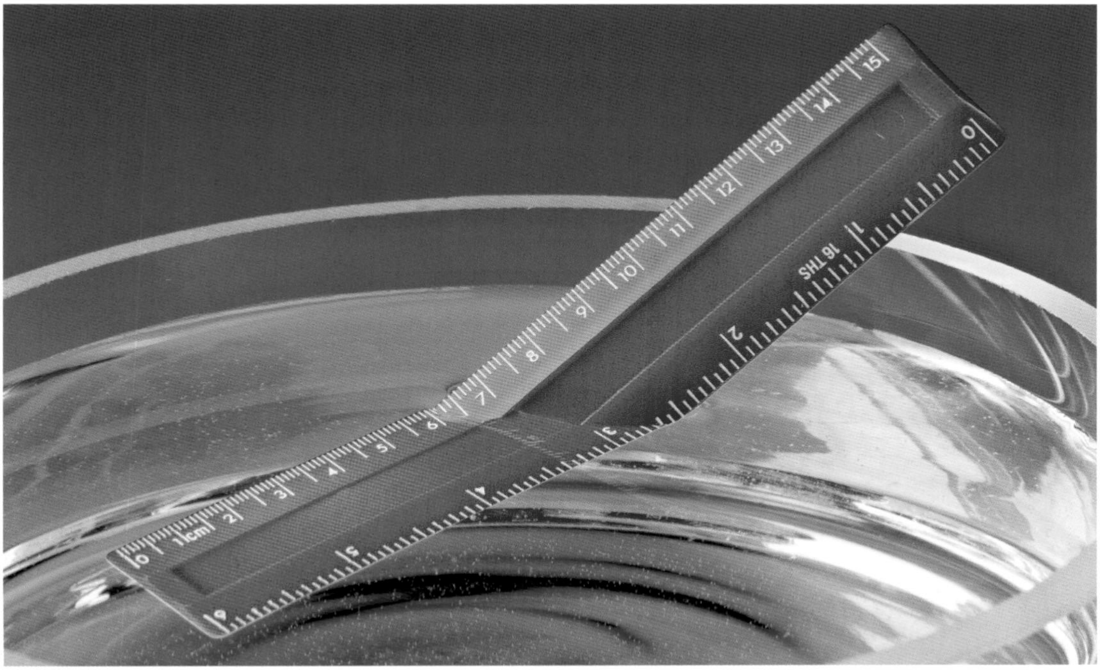

Minimum effort

Maupertuis wondered whether a similar principle could govern the behaviour not just of light beams, but also of moving objects. He formulated a principle whereby the total energy of motion of an object, added up along its trajectory, must be as small as it can be. Adding up the energy of motion along every possible trajectory and picking the one for which it is minimized then tells you which trajectory the object actually takes.

In the year 1760, a similar principle was adopted by the Italian mathematician Joseph-Louis Lagrange – and later by the Irish physicist William Rowan Hamilton – to formulate the laws of Lagrangian dynamics. This is a convenient method for studying the motion of complex physical systems that have many variables – an example being three or more stars all orbiting in each other's gravitational field.

'Maupertuis formulated a principle whereby the total energy of motion of an object, added up along its trajectory, must take the minimum value possible.'

The procedure involves calculating the Lagrangian of the system, which is given by its total energy of motion minus its 'potential energy' – the stored-up energy in the system (for example, the energy in a stretched spring). The Lagrangian is added up along all possible future trajectories of the system – to give a quantity known as the 'action'. The system follows the trajectory for which the action is minimized – hence 'the principle of least action'.

Lagrangian dynamics reproduced Newton's laws of motion. Its systematic approach to complex problems has seen it applied to all the major branches of physics including electromagnetism, general relativity and quantum theory.

Path integrals

Nobel-prize-winning physicist Richard Feynman used Lagrangian dynamics in 1948 to derive his path integral formulation of quantum theory. Quantum theory is not a deterministic science, but deals instead with the probability of finding particles at different points in space. Path integrals use the principle of least action to calculate particle probabilities by summing over all possible behaviours that the system can exhibit.

Feynman was able to show that during a quantum process – say, a particle travelling from A to B – all the indirect routes interfere with one another so as to effectively cancel out, leaving the true path that the particle takes. The path integral formulation is also useful to physicists because it shows explicitly the connection between quantum and non-quantum behaviour.

The principle of least action has become regarded as one of the most fundamental concepts in science. Scientists working at the cutting edge of theoretical physics are today using it to investigate new ideas in particle physics, astrophysics and theories of gravity.

17 Galaxies

DEFINITION DISTANT COLLECTIONS OF HUNDREDS OF BILLIONS OF STARS GROUPED TOGETHER UNDER GRAVITY

DISCOVERY THE EXISTENCE OF GALAXIES WAS FIRST PROPOSED BY THE ENGLISH ASTRONOMER THOMAS WRIGHT IN 1750

KEY BREAKTHROUGH REALIZING THE MILKY WAY IS A DISK OF STARS, WRIGHT SUGGESTED SPACE IS LITTERED WITH SIMILAR SYSTEMS

IMPORTANCE THE DISCOVERY OF GALAXIES WAS AN ESSENTIAL STEP IN UNDERSTANDING HOW THE UNIVERSE WORKS

The stars of our universe are gathered together into islands known as galaxies, separated by vast oceans of empty space. The first galaxy to be recognized as such was our own Milky Way, in 1750. Now the space beyond is known to be teeming with billions of them, and as telescopes become more powerful, so ever more of these distant galaxies keep rolling over the horizon and into view.

When Galileo turned the first astronomical telescope upwards to the heavens in 1610, one of the first things he looked at was the Milky Way – a bright band of light that can be seen carving the night sky in half on a clear evening. Before Galileo's observations, astronomers had believed the Milky Way to be simply clouds of cosmic gas. But Galileo's telescope revealed otherwise – the Milky Way was, in fact, a swarm of thousands upon thousands of stars.

The idea had been first suggested by Arab astronomers during the Middle Ages. And yet even after Galileo demonstrated its veracity, it would take another 140 years before someone grasped the implications of the discovery.

Englishman Thomas Wright, born in the year 1711, was variously an astronomer, mathematician and garden designer. In 1750, he published a book called *An Original Theory of the Universe* in which he gave his interpretation of the true nature of the Milky Way. He believed that the bright band he could see across the night sky was actually a disk of stars, within which both our Sun and solar system are embedded. Wright was absolutely correct. The dark areas of sky visible on either side of the band correspond to directions in which we are looking out of the disk. The bright band itself is what we

LEFT The Milky Way can be seen overhead on a clear summer's night. It is a disk-like galaxy of stars, within which our Sun and solar system are embedded. The brighter regions are the diffuse glow of hundreds of billions of stars, while the darker areas are a band of dust that sits in the galaxy's mid-plane.

see as we look into the disk. It looks brightest during the summer months, as we stare towards the Milky Way's centre. Dark streaks of dust can also be seen crossing the disk. But as well as fathoming the nature of our own place in the universe, Wright added an insightful corollary to his theory – supposing that there could be other, similar agglomerations of stars lying at great distances from the Milky Way. German natural philosopher Immanuel Kant read Wright's book and agreed with him. Kant called the Milky Way and these other distant star systems 'island universes' and suggested that they might be rotating, like our solar system.

Kant's view was far from universally held within the scientific community, even when astronomers began to discover mysterious faint, fuzzy patches on the night sky that defied explanation. The first were found by French astronomer Charles Messier. His speciality was discovering comets – great chunks of ice that circle the solar system, developing a hazy appearance as they draw near to the Sun and its heat boils water vapour from their surfaces. Normally, a comet is seen to shift its position from night to night as it moves along in its orbit. But between the years 1760 and 1784, Messier discovered over 100 more objects that looked like comets but which remained stock still against the night sky.

Colour-splitting spectroscopy

It was not until 100 years later, during the 1860s, that the nature of these objects became more clear – as a result of the new experimental science of spectroscopy. This involves splitting up light into its constituent colours and measuring the brightness of each individual colour. Certain chemical elements emit or absorb light at particular colours, creating peaks or dips in brightness. Detecting the same peaks and/or dips in the brightness of these colours within the light from a celestial object indicates that the object contains those same elements.

'The Sun and solar system lie about 26,000 lightyears from the galactic centre on the inner edge of the Orion spiral arm, one of several making up the Milky Way's spiral pattern.'

When astronomers applied this theory to Messier's 'nebulae' (so named after the Latin word for 'cloud' because of their fuzzy appearance) they discovered patterns of peaks and dips characteristic of the chemical elements normally found in stars. Many of these groupings also displayed spiral structure, suggesting that they are rotating – just as Kant had supposed.

Cosmic oases

The major question remaining to be answered was where these swirling nebulae are actually located. Do they lie within the bounds of the Milky Way – or are they, as Wright had speculated, far beyond? This final piece of the puzzle was slotted into place by American astronomer Edwin Hubble during his research in the 1920s. Hubble found a method of measuring the distances to these starry nebulae (which would later prove pivotal in his

discovery that the universe is expanding), and this put the nebulae firmly outside the Milky Way. Fellow American astronomer Harlow Shapley had deduced in 1918 that the Milky Way is about 100,000 lightyears in diameter. By comparison, the starry nebulae were typically millions of lightyears distant. Their similarity to the Milky Way led them to be called 'galaxies', from the Greek word *galaxias*, meaning 'milky'.

ABOVE The Hubble Space Telescope returns stunning images of galaxies from across the universe, including the Sombrero Galaxy located 30 million lightyears away in the constellation of Virgo. It measures 50,000 lightyears across and is etched with dark lanes of dust.

Home sweet home

More detailed observations of the Milky Way now reveal it to have a spiral appearance as well. The Sun and solar system lie about 26,000 lightyears away from the galactic centre on the inner edge of the Orion spiral arm, one of several arms making up the Milky Way's spiral pattern. The spiral arms orbit the galaxy once every 250 million years while the Milky Way as a whole is home to as many as 400 billion other stars.

And there are hundreds of billions more galaxies, just like the Milky Way, out there in deep space. Deep-field images of the universe relayed by the super-powerful Hubble Space Telescope show a haze of faint dots – almost as numerous as the explosion of stars within the Milky Way's bright band – and each of them is a galaxy in its own right. In October 2010, astronomers at the European Southern Observatory found the most distant galaxy ever observed – lying an incredible 13.1 billion lightyears away from Earth.

Thomas Wright probably thought he was pushing the boundaries of astronomy when he announced his theory of the Milky Way in 1750. Now it seems the cosmos is bigger and packed with more of these starry islands than even he dared consider.

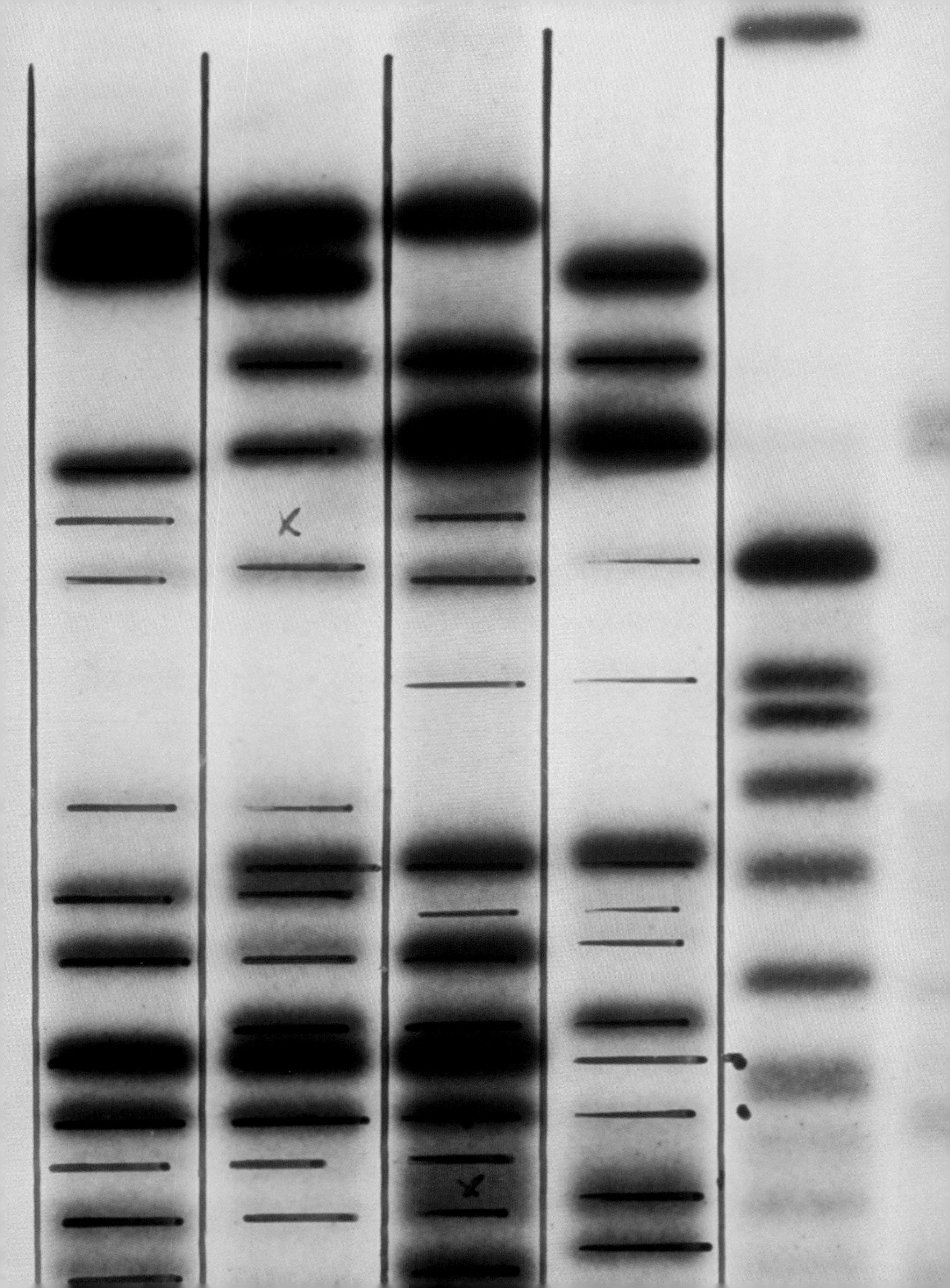

18 Bayes's theorem

DEFINITION FORMULA FOR CALCULATING PROBABILITY OF AN EVENT HAPPENING, GIVEN OTHER EVENTS THAT HAVE ALREADY HAPPENED

DISCOVERY FORMULATED BY ENGLISH CLERGYMAN THOMAS BAYES AND PUBLISHED POSTHUMOUSLY IN 1763

KEY BREAKTHROUGH BAYES WAS LED TO THE THEOREM AS A WAY TO MAKE INFERENCES FROM STATISTICAL DATA

IMPORTANCE CRUCIAL FOR INTERPRETING EVERYTHING FROM LEGAL EVIDENCE AND POLITICAL EXIT POLLS TO MEDICAL STATISTICS

Along with his religious duties, Presbyterian minister Thomas Bayes had a passion for mathematics, in particular the theory of probability. In 1763, two years after his death, a theorem he had derived for computing 'conditional probabilities' – for instance, the probability of it raining given the condition that it is already cloudy – was published on his behalf by a friend. It remains today one of the single most important ideas in probability theory.

Probability is a measure used by mathematicians to gauge the likelihood of certain events happening. It is quantified by a number ranging between 0 (meaning the event will never happen) and 1 (the event will always happen).

Calculating the probability of an event means taking the number of outcomes in which the event happens and dividing it by the total number of all possible outcomes. For example, given a bucket containing four red balls and six blue balls, the probability of randomly drawing a red ball is 4 (the number of outcomes in which the event happens) divided by 10 (total number of possible outcomes) – which is 2/5. This is written as P(red ball) = 2/5.

Bayes's theorem centres on 'conditional probability'. This is the probability of an event happening given that another event has already taken place. With the bucket of balls, what is the probability of the second ball you draw being red given that the first ball was red? In this case, there are three red balls left and nine balls left in total, so this is just 3/9 = 1/3. Mathematicians write this as P(2nd ball red|1st ball red) = 1/3. Bayes's theorem then gives a relationship between a conditional probability, such as the probability of event A given event B, P(A|B), and also the reverse situation, the probability

LEFT Audiographs provide a 'genetic fingerprint' that can be used to place criminal suspects at a crime scene. Prosecutors cite the tiny probability of two people having the same genetic fingerprint, and equate this to the probability of a person's guilt or innocence. But Bayes's theorem reveals that this approach is used in error and could lead to wrongful convictions.

of B given A, P(B|A) in terms of the individual unconditional probabilities of A and B happening – P(A) and P(B). In mathematical terms, the theorem states that $P(A|B) = P(B|A) \cdot P(A) / P(B)$, with '•' symbolizing multiplication.

It looks innocuous but it is a powerful theorem that can throw up some very surprising results. For example, let us say you are being tested for a rare illness that affects just 0.01 percent of the population. You are told that the test you are taking is 99 percent accurate. To your horror, you test positive – what are the chances that you actually have the illness? At first sight, you might think it is 99 percent – a probability of 0.99. But a closer examination with Bayes's theorem reveals very much otherwise.

What you want to know is the conditional probability of having the illness given a positive test result. Referring back to Bayes's theorem, outcome 'A' is having the illness while outcome 'B' is getting a positive test result. You want to know P(A|B). You know that the probability of B given A, getting a positive test given you have the illness, $P(B|A) = 0.99$ – because that is the accuracy of the test. And you know that the unconditional probability of having the illness in the first place is $P(A) = 0.0001$ (0.01 percent expressed as a decimal).

'Bayes's theorem tells you that your probability of being ill is 0.0098, or about one percent – very different to the 99 percent you feared.'

The unconditional probability of B (getting a positive result) is a little trickier to work out. There are two possible cases – you could either have the illness and get an accurate test (probability $0.0001 \times 0.99 = 0.000099$) or you could be fit and well but have an inaccurate test ($0.9999 \times 0.01 = 0.009999$). By simply adding these two numbers together gives the total probability of receiving a positive result, $P(B) = 0.010098$. Bayes's theorem then tells you that your actual mathematical probability of being ill is 0.0098, or about 1 percent – very different to the 99 percent you initially feared.

What's going on? The discrepancy stems from the fact that the test is not accurate enough to detect such a rare illness. If 10,000 people took this test then, on average, only one of them would actually be ill (0.01 percent). The test, however, gets it wrong one in 100 times, meaning that it gives roughly 100 people positive results. If you are one of those 100, then your conditional probability of being the one who is sick is one in 100, or 1 percent.

Statistical inference

Bayes's theorem has a raft of applications today. It is used in email spam filters – where messages containing certain words, like 'Viagra', are blocked according to the conditional probability of them being spam given that they contain those words. 'Bayesian inference' is a specialized use of Bayes's theorem, whereby a result from a small statistical sample (perhaps an exit poll on election day) is used to infer the probability that the same trend will be seen in a larger sample (such as the rest of the voting population).

LEFT What is the chance of hitting three cherries on a fruit machine, having already got two? If the initial probability of getting two cherries is $P(2) = 0.01$ (1 percent) and the initial probability of getting three is $P(3) = 0.001$ (0.1 percent), then the probability of getting a third cherry on the machine, given that you already have two, is $P(3) / P(2) = 0.1$, which is 10 percent.

Legal application

Bayes's theorem is also of crucial importance in the interpretation of statistical evidence. This could be data gathered in clinical trials of new drugs – or evidence presented in the law courts. A poignant demonstration of this took place in the late 1990s, when English lawyer Sally Clark was accused of murdering her two baby sons. Her defence claimed that they had been victims of cot death. The prosecution cited the probability of two children in the same household dying from cot death as being 1 in 73 million and then equated this figure to the probability of Clark's innocence.

Wrong diagnosis

But this was a probabilistic blunder. The prosecution had not realized that double fatalities of any sort are unlikely – and that this is the main reason why the chance of a double cot death is so small. Similarly, the chance of two children in the same home dying in a double murder is also vanishingly small. The statistic the prosecution should have used is the conditional probability of cot death being the cause, given that a double fatality had already occurred. This is much bigger – as a calculation using Bayes's theorem would have revealed. Despite a press release by the Royal Statistical Society pointing out the error, in 1999 Clark was convicted of the murders.

In 2002, a mathematician at Salford University worked out Sally Clark's true probability of innocence, finding it to be in excess of 0.8. Remember, probability ranges between 0 (impossible) and 1 (a certainty). Given that the two boys had died, the probability that their deaths were caused by cot death was much greater than the probability that she had murdered them. Her conviction was overturned in 2003. But it was too late. Losing two children to illness and then being accused of their murder had already destroyed Clark – she died of alcoholic poisoning in 2007. This case is a tragic reminder of the subtleties involved in making inferences from probability theory – subtleties that Thomas Bayes had warned us about two-and-a-half centuries before.

19 Combustion

DEFINITION A CHEMICAL REACTION IN WHICH FUEL AND OXYGEN COMBINE TO PRODUCE HEAT

DISCOVERY THE THEORY OF COMBUSTION WAS MADE POSSIBLE BY THE DISCOVERY OF OXYGEN IN 1772 BY CARL WILHELM SCHEELE

KEY BREAKTHROUGH SCHEELE OBSERVED CHEMICAL REACTIONS PRODUCING A GAS THAT MADE FLAMES BURN BRIGHTER

IMPORTANCE THE THEORY OF COMBUSTION PAVED THE WAY FOR AUTOMOTIVE ENGINES, SPACE ROCKETS AND MODERN EXPLOSIVES

Of all the chemical reactions, fire is perhaps the one that has been used by humans the longest – for heat, light, cooking and as a weapon. Yet we only came to understand it relatively recently, following the discovery of oxygen by Carl Wilhelm Scheele. Scheele enabled others to enumerate fire as a chemical process, one that underpins everything from rockets to living cells.

The first scientific theory for how things burn involved a fire element known as 'phlogiston'. The theory – which harks back to the ancient alchemical view that the world is composed of four elements called earth, air, water and fire – was put forward in 1667 by the German alchemist Johann Joachim Becher. In essence, the theory says that flammable materials are rich in phlogiston. When they burn they gradually release the phlogiston, and they continue to burn until all the phlogiston is exhausted, at which point they become 'dephlogisticated'.

Scientists began to question the theory in the 1750s, when it was noticed that some metals actually become heavier after they have been burned. If phlogiston was being depleted in the metals then they should get lighter – instead, the observations were suggesting that something was being added to the metal by the burning process.

LEFT Fireworks rely on the chemical reaction of combustion – where oxygen, fuel and heat react to produce gas, solid particles and more heat. Different colours are produced by the combination of chemicals packed into the firework – for example, barium produces green flames and lithium burns red.

That something was discovered in 1772, by a German-Swedish chemist called Carl Wilhelm Scheele. Born in 1742, he had a keen interest in science and read widely – giving himself the education that his family could not afford. At the tender age of 14, Scheele became an apprentice pharmacist, working for the firm Martin Anders Bausch in Gothenburg. His interest in science, coupled to the plentiful supply of chemicals, soon led him to start experimenting. In a number of experiments carried out in around 1772, Scheele was able to produce a gas that he referred to as 'fire air' – because it enhanced the burning of other materials. Having studied the phlogiston theory, he believed that fire air burned by attracting phlogiston. He also discovered a second gas, which he named 'foul air', and correctly deduced that the air we breathe is a mixture of these two gases. In fact, what Scheele had discovered were the gases now known as oxygen and nitrogen.

Dephlogisticated air

In 1774, English philosopher Joseph Priestley replicated Scheele's results, referring to oxygen as 'dephlogisticated air'. He found that not only do flames burn more intensely in the presence of this gas, but that a mouse in a sealed chamber would live longer if the chamber was filled with this gas rather than ordinary air.

It was the French chemist Antoine Lavoisier who would make the key connections. He showed that combustion was not due to the depletion of phlogiston, but rather the consumption of this new gas. In 1777, he named the gas oxygen – from the Greek word *oxys*, meaning 'acid' (Lavoisier mistakenly believed oxygen to be a component of all acids). His work would consign the misguided phlogiston theory to the scientific scrap heap, where it has rightly remained ever since.

'In a number of experiments carried out in around 1772, Scheele was able to produce a gas that he referred to as "fire air" – because it enhanced the burning of other materials.'

Scheele's contribution in all this was fundamental: most historians agree that he was the first to discover oxygen, even though Priestley was the first to actually publish his findings. Scheele was also the first to identify a number of other chemicals, including chlorine, but was again beaten in getting the discovery published – this time by the English chemist Humphry Davy. This led the science fiction author Isaac Asimov to dub him 'hard-luck Scheele'.

Farewell, phlogiston

Thanks to the work of Scheele, Priestley and Lavoisier, combustion is now understood as a chemical reaction involving fuel, oxygen and heat. The reaction is 'exothermic' – that is, once it is under way it produces its own heat that sustains the reaction until it either runs out of oxygen or fuel.

For example, burning the gas butane – each molecule of which is made of four carbon atoms and ten hydrogen atoms (chemical formula C_4H_{10}) – proceeds via a chemical reaction $2C_4H_{10} + 13O_2 \Rightarrow 8CO_2 + 10H_2O$. In other words, each reaction combines two butane molecules and 13 dioxygen molecules (dioxygen is the most stable form of oxygen, made by joining together two oxygen atoms) to produce eight carbon dioxide molecules and ten molecules of water vapour (steam).

This is a clean reaction that produces no smoke particles whatsoever – just gases. Other combustion reactions are not so clean. For example, burning aluminium powder (chemical symbol Al) in a hot flame is governed by the chemical reaction $4Al + 3O_2 \Rightarrow 2Al_2O_3$. The chemical compound produced is aluminium oxide, which is given off as particles of smoke.

The exothermic nature of combustion is what makes it an excellent source of power. For example, in a car's internal combustion engine a mixture of air and fuel is burned together inside a piston. The rapid rise in temperature makes the pressure shoot up, causing the piston head to move and drive a crank shaft that turns the wheels. The optimum air-fuel mixture for a petrol (gas-powered) engine, which ensures there is enough oxygen to burn all the fuel molecules, is 14.7:1. And this oxygen all comes from the atmosphere.

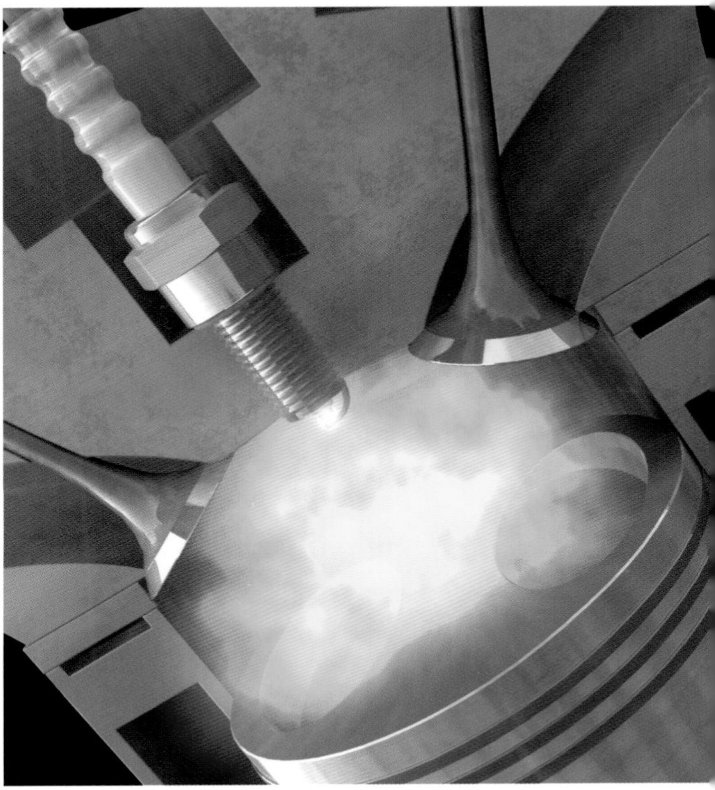

ABOVE An internal combustion engine uses the expansion of burning gases to drive pistons and generate work. The fuel and oxygen must be mixed in exactly the right proportions by the engine's carburettor, which blends the two together. For a petrol (gas-powered) engine, this corresponds to an air-fuel ratio of 14.7:1.

Bring your own oxygen

But what if there is not always oxygen available? A rocket engine, for example, has to operate in space where there is no atmosphere and no oxygen. This is why space rockets have to carry not just fuel (usually in the form of kerosene or liquid hydrogen) but also tanks of liquid oxygen with which to burn it. Similarly, gunpowder carries its own solid source of oxygen in the form of potassium nitrate, which when mixed with a fuel of charcoal and sulphur creates a powder that burns extremely rapidly. If the powder is contained within a vessel and ignited, the result is an explosion.

Another form of combustion is even responsible for the energy-generating chemical reactions that take place within our bodies. Here, oxygen combines with nutrients to generate energy in a process known as 'cellular respiration'. It is a lot slower than the burning of gunpowder, but it is still a form of combustion nonetheless.

From the exploration of space to the way we power our bodies – our understanding of combustion all derives from the work of 'hard-luck Scheele' and his ground-breaking discovery of oxygen. Little good it did him, though. Scheele's chemical experiments – often using hazardous chemicals ranging from mercury to cyanide – led him to an early grave in 1786, aged just 43.

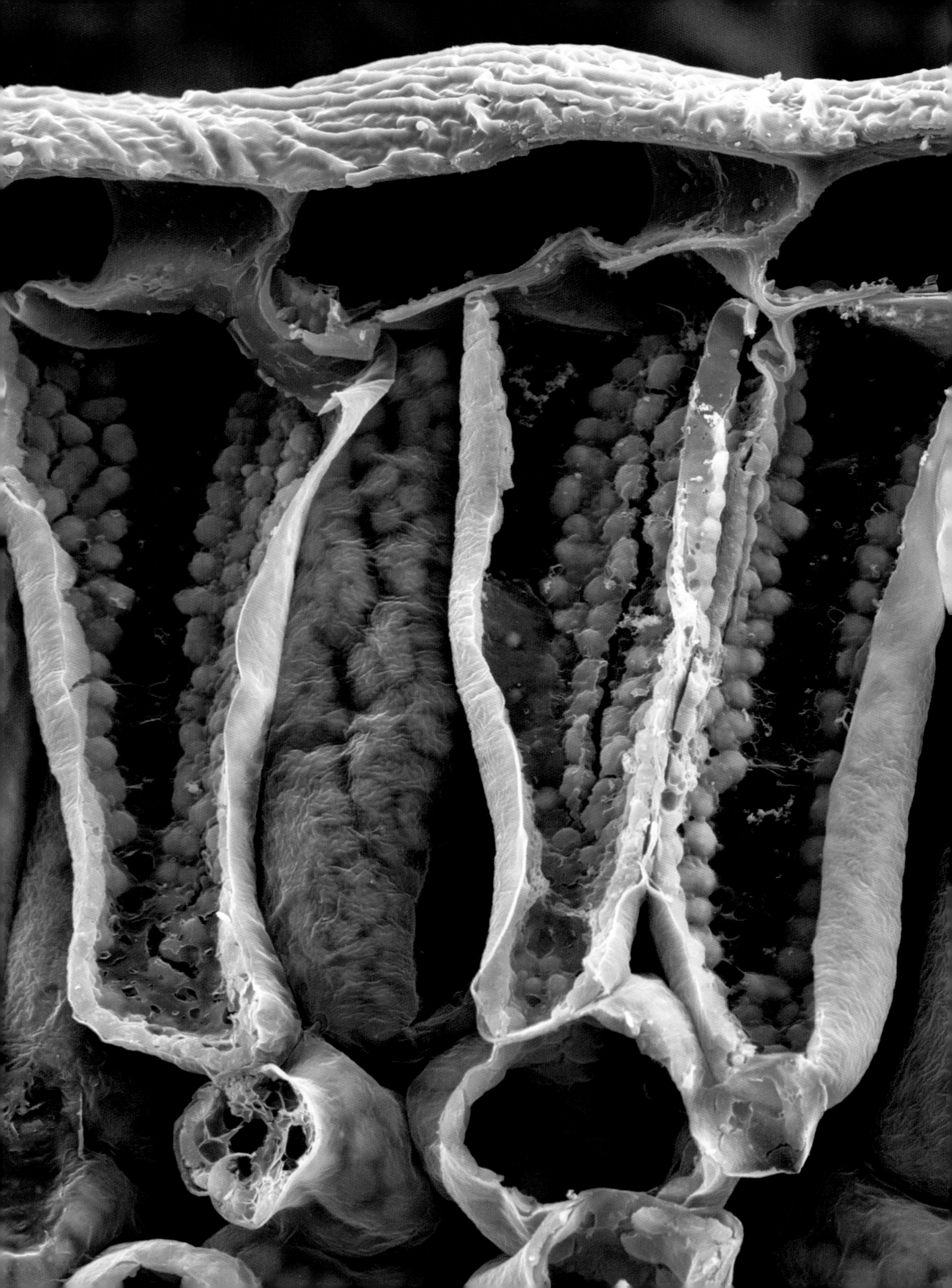

20 Photosynthesis

DEFINITION A CHEMICAL REACTION IN WHICH PLANTS CONVERT SUNLIGHT, WATER AND CARBON DIOXIDE INTO FOOD

DISCOVERY THE EXACT NATURE OF THE PROCESS WAS FIRST REALIZED BY DUTCH SCIENTIST JAN INGENHOUSZ IN 1779

KEY BREAKTHROUGH INGENHOUSZ SAW BUBBLES EMITTED BY UNDERWATER PLANTS, WHICH HE IDENTIFIED AS OXYGEN

IMPORTANCE PHOTOSYNTHESIS IS THE MOST IMPORTANT CHEMICAL PROCESS ON EARTH – SUPPORTING ALL LIFE ON THE PLANET

Next time you take a breath of fresh air, thank the plants for generating the oxygen that you have just inhaled. Plants extract carbon dioxide from the atmosphere and, in return, give out oxygen in a process known as photosynthesis. It is probably the most important chemical reaction on Earth – creating plant matter for the food chain as well as oxygen to sustain all animal life. Photosynthesis was discovered by a Dutch scientist in 1779.

Photosynthesis is the basis for all life on our planet. It takes place in green plants, algae and some bacteria species and it gives these organisms the food (in the form of sugars) that they need in order to survive. In turn, plants and algae form the first link in the food chain – without this there would be nothing for higher forms of life to eat.

But photosynthesis has another role that is even more important. A by-product from the manufacture of plant food is the oxygen that we need to breathe. It is sometimes said that plants are the lungs of the planet – and with good reason. For if there was no plant life here on Earth then there could be no animals either – quite simply because they would not be able to breathe.

LEFT Plants make their own energy by combining carbon dioxide, water and sunlight through photosynthesis. Chloroplasts are the bright green cells within which the vital work is carried out, as shown on this section of a Christmas rose leaf (*Helleborus niger*), magnified 750 times.

Oxygen is an essential requirement for animal life, combining inside cells with nutrients from food to generate energy – and doing so much faster than energy can be produced in plants. This is why animals are able to move around, hunt prey and perform other activities, while plants – with their lower energy reserves – are more or less stationary. No surprise, then, that the first life forms to arise on Earth generated their energy by photosynthesis. These were primitive bacteria that existed around two billion years ago, later joined by photosynthetic algae. The oxygen they produced is what allowed animal life on Earth to get going – first emerging as single-celled life forms called 'flagella', which had a tail enabling them to propel themselves around.

Up until this time, Earth's atmosphere was choked with carbon dioxide (CO_2) that had been spewed out by the many active volcanoes peppering the planet's surface. Photosynthesis converted this carbon dioxide smog to give the planet its current atmospheric composition of 21 percent oxygen and 78 percent nitrogen (and just a trace of carbon dioxide).

Busy chloroplasts

In green plants today, photosynthesis takes place in specialized cells known as 'chloroplasts'. Green pigment in the plant – called chlorophyll – harnesses the energy in light from the Sun to combine carbon dioxide from the atmosphere with water drawn up through the plant's roots. This process creates carbohydrate (sugars upon which the plant feeds) plus oxygen – which is released back to the atmosphere through pores on the surface of the plant's leaves, known as 'stomata'.

'The first life forms to arise on Earth generated their energy by photosynthesis. These were primitive bacteria that existed around two billion years ago.'

The process divides into two separate chemical reactions. The first, known as the 'light-dependent reaction', happens during the plant's exposure to sunlight. It uses the energy of the light to split water molecules into hydrogen plus oxygen and free electrons (negatively charged particles that orbit into the outer layers of atoms). There then follows a series of 'light-independent reactions' in which these reaction products are combined with carbon dioxide to form oxygen and carbohydrate.

Oxygen source

Photosynthesis was initially discovered by scientists during the late 18th century. The first hint of its existence came during the discovery of the gas oxygen. English chemist Joseph Priestley made the first published discovery of oxygen, even though, in reality, he was just beaten by German-Swede Carl Wilhelm Scheele. In one of his experiments, Priestley placed a lit candle and a plant together in a sealed glass enclosure. When the candle had burnt all the oxygen inside the enclosure, it went out. But some time later, Priestley was able to relight the candle (using magnified sunlight) – the implication being that the air in the enclosure had had its capacity to support the combustion of the candle wick restored.

In 1779, Dutch biologist Jan Ingenhousz explained what was actually going on in this experiment. He put plants into transparent containers underwater. When he did this, and allowed sunlight to fall on them, he noticed bubbles forming on the undersides of their leaves. He collected the gas that was being produced and showed that when it was introduced to a flame, the flame burned brighter – the gas being produced was oxygen.

LEFT When Ingenhousz submerged plants underwater and let light fall onto them, he observed bubbles of gas forming under the leaves (shown here on *Elodea* pondweed). The bubbles are oxygen, which is the gas expelled during photosynthesis. Carbon dioxide is absorbed during the same process.

Carbon consumption

In 1796, Swiss botanist Jean Senebier showed that while plants are producing oxygen they also consume carbon dioxide. This is why planting trees is currently being pursued as a method to remove carbon dioxide from the atmosphere and thus combat the negative impact of the greenhouse effect.

Other researchers showed soon after that the difference in mass between the carbon dioxide taken in by plants and the oxygen emitted is not enough to explain the observed rate of growth in plants. It was then that scientists understood the role played by water in the growing process, which is taken in by plants via their roots from the surrounding soil.

The precise nature of photosynthesis was not deduced until much later. In the 1940s, American researchers Samuel Ruben and Martin Kamen used a radioactive isotope to track oxygen movement through plants – showing that the oxygen released comes from water absorbed by the roots. In the 1950s, a team led by American biochemist Melvin Calvin used a radioactive carbon isotope to trace the movement of the carbon in plants. His work revealed the chemical process by which carbon gets extracted from the input gases and incorporated into the plant: in 1961, he won the Nobel Prize for Chemistry. Calvin had slotted into place the final piece of the puzzle needed to explain the greatest chemical reaction on Earth – completing the pioneering research started by Ingenhousz nearly 200 years earlier.

21 Deep time

DEFINITION THE IDEA THAT EARTH'S GEOLOGICAL HISTORY SPANS NOT THOUSANDS BUT MANY MILLIONS OF YEARS

DISCOVERY FIRST HYPOTHESIZED BY THE SCOTTISH SCIENTIST JAMES HUTTON IN 1785

KEY BREAKTHROUGH HUTTON FOUND THAT ROCK STRUCTURES CALLED 'UNCONFORMITIES' REQUIRED AN OLD EARTH TO FORM

IMPORTANCE HUTTON'S IDEAS FORMED THE BASIS FOR THE MODERN VIEW OF OUR PLANET'S GEOLOGY

'We find no vestige of a beginning, no prospect of an end,' wrote geologist James Hutton in one of his key scientific papers on the geology of the Earth. The meaning was simple – the planet has been here for a very long time indeed, and may remain for even longer. It was a radical, yet correct, theory that overturned existing ideas that the Earth is just a few thousand years old. These findings established Hutton as the father of the science of geology.

Up until the end of the 18th century, the Earth was generally believed to be no more than a few thousand years old. That was the view of the planet's provenance put forward by the Judeo-Christian faith traditions.

But in 1785, that established school of thought was shaken to the core by the work of a Scottish geologist called James Hutton. Born in 1726, Hutton initially studied classics at the University of Edinburgh and began an apprenticeship as a lawyer, before realizing that his true interests lay in science. He went back to university to study medicine, gaining his doctorate from Leiden University in 1749.

In the 1750s, Hutton and a friend developed a way to produce ammonia salts from soot. This led to a profitable business. That, along with the money he earned renting out properties in Edinburgh, earned him a comfortable income – allowing him to focus more time on his scientific interests.

LEFT In this example of a rock unconformity from Kilkenny Bay, England, the upper rock formation was created by layers of particles compressed over time. Beneath lies igneous rock, tilted by geological forces and left at an angle.

Shortly after, Hutton inherited his father's farm near the Scottish borders and moved there. He took a scientific approach to agriculture. Working with the land is thought to be what first ignited his interest in geology, and he

ABOVE Granite outcrops, such as these at La Digue in the Seychelles, are examples of volcanic rock formed out of molten magma, then eroded by the action of ocean waves.

began to formulate various ideas about rock formations and how they were created. But geology was not a pursuit he rushed into with any urgency. Hutton spent around 25 years studying and refining his ideas about Earth's geological history before going public with any of his findings.

Hutton's central thesis was an idea that became known as 'uniformitarianism' (the term itself was not coined until later, in 1830, by the scientist William Whewell). It essentially states that the same processes that have shaped the planet in the geological past are still shaping it today. This idea is in contrast to the idea of 'catastrophism' that was popular at the time, and which held that rather than slow changes, Earth was being battered by sudden events.

Rock of ages

Hutton's theory of slow geological change, coupled to the rich geological structures that he examined, led him to the idea of geologic time, or 'deep time'. He argued that in order for these structures to form, the Earth must be very much older than the few thousand years that the text of the Bible would have us believe.

This deep time view was backed up by geological sites he found called 'unconformities'. Outcrops and cliff faces often showed layers of sedimentary rock, which Hutton believed (correctly) to have been formed under the sea by the gradual deposition of sediments that were later compacted into rock. The sediments had come from the land and were washed into the sea by rivers. Hutton's unconformities appeared to be made of two sections: a lower section of layered sedimentary rock in which the layers form stripes that are almost vertical, overlain by an upper section of younger sedimentary rock in which the layering is horizontal.

Vertical ripples

Hutton saw these formations as supporting evidence for his notion of gradual geological change taking place over not thousands but millions of years. Ripple marks in the older, vertical layers of rock confirmed that they had been lain down as ocean sediment and then turned into rock over many millions of years. Gradual movement of the land had then tilted this rock bed to make the layers vertical before further sediments were laid down, to form the newer, horizontal rock layers at the top. The process must have taken, quite literally, ages.

'Ripple marks in the older, vertical layers of rock confirmed that they had been lain down as ocean sediment and then turned into rock over many millions of years.'

Hutton also advocated an idea known as 'plutonism' – that much of the rock making up the Earth's surface originated as molten magma from deep underground. He was led to this conclusion by studies of granite formations that made it clear the granite had been molten at the time of formation. This was in contrast to another theory in existence at the time called 'neptunism', which held that all the land on Earth crystallized out of the seas.

Hutton envisaged plutonism as driving the formation of new rock beds at sea, which then grow by sedimentation. The new rock beds can undergo further undulations that may push them up to form new continents – and even tilt them up on end, as was seen at the sites of unconformities.

Modern geology

Hutton's view is more or less the one adopted by geologists today. New rock is known to spread out from underwater oceanic ridges – where subsurface magma bubbles up and solidifies to form new crust. Similarly, at other sites – called oceanic trenches – old crust is getting pulled down beneath the planet's surface in a process known as subduction.

Hutton presented his findings first as two lectures to the Royal Society of Edinburgh, in 1785. He later published a scientific paper, 'A Theory of the Earth; or an Investigation of the Laws Observable in the Composition, Dissolution, and Restoration of Land upon the Globe', in 1788. In 1794, he expanded on the theory in a book, *An Investigation of the Principles of Knowledge and of the Progress of Reason, from Sense to Science and Philosophy*, which, with over 2000 pages of dense prose, was largely regarded as utterly unreadable.

It was Scottish geologist Charles Lyell's 1830 book *Principles of Geology* that ultimately popularized Hutton's ideas, bringing them the widespread acceptance they deserved. Hutton's notion of deep time was ultimately vindicated in the early 20th century when estimates for the age of the Earth based on the decay of radioactive rocks yielded figures in excess of a billion years. The final age of the Earth, obtained in the 1950s, revealed that our planet has been circling the Sun for approximately 4.54 billion years. Geologic time, it seemed, ran deeper than even Hutton had suspected.

SLEEVES

22 Conservation laws

DEFINITION PRINCIPLES IN PHYSICS DICTATING THAT QUANTITIES, INCLUDING ENERGY AND MASS, DO NOT CHANGE WITH TIME

DISCOVERY FRENCH CHEMIST ANTOINE LAVOISIER FIRST PROPOSED CONSERVATION OF MASS IN 1789

KEY BREAKTHROUGH LAVOISIER WAS ABLE TO MEASURE THE TOTAL MASS OF CHEMICALS BEFORE AND AFTER A CHEMICAL REACTION

IMPORTANCE CONSERVATION LAWS PROVED ESSENTIAL IN THE DEVELOPMENT OF MODERN PARTICLE PHYSICS

Energy can neither be created nor destroyed – the principle of physics known as energy conservation. It explains various phenomena, from the operation of steam engines to the behaviour of balls rolling downhill. Energy is just one of a number of conserved quantities in physics that include mass, momentum and electric charge. Conservation laws are now a fundamental guiding principle in the formulation of new theories in physics.

The idea that it is impossible to create something from nothing was first noted by the Ancient Greeks in the fifth century BC. Yet it would take another two millennia for the idea to be framed in exact scientific terms. That happened in 1789, when the French chemist Antoine Lavoisier put forward the principle of the conservation of mass. He had found that when a chemical reaction occurs between two or more substances then the total mass that goes in adds up to the total mass that comes out – mass is conserved.

That might sound like an obvious statement, but it was one that was surprisingly difficult to prove because many chemical reactions involve the release of gases. Unless these gases could be captured and weighed, then it was impossible to determine accurately the total mass of material after the reaction. For example, when a lump of coal is burned a good deal of matter is lost in the form of gases such as carbon dioxide, as well as smoke particles.

With the rise of kinetic theory and thermodynamics, interest was growing in the idea that heat could be used to do mechanical work (that is, to move objects around). For example, in the 1770s, Scottish inventor James Watt had built his first steam engines: devices that generated mechanical work

LEFT In the collision of a racquet and ball, chemical energy is liberated in the muscles of the player and transformed into the racquet motion. This kinetic energy is transferred to the ball, causing it to change direction and shape, and to increase in speed. It also stretches the racquet strings.

(for example, turning a wheel) by heating water. This led English scientist Thomas Young to propose the concept of 'energy' – a quantity that could exist in the form of heat and which has the potential to carry out work, and in so doing get converted into energy of motion, thus explaining the physics behind Watt's engines.

Energy conservation

In 1837, German chemist Karl Friedrich Mohr was the first to suggest that, like mass, energy might be conserved. This was made explicit for the science of heat engines when, in 1850, Rudolf Clausius proposed the 'first law of thermodynamics'. It stated mathematically that the energy put into a system was equal to the increase in heat energy plus the work that the system then does.

'When a chemical reaction occurs between two or more substances then the total mass that goes in adds up to the total mass that comes out – mass is conserved.'

Heat is just one form of energy. Other energies exist, including potential energy. This is the energy, for example, stored in a stretched spring or the energy gained from climbing to the top of a hill against gravity. Of course, you have not really gained energy by climbing the hill – you have merely converted it from one form (in this case, chemical energy stored in your muscles) into work (moving from the bottom of the hill to the top), which manifests itself as potential energy once you are at the top. If you rolled down to the bottom of the hill, this potential energy would be converted into energy of motion – or kinetic energy. These are all examples of energy conservation.

Imperfect world

In a perfect world, the kinetic energy you have after rolling to the bottom of the hill equals your potential energy when stood still at the top, which in turn is exactly equal to the chemical energy that you burned in your muscles by climbing. But in actual fact, the world is imperfect and it is impossible to convert stored energy into work with 100 percent efficiency. This does not mean that the conservation of energy is violated – simply that energy is converted into other forms as well. So as you climb the hill, in addition to turning chemical energy into work, some of it is converted into heat, which is why you get hot from the effort. Similarly, as you roll down, not all of the potential energy is turned into motion – some is lost as heat and sound.

Soon other quantities in physics were found that also obeyed conservation laws. The momentum of a moving body is defined by its mass multiplied by its speed. When two (or more) objects, such as billiard balls, collide then the total momentum before the collision is equal to the total momentum after. So if a moving ball, A, collides with a stationary ball, B, and is brought to rest by the collision then ball B must end up moving with the same momentum with which ball A began. This effect is also illustrated in the desktop toy known as Newton's cradle, where a row of ball bearings suspended by threads transfer all their momentum from one ball to the next.

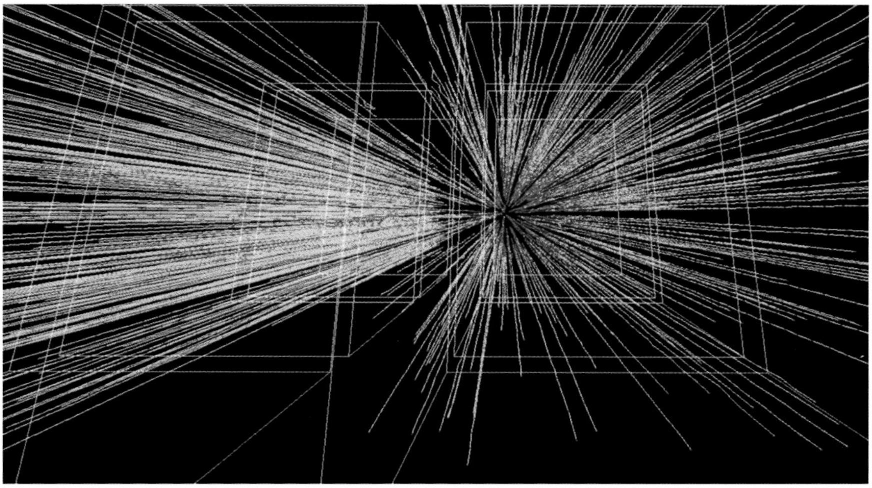

LEFT These computer-generated subatomic particle tracks give a greater understanding of the early universe. They demonstrate the 'colour charge' of the strong nuclear force – a conserved quantity carried between quark particles by smaller particles called gluons. The laws of conservation of energy apply to a process such as this in quantum physics.

This is known as the conservation of linear momentum – 'linear' because it is a property of linear, or straight-line, motion. Rotating bodies also have what is called angular momentum, and this is conserved too. Angular momentum is defined not just by the mass and the rate of spin but also by the direction in which the spin axis points. Conservation of this quantity explains the apparently bizarre behaviour of a spinning gyroscope – a heavy disk mounted on a spindle which, when the disk is rotating fast enough, can balance on end at seemingly impossible angles. It does not topple over because conservation of angular momentum holds its rotation axis in place.

Noether's theorem

In the 20th century, German mathematician Emmy Noether uncovered connections between conservation laws and symmetries in physics. A symmetry is just a transformation that leaves a particular law of physics unchanged. For example, all physical laws are invariant under time transformations – if you conduct a physics experiment today then the results will be the same if you repeat it next week, provided that you replicate the conditions. Noether found that the conservation of energy is a consequence of this time symmetry. Similarly, conservation of momentum arises because physics is also symmetric under transformations through space.

Her discovery – known as Noether's theorem – proved invaluable in developing subatomic particle physics. It is easy to apply intuition to everyday phenomena, such as billiard balls. But with the particle world, physicists had to fly entirely by their instruments – in this case, mathematics. Noether's theorem revealed quantities that are conserved during particle physics interactions, enabling scientists to unpick the physics of the particle world. These quantities included electric charge, but also more esoteric kinds of charge that govern particle processes inside the cores of atoms.

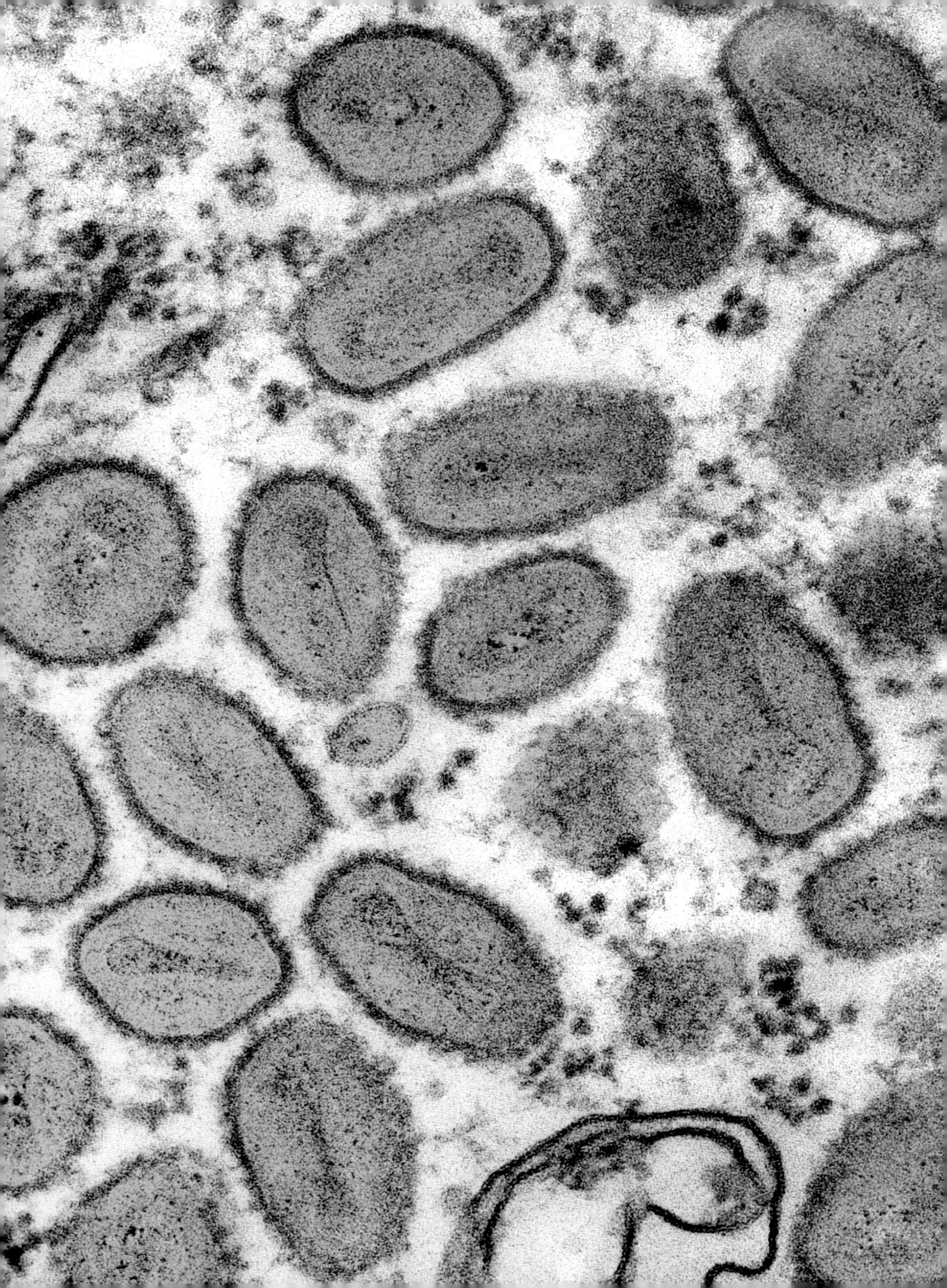

23 Vaccination

DEFINITION A WAY TO BOOST IMMUNITY TO A DISEASE BY INJECTING A WEAKENED FORM OF THE VIRUS OR BACTERIA THAT CAUSES IT

DISCOVERY VACCINATION WAS FIRST DEMONSTRATED SCIENTIFICALLY BY ENGLISH PHYSICIAN EDWARD JENNER IN 1796

KEY BREAKTHROUGH JENNER CARRIED OUT AN INVESTIGATION INTO CLAIMS THAT THE COWPOX VIRUS OFFERS IMMUNITY TO SMALLPOX

IMPORTANCE VACCINES HAVE PROBABLY SAVED MORE LIVES THAN ANY OTHER MEDICAL DISCOVERY

If you had to name the single breakthrough in medical science that has saved the most human lives, then vaccination would almost certainly be it. The field was pioneered in 1796 by English physician Edward Jenner, who developed a vaccine for smallpox. Since then, vaccination has been developed to prevent diseases ranging from rabies to hepatitis to swine flu.

Vaccination works by introducing a mild dose of pathogen – the agent that causes a particular disease or illness – into a patient's body. The idea is that if the patient is subsequently exposed to a full dose of the pathogen in the future then the vaccination will have trained their natural immune system to attack and destroy it.

Vaccination was pioneered by English physician Edward Jenner in the late 18th century. Jenner was born in 1749, in the English county of Gloucestershire. He trained in medicine during his youth, becoming a general practitioner in 1773. At the time, nationwide outbreaks of smallpox were common. It was a deadly disease: untreated, one-third of the people who contracted it would die. And even those who survived were often left horribly scarred by the blisters that would cover their skin.

There was a preventative treatment for smallpox already in existence but it was far from a guarantee of immunity and, if you were really unlucky, it could actually kill you. Named 'variolation', the treatment was introduced to Britain by Lady Mary Wortley Montagu, who had observed the practice in Constantinople, in the Ottoman Empire (modern-day Turkey). Smallpox is caused by two different viruses: *Variola major* and *Variola minor*. As the name

LEFT The vaccinia virus, particles of which are shown here in a transmission electron micrograph (TEM), causes the cattle disease cowpox. Vaccinia prompted Jenner's search for a protective vaccine against the more deadly human disease, smallpox.

suggests, *Variola minor* is the less virulent of the two forms. Surviving infection from this weaker type of smallpox was found to fortify the immune system against both varieties.

The idea behind variolation, then, was to introduce *Variola minor* through a single infection point – usually a scratch in the skin. Using just a small quantity of the weaker virus usually meant that the internal immune system could fight off the infection before it spread, resulting in just a few blisters around the infection site – after which the patient would start to make a full recovery. The effect of variolation was dramatic, reducing the mortality rate of any later *Variola major* infection from 30 percent to less than two percent.

'Jenner decided to investigate a popular belief – yet one that was unproven scientifically – that people who had contracted cowpox subsequently became immune to the virus causing smallpox.'

On the face of it, this was impressive. However, roughly four percent of people would die from the variolation procedure itself. Despite its benefits, Jenner disliked the fact that healthy people were having to be infected with a potentially lethal virus. He believed he could do better.

Human trials

Jenner decided to investigate a popular belief – yet one that was unproven scientifically – that individuals who had contracted cowpox subsequently became immune to the virus causing smallpox. Cowpox, which was a disease common among rural dairy workers, produced red blisters on the skin but was not life-threatening.

In 1796, Jenner conducted a controlled test on James Phipps, the eight-year-old son of his gardener. He innoculated Phipps with the cowpox virus in the form of pus extracted from blisters on the hand of an infected milkmaid. Once the boy had recovered from the resulting bout of cowpox, Jenner then infected him with *Variola minor*, as he would in a standard variolation procedure. The boy showed no sign of developing the smallpox disease – not even any blisters around the infection site. Jenner repeated the process on a total of 23 people and found the same result in all of them. Jenner had found a safe method of guarding against the deadly smallpox virus. Because cowpox is spread by the vaccinia virus, Jenner's new treatment then became known as 'vaccination'.

New vaccines

It is unlikely that Jenner was the first person to carry out the procedure. Dorset farmer Benjamin Jesty is believed to have immunized his own family in this way during a smallpox epidemic in the 1770s – and probably saved their lives. However, Jenner was the first to conduct a careful scientific study proving that the vaccination procedure works – and to then publish his results. In 1840, based on his findings, the English civic authorities banned variolation and replaced the procedure with Jenner's vaccinations using the cowpox vaccinia virus.

During the late 19th century, the French microbiologist Louis Pasteur built upon Jenner's work. While investigating chicken cholera, Pasteur infected some chickens in his lab with samples of the cholera bacteria that had spoiled with age. Because they had been given weakened bacteria the birds survived and, as with Jenner's cowpox vaccinations, they gained immunity to the full-blown illness. But whereas Jenner had used a naturally occurring virus as his vaccine, Pasteur had effectively created his own supply – by removing the harmful elements from a pathogen (even if, in this case, it was inadvertent).

This line of reasoning led to a method of immunizing cattle against the disease anthrax – using chemicals to neutralize the lethal components in the bacteria and then vaccinating animals with these weakened pathogens. In the 1880s, Pasteur used a similar technique to develop a rabies vaccine.

Infant immunization

Pasteur's method allowed vaccines to be developed for a range of diseases, including polio, hepatitis, diphtheria, typhoid and many others. But the benefits have not always been clear-cut in the public eye. In 1998, controversy erupted in Britain with claims from within the medical community of a link between autism and the MMR vaccine – a combined immunization against measles, mumps and rubella that was universally administered to children.

However, follow-up studies have failed to confirm such a link, and in Japan – where MMR is no longer administered – autism rates continue to climb. Meanwhile in the UK, the controversy has deterred some parents from using the MMR vaccine, and the mortality rate from measles has begun to rise.

Today, with the ever-present hazard of deadly influenza viruses spreading from animals to humans (notably bird flu and swine flu), vaccines continue to be of paramount importance in global health care. Efforts also continue to pin down vaccines for serious conditions such as AIDS and cancer and, if historical precedents are anything to go by, it may only be a matter of time before these are found and trialled successfully. In 1979, a little under 200 years after Jenner's pioneering work, smallpox was finally eradicated from the natural world.

BELOW Vaccines today are administered with hypodermic needles, but in Jenner's day, a double-tipped bifurcated needle was used. Drops of vaccine were held in the tip of the needle, which was then jabbed into the patient's skin.

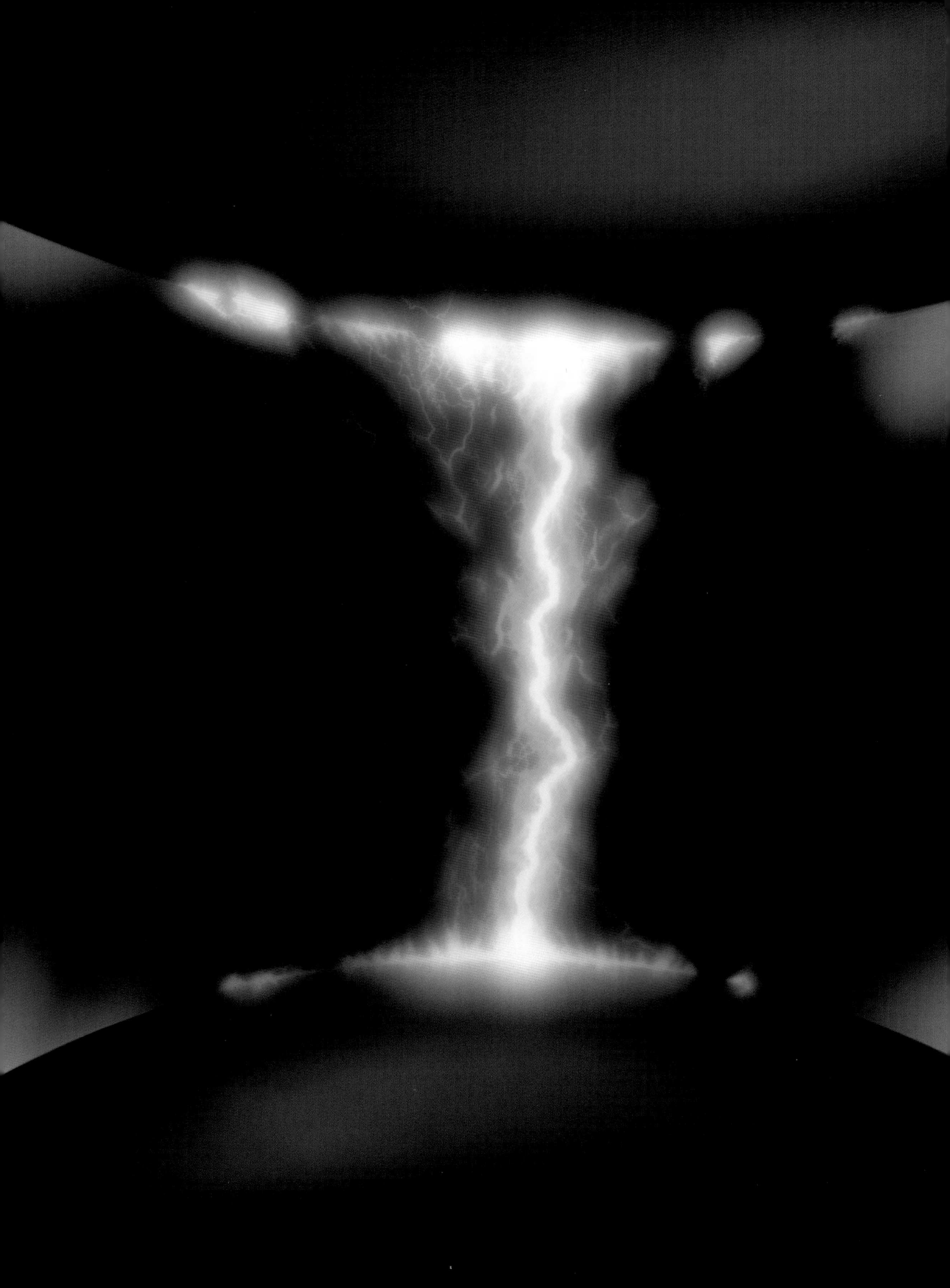

24 Electricity

DEFINITION ELECTRIC CURRENT IS A FLOW OF ELECTRICALLY CHARGED PARTICLES – USUALLY NEGATIVELY CHARGED ELECTRONS

DISCOVERY THE FIRST SOURCE OF STEADY ELECTRIC CURRENT WAS THE BATTERY, INVENTED BY ALESSANDRO VOLTA IN 1800

KEY BREAKTHROUGH VOLTA COPIED THE SPECIALIZED ORGANS IN THE TORPEDO FISH THAT GENERATE ELECTRICITY TO STUN PREY

IMPORTANCE IT WAS THE FIRST STEP TOWARDS ESTABLISHING ELECTRICITY AS THE POWER SOURCE OF OUR CIVILIZATION

Electricity was first produced in a form that could be harnessed in 1800 when Italian physicist Alessandro Volta built the world's first battery, by stacking up zinc and copper plates within acid. His discovery paved the way for electric light, television, communications technology, and all the other electrical devices upon which we now rely.

Electricity, or more specifically electric current, is a flow of electric charge – normally in the form of negatively charged particles called electrons. Just as scientists figured out how to harness flowing water to perform useful tasks, so in the 18th and 19th centuries they gained the understanding needed to generate and harness electricity.

Much scientific research in the 18th century involved electrostatics – the science of stationary electric charge. In the 1740s, physicists developed the 'Leyden jar' (so named because it was invented at the University of Leyden) – a device that could store a static charge and create an impressive spark when it was released. But the turning point in the development of electricity as a usable technology came when Alessandro Volta built the first device that could supply a steady electrical current – the battery.

Volta was born in the city of Como in 1745. In 1774, he became professor of physics at the city's Royal School, where he made a name for himself through work on the 'electrophorus', a device used for generating static electricity. In 1779, he moved to the nearby University of Pavia, and it was here that he began to pursue his real interest – finding a way to generate a continuous electric current.

LEFT These two electrically charged metal spheres demonstrate the action of electrostatic discharge (ESD), creating a sudden and momentary spark. In the real world, unregulated electrical fields caused by ESD can damage electronic equipment.

Volta was inspired by the work of his countryman Luigi Galvani who had investigated electricity within the context of biology. Galvani realized that electricity is what activates the muscles in animals. In one experiment, while dissecting a frog, his assistant touched an exposed nerve with a metal scalpel that had picked up an electrostatic charge – causing the frog's leg to kick.

Stacks of electric charge

Moreover, Volta had read about the torpedo fish, a kind of ray that generates electric currents in order to stun prey. Dissection of these fish showed that the current is generated by two organs on either side of the head. Each is made up of columns consisting of millions of gelatinous plates stacked up.

Volta sought to mimic the design of this stack of plates. But from what would he make his plates? In 1800, he carried out a series of experiments to find the answer. Volta dipped two plates made of different metals into a glass of salt water, finding that electric current would indeed flow through a conductor placed across them. He repeated the experiment many times, finding that the metals producing the greatest current were zinc and silver. However, zinc and copper proved almost as effective and a lot more affordable. He also tried different liquids – known as the 'electrolyte' solution – finding dilute sulphuric acid to be much more effective than salt water.

'Volta dipped two plates made of different metals into a glass of salt water, finding that electric current would indeed flow through a conductor placed across them.'

The device was working because of an electrochemical reaction that was taking place between the two metals and the electrolyte. Sulphuric acid is a chemical compound formed from ions – electrically charged atoms and molecules. It is made when two positively charged hydrogen ions and a negatively charged sulphate ion stick together – because of their opposite electric charge (opposite charges attract).

In Volta's primitive battery, the positively charged hydrogen in the acid was grabbing negatively charged electrons from the zinc, leaving the zinc with an overall positive charge. Meanwhile, the copper was gaining electrons from the negatively charged sulphate ions, leaving it negatively charged. Connecting the two metals up to one another with a conductor then allowed electrons from the negatively charged copper to flow towards the positively charged zinc – the chemical reaction had given rise to an electric current.

Voltaic piles

Like the jelly plates in the torpedo fish, Volta then stacked up his metal plates in an alternating sequence, with pairs of plates separated by a layer of leather that had been soaked in sulphuric acid. The device, which was called a 'Voltaic pile' because it was literally a pile of plates and electrolyte layers, generated an impressive amount of electricity – or rather, electromotive force (emf), the ability of a battery to drive a current through a circuit. (In contrast, electric current was defined as the rate of flow of electric charge through a conductor that the emf produced.) Volta's pile produced an emf of 50 volts, as measured in the units that now bear his name.

In 1801, Emperor Napoleon summoned Volta to give a demonstration of the device and was so impressed that in honour of the discovery he enobled Volta with the title of count. This social elevation was more or less the end of Volta's productive life as a scientist.

ABOVE Half-cells can create electricity. The copper half-cell (right), comprised of a copper rod placed in a salt-copper-acid solution, is linked by a salt 'bridge' to a manganese half-cell (left) that is created in a similar way. Reactions inside the cells create electrons and, since the copper cell makes more electrons, a current flows when the rods are linked by a wire.

Ohm's law

After Volta, other researchers took up the baton. In 1827, the German physicist Georg Ohm figured out his famous law linking the voltage supplied by a battery to the current it set up in a conductor, via a quantity called resistance – a property of materials specifying how easy it is for a current to flow through them. Later that century, American inventor Thomas Edison and others would use this discovery to calculate how much a wire heats up as a current gets forced through it – causing the wire to emit light. It enabled them to build the first filament light bulbs. Meanwhile, electric telegraphy was also beginning to take off around the world.

Ohm's law was the beginning of the scientific study of the flow of electricity. This was developed further by the German physicist Gustav Kirchhoff who, in 1845, put forward two laws for analyzing the flow of current through the wires making up an electrical circuit.

Magnetic motors

But even more interesting developments were afoot. Physicists were noticing that electric currents tended to create magnetic fields, and vice versa. English scientist Michael Faraday used this to build electric motors and generators, devices that used moving coils and magnetic fields to turn physical movement into electric current.

Faraday's innovative work was to have deep ramifications for fundamental physics – pointing the way to a unified theory that pulled together electricity and magnetism as different facets of the same underlying entity: electromagnetism. For everyone else, it marked the beginning of the mass distribution of centrally generated electricity through the networks upon which we all now rely to run our lights, computers, TVs, and to make us hot drinks in the morning.

25 Wave theory of light

DEFINITION A DESCRIPTION OF LIGHT RAYS AS WAVES OF ENERGY TRAVELLING THROUGH SPACE

DISCOVERY THE THEORY WAS PROVED IN A LANDMARK EXPERIMENT CONDUCTED BY ENGLISH PHYSICIST THOMAS YOUNG IN 1801

KEY BREAKTHROUGH YOUNG DEMONSTRATED A WAVE EFFECT CALLED 'INTERFERENCE' TAKING PLACE BETWEEN LIGHT BEAMS

IMPORTANCE ACCOUNTED FOR LIGHT BENDING THROUGH GLASS AND EXPLAINED COLOURS AS DIFFERENT WAVELENGTHS OF LIGHT

Light is something we take for granted but it has been remarkably unwilling to give up its secrets. In the 17th century, Isaac Newton believed light to be made of tiny particles, but this view was hard to reconcile with experiments. In the early 19th century, physicist Thomas Young performed an experiment proving that light is, in fact, a wave, and the discovery greatly enriched the science of optics.

Scientists have spent centuries trying to fathom the nature of light. In the third century BC, Greek philosopher Euclid speculated that light moves in straight lines, which is broadly correct – and he used the idea to study the phenomenon of reflection. He also proposed the (incorrect) idea that we see by emitting beams from our eyes – which, in order to see the distant stars, must travel through space infinitely fast.

The first estimate of the speed of light was made in 1676 by Danish physicist Ole Rømer. He used discrepancies in the eclipse timings of Jupiter's innermost large moon Io, made at different times of the year, to work out that it takes light about 22 minutes to cross the diameter of Earth's orbit around the Sun. That meant that light moves at around 220,000,000 metres per second (240,594,926 yards per second) – very fast, but not infinite. Rømer's estimate was the right order of magnitude – the best measurements today yield 299,792,458m/s (327,857,019 y/s).

This information still said nothing about what light is actually made of. However, English physicist Isaac Newton was already searching. In 1675, he had put forward his own theory of light, in which he supposed that it is

LEFT A demonstration of an interference pattern created using water waves in a ripple tank. Like water, light rays also exhibit 'interference', with peaks and dips in wave patterns reinforcing one another and cancelling each other out.

made up of very many tiny particles – or 'corpuscles'. Newton's theory was thus known as the corpuscular theory of light, and was published formally in his 1704 book, *Opticks*. The theory explained the reflection of light, but struggled with more complex phenomena, such as refraction – the way a light ray entering an optically denser medium at an angle bends towards the denser medium. The book was not particularly successful.

Luminiferous ether

The alternative to Newton's approach was a wave theory of light. This idea had been put forward by Newton's rival Robert Hooke and by the Dutch physicist and astronomer Christiaan Huygens. The basic idea was that rather than a hail of tiny particles, light was made up of ripples travelling through a medium that pervaded the whole of space – known as the 'luminiferous ether'. Light waves in the ether could then be thought of as rather like water waves on the surface of a pond.

The wave model had no trouble explaining refraction. According to the theory, the light-bending effect is due to the light ray slowing down in the denser medium – rather like the way that applying the brakes on just one side of a car causes it to veer in that direction.

But it was a key experiment carried out in 1801 by the English physicist Thomas Young that finally put the wave theory on the map. Young shone a beam of light onto a screen that had a pair of narrow slits cut into it. The light passed through the slits to fall on a second screen, this time with no slits. But the light did not illuminate the second screen evenly: instead, it formed a pattern of bright and dark bands known as 'interference fringes'. As Young showed, these could only be present if light was a wave.

'It was a key experiment carried out in 1801 by the English physicist Thomas Young that finally put the wave theory on the map.'

The situation is rather like what happens when two stones are thrown into a pond next to each other. The ripples from the two impact points spread out and eventually overlap. When this happens, the two waveforms add together in an effect called 'superposition'. Where two peaks in the ripple pattern from each stone meet, they add to form a large peak (this is called 'constructive interference'). Similarly, where two dips coincide, the result is a large dip (constructive interference again). But where a dip and a peak coincide they cancel one another out (this is called 'destructive interference').

Interference bands

The waveforms of the light from the two slits in Young's experiment were behaving in much the same way as the ripples in the pond. Where two peaks or two dips in the waveforms overlapped the result was constructive interference and a bright band on the screen. On the other hand, where a peak and dip coincided they interfered destructively, cancelling one another out to form a dark band. And this accounted for exactly the alternating pattern of bright and dark fringes on the screen that Young saw. The evidence was there, as plain as day: light was a wave. Young also correctly deduced that different colours of light correspond to different wavelengths – the distance from one wave crest to the next.

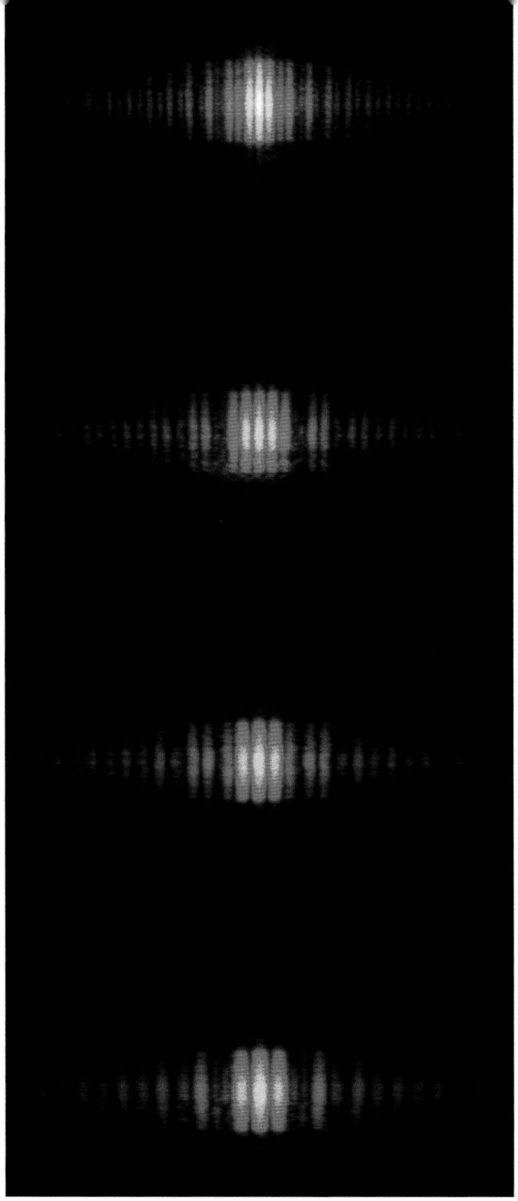

But that was not the end of the story. Another property of light waves is that they can be polarized. Thinking of waves on water again is a good analogy. The surface of the water undulates up and down in the vertical direction only as a wave passes – the waves are 'vertically polarized'. In general, a light wave moving in three-dimensional space will vibrate in all directions perpendicular to its direction of motion. But it too can be polarized to vibrate in just one direction using special filters. In 1845, English physicist Michael Faraday showed that the plane in which light is polarized can be rotated by a magnetic field. The result suggested that light waves have something to do with magnetism. But what?

Electromagnetic waves

All became clear following the publication in the 1860s of James Clerk Maxwell's seminal theory of electromagnetism (see page 153). The theory pulled together electricity and magnetism into a single unified entity. Moreover, it predicted the existence of waves of electrical and magnetic energy moving through space at the speed of light – and exhibiting all the other properties of light that scientists had been trying to explain over the years, including Faraday's rotation effect. Light, Maxwell had shown, is an electromagnetic wave.

ABOVE Young's double-slit experiment created a characteristic pattern of bright and dark interference fringes. At the time, this was regarded as conclusive proof that light is a wave. But quantum theory emerged a few decades later to fog the issue once again.

But light accounts for just one portion of the electromagnetic spectrum, occupying a narrow wavelength band. Other wavelengths correspond to different types of electromagnetic waves, such as radio, infrared, ultraviolet and X-rays. With this realization there was no need for the luminiferous ether as a medium for light to move through, and the concept was soon dropped. Maxwell had discovered the true wave nature of light. But the celebrations were short-lived because within 50 years, scientists discovered quantum theory (see page 237) – and then everything changed again.

26 Atomic theory

DEFINITION THE IDEA THAT MATTER IS NOT SMOOTH, EVEN ON THE SMALLEST SCALE, BUT MADE OF MANY TINY, INDIVISIBLE CHUNKS

DISCOVERY FIRST PUT FORWARD IN A SCIENTIFIC PAPER BY THE ENGLISH CHEMIST JOHN DALTON, PUBLISHED IN 1805

KEY BREAKTHROUGH REALIZATION THAT ATOMIC THEORY EXPLAINED THE PROPORTIONS IN WHICH CHEMICALS UNDERGO REACTIONS

IMPORTANCE PAVED THE WAY FOR THE QUANTUM THEORY OF MATTER AND THE REALM OF NUCLEAR PHYSICS

It was an idea first put forward nearly two and a half thousand years ago by the Ancient Greeks – that matter is made up of indivisible chunks called atoms. But, alas, they had no way to carry out the experiments needed to prove the theory. In the early 19th century, John Dalton reintroduced the theory in a modern scientific context. Today, it forms the basis for our understanding of chemistry and fundamental physics.

Atomic theory says that matter is not continuous. Keep chopping a lump of material in half over and over again and eventually you get to a point where you cannot divide it any further. What you have left is an 'atom' – the smallest possible piece of any given chemical element. The name comes from the Greek word *atomos*, meaning 'indivisible'.

The idea of atoms was first put forward by the Greek philosopher Democritus, in the fifth century BC. He believed that the nature of a substance was derived from that of its atoms – so ice has round slippery atoms, iron has strong and rigidly bonded atoms, while fire has weird-looking thorny atoms that somehow create the sensation of heat. It was a very whimsical theory, and the finer details that Democritus put forward were way off beam, but the general gist of what he was saying was correct. Matter really is composed of indivisible units – many trillions of them.

However, it took until the 19th century for this to be proposed on a scientific footing – and another 100 years for that proposal to be verified experimentally. The proposal itself was made by an English chemist called John Dalton. He had been born to a Quaker family in the county of Cumberland in 1766. As a

LEFT Each atom is composed of a heavy central nucleus containing positively charged protons and electrically neutral neutrons. In the outer layers orbit negatively charged electrons, usually equal to the number of protons, so that the atom remains, overall, uncharged. The electrons determine each atom's chemical properties.

young man, he was keen to study science, but his religion made this difficult – at the time anyone following a Christian denomination not conforming with the Church of England was a 'Dissenter' and prohibited from attending university. Instead, between 1790 and 1793, Dalton was tutored privately. Following this he taught for a time at New College in Manchester, which was sympathetic to Dissenters, before becoming a private tutor himself and using this income to fund his own scientific research.

Dalton carried out much research into meteorology, but around the turn of the 19th century he became interested in something entirely different. A number of experimental observations were emerging in chemistry, suggesting that the proportions in which elements combined during chemical reactions obeyed some kind of hitherto undiscovered law.

'Dalton had noticed with interest a result obtained by the French chemist Joseph Louis Proust, showing that substances tended to combine in whole-number ratios.'

For example, Dalton had noticed with interest a result obtained by the French chemist Joseph Louis Proust, showing that substances tended to combine together in whole-number ratios. In one experiment Proust had shown that, in the formation of tin oxide, a piece of tin will combine with either 13.5 percent or 27 percent of its own weight in oxygen. There was no middle ground – oxygen had to be added in whole-number increments of 13.5 percent of the mass of the tin. Atomic theory could explain this neatly – each tin atom can only combine with a whole number of oxygen atoms depending on the type of oxide formed, and each oxygen atom must therefore weigh 13.5 percent the mass of a tin atom.

Atomic axioms

Using these results, Dalton formulated his atomic theory as five axioms:

1. All chemical elements consist of particles called atoms.
2. All atoms of a particular chemical element are identical.
3. All atoms of any given chemical element are different from those of other elements.
4. Atoms of one element can combine with atoms of other elements to create compounds, and when they do this they combine in equal amounts.
5. Atoms cannot be created, divided or destroyed.

It is now known that the second axiom can be violated by the existence of 'isotopes' – atoms of the same chemical element that have different mass. This is due to different numbers of neutron particles inside the atom. In addition, the second part of the fourth axiom is also known to be wrong – atoms can, in fact, join to form molecules in all sorts of complicated ways. And the final axiom has been known to be totally wrong since the development of nuclear physics, which enables physicists to transmute atoms of one chemical element into those of another. However, the first and third points still hold up today – and, happily, these are the most important parts of the whole theory.

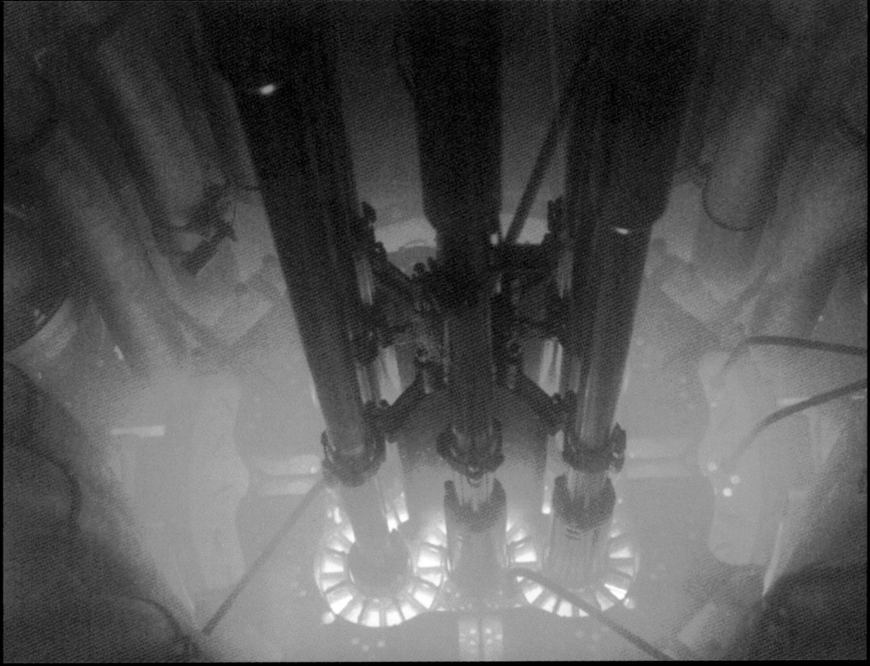

LEFT The Advanced Test Reactor at the Idaho National Engineering and Environmental Laboratory, USA, operates with a power of 250,000 kilowatts. The blue glow is radiation emitted when charged particles travel faster through cooling water than light does. Atoms became a new power source after physicists learned how to open the atomic nucleus and extract energy.

Dalton presented his theory to the Manchester Literary and Philosophical Society in 1803, followed by a scientific paper in 1805. He expanded upon it further in his book *A New System of Chemical Philosophy*, first published in 1808.

Experimental proof

Verification of the existence of atoms came in 1905 when the young Albert Einstein published a paper interpreting an observation that had been made by the Scottish botanist Robert Brown in 1827. Brown observed that tiny particles of pollen, suspended in a gas or fluid, exhibit a seemingly random jittering motion – as if they were being knocked about. Einstein interpreted the movements of the particles as recoils from multiple impacts with atoms in the gas or fluid. He supported his theory with calculations, the results of which matched with Brown's results. Dalton's theory had been proven.

Later research would refine the theory. We now know that atoms are made of positively charged proton particles bunched together with electrically neutral neutrons, to form the nucleus of the atom, around which orbit negatively charged electrons. The structure of the electron orbit determines the chemical properties of an atom – crucial in determining how it can bond together with other atoms to produce molecules. As physicists discovered in the 1920s, the electron structure is governed by quantum theory – given by the solutions to Schrödinger's equation for the atom (see page 241).

From materials science to nuclear reactions, building upon Dalton's work in understanding the atom has brought some of the biggest advances in physics and chemistry. And although quite different to how Dalton originally envisaged it, the atom today is, quite rightly, a scientific icon.

27 Molecules

DEFINITION THE SMALLEST POSSIBLE UNITS OF CHEMICAL COMPOUNDS, MADE BY JOINING TOGETHER ONE OR MORE ATOMS

DISCOVERY MOLECULES WERE PUT FORWARD AS ENTITIES DISTINCT FROM ATOMS BY ITALIAN SCIENTIST AMEDEO AVOGADRO IN 1811

KEY BREAKTHROUGH AVOGADRO REALIZED THAT THE ATOMS IN DALTON'S ATOMIC THEORY COULD BE JOINED TO ONE ANOTHER

IMPORTANCE INVESTIGATING MOLECULES AND THEIR PROPERTIES WAS FUNDAMENTAL TO THE DEVELOPMENT OF CHEMISTRY

Amedeo Avogadro expanded the theory of atoms that had been put forward by English scientist John Dalton. Avogadro realized that atoms could bond together to form new compounds – and when this happens the smallest unit of the new compound is called a molecule. His work cleared up the confusion in defining the difference between atoms and molecules, which had muddied the waters of research for nearly 200 years. These developments allowed chemistry to evolve into a modern science.

Whereas atoms are the simplest, smallest possible units of any given chemical element, a molecule is the smallest unit of a given chemical compound. A compound is made by bonding together atoms of different elements. One of the simplest compounds we encounter every day is water, made by bonding two atoms of hydrogen (H) to a single atom of oxygen (O). Chemical compounds are denoted as a written formula, with a subscript indicating the number of each atomic species present. Water molecules are written as H_2O.

The word 'molecule' comes from the Latin-derived French words for 'mass' (*mole*) with a suffix meaning 'little' (*cule*), indicating the smallest mass of a substance. The word was first put forward in the 17th century by French philosopher René Descartes, but until the 19th century there was confusion over its precise meaning – with the words 'atom' and 'molecule' used interchangeably. The distinction was finally set out by Amedeo Avogadro.

Avogadro was born into Italian nobility on 9 August 1776 in the town of Turin. After initially studying canon law, he switched to natural philosophy instead. Much like John Dalton, who formulated the atomic theory, Avogadro

LEFT When mercury is covered by water, the boundaries between layers form a 'meniscus' caused by forces between molecules. The force between the mercury and the glass tube is less than between the mercury molecules themselves, forming a convex meniscus. The force between the water molecules and the glass is greater, pulling the water into a concave meniscus.

became a skilled tutor in these scientific subjects – but carried out private research in his spare time. Avogadro's seminal breakthrough was proposing the idea that atoms are the indivisible units of the fundamental chemical elements, and that molecules are made by bolting atoms together.

Elementary molecules

In 1811, Avogadro published his ground-breaking idea in the somewhat verbosely titled essay '*Essai d'une manière de déterminer les masses relatives des molécules élémentaires des corps, et les proportions selon lesquelles elles entrent dans ces combinaisons*' ('Essay on Determining the Relative Masses of the Elementary Molecules of Bodies and the Proportions by Which They Enter These Combinations'). The essay was written in French because, at the time, Italy was part of the Napoleonic empire.

Also in the essay was the first formal statement of what has become known as Avogadro's law. This essentially states that at equal temperature and pressure, equal volumes of different gases contain the same number of molecules. Technically, Avogadro's law only applies to an 'ideal gas' – a gas in which the molecules exert no forces upon one another. In reality, intermolecular forces always exist meaning that gases are not ideal and create deviations from Avogadro's law – though it remains a good approximation and a powerful scientific principle.

Molecular geometry

Intermolecular forces are, as their name suggests, the powerful forces that operate between groups of molecules. They are responsible for forming the rigid lattices of molecules that lock matter together in its solid form. The bonds within this lattice naturally break up when the solid material is heated (see Kinetic theory on page 65), leading first to a liquid state and finally to a gas, where the molecules are able to drift away.

'Avogadro's seminal breakthrough was the idea that atoms are the indivisible units of the fundamental chemical elements, and that molecules are made by bolting atoms together.'

Intermolecular forces come in two principal types. The first type are Van der Waals forces, which are a complex kind of electrical interaction between molecules; the second type are hydrogen bonds, where a hydrogen atom within a molecule tries to bond with other nearby atoms.

Although a molecule can often be specified purely by its chemical formula, there are a number of cases where this is not possible. For example, the two chemicals ethanol and dimethyl ether have the same chemical formula – C_2H_6O – but possess very different properties. These differences are due to the way the individual atoms are arranged – that is, the unique geometry of the molecule. In each case, the atoms are bolted together into three groups – in ethanol it is CH_3–CH_2–OH while in dimethyl ether it is CH_3–O–CH_3. Two compounds that have the same chemical formula but different molecular geometry are known as 'isomers'.

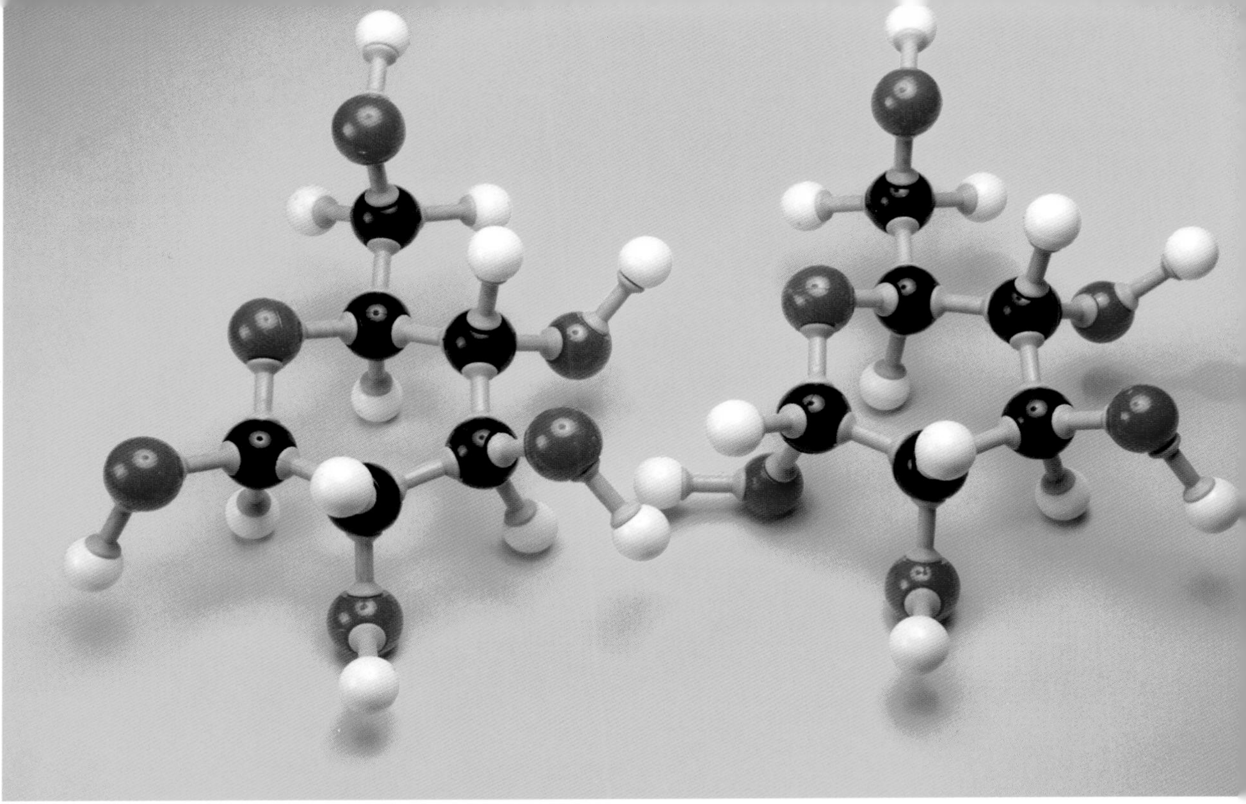

ABOVE Stick-and-ball models are used to teach molecular structure, with different atoms represented by coloured balls. Here, black is carbon, red is oxygen and white is hydrogen. The grey sticks represent the bonds between them. The molecules here represent forms of glucose ($C_6H_{12}O_6$).

Avogadro's name remains today attached with the theory of molecules in chemistry and physics, via a number known as Avogadro's constant. The number relates to the idea of molecular mass, which is the mass of a molecule that you get by adding up the atomic masses of all its constituent atoms. The atomic mass of an atom is given by adding up all the heavy particles (protons and neutrons) in its nucleus. For example, hydrogen has an atomic mass of 1 and oxygen has an atomic mass of 16 – meaning a water molecule, H_2O, then has a molecular mass of 18.

If you have a sample of a substance with a mass in grams equal to the substance's molecular mass, then the number of molecules in your sample is equal to Avogadro's constant. It takes the value 6.022×10^{23} – that is, 6.022 multiplied by a 1 with 23 zeroes after it. This number of molecules is sometimes referred to by chemists as a 'mole' of a substance. So 18 grams of water contains one mole of water molecules.

Avogadro's constant was not deduced by the great man himself, but was figured out later. In 1909, French physicist Jean Perrin suggested that the number be named after Avogadro in honour of his momentous contributions to molecular theory. Perrin himself took some fairly big strides, making the most accurate determinations of Avogadro's constant and proving beyond doubt that molecules really exist just as Avogadro had forecast 100 years earlier – a result for which he received the 1926 Nobel Prize in Physics.

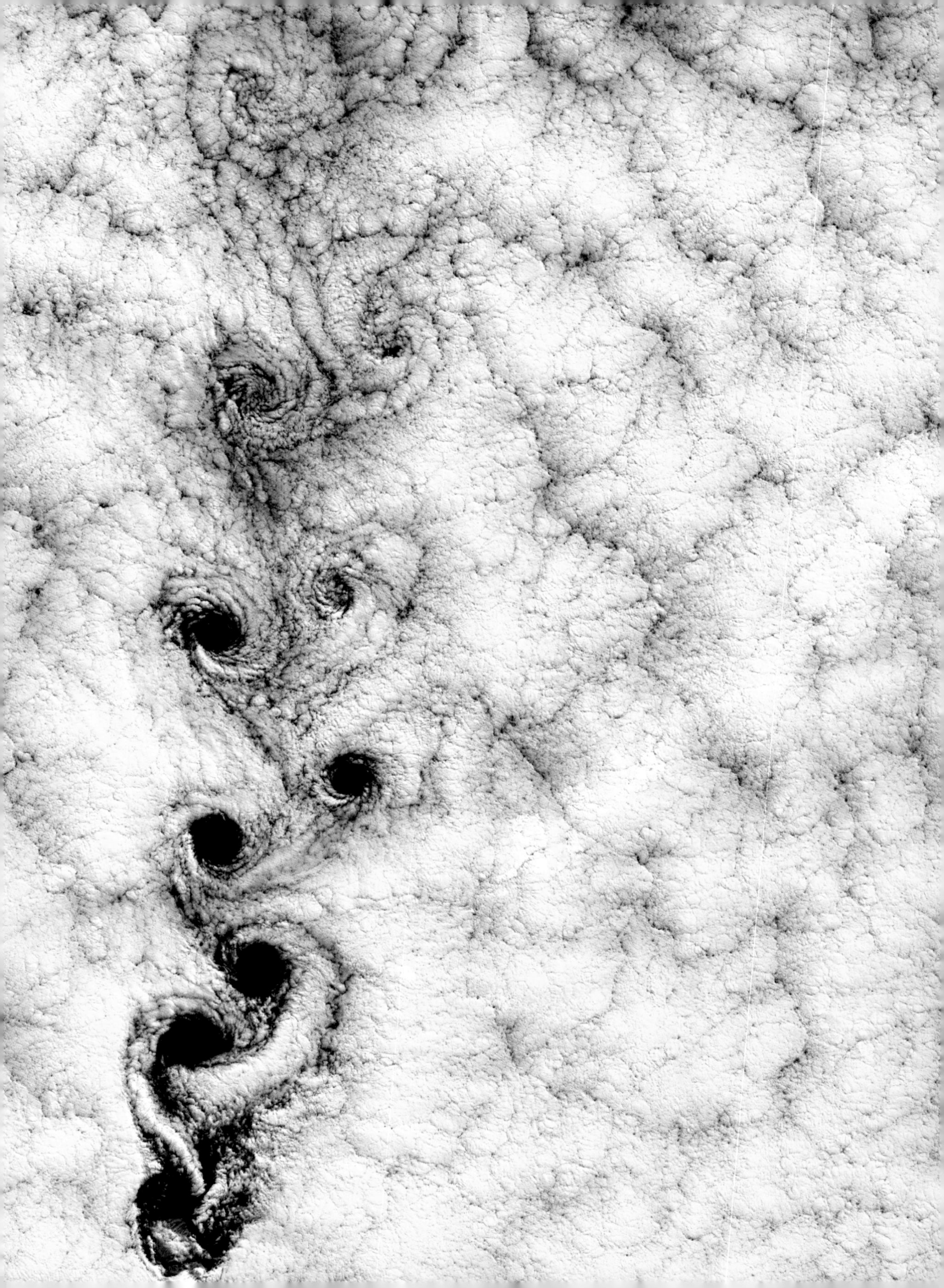

28 Navier-Stokes equations

DEFINITION A SET OF THREE MATHEMATICAL EQUATIONS THAT GOVERN THE BEHAVIOUR OF FLUIDS

DISCOVERY THE BASIC EQUATIONS WERE PUT FORWARD BY CLAUDE NAVIER IN 1822; THEY WERE REFINED BY GEORGE STOKES IN 1845

KEY BREAKTHROUGH FINDING A NEW WAY TO MODEL VISCOSITY AND HOW TO TREAT FLUIDS THAT ARE COMPRESSIBLE

IMPORTANCE USED IN CLIMATE SCIENCE, AIRCRAFT DESIGN, MARITIME ENGINEERING AND ASTROPHYSICS

The Navier-Stokes equations are some of the most fiendishly complex mathematical formulae. How exasperating, then, that they are also some of the most crucial – affecting the design of ships and aircraft, and the day-to-day modelling of how the planet's weather systems operate. Scientists now use powerful computers to try and solve the equations.

In 1759, Swiss mathematician Leonhard Euler formulated a set of equations giving a description of fluid flow. In mathematical terms, a fluid is any continuous, non-rigid form of matter – that can be a liquid, such as water flowing over the hull of a ship, or a gas such as air flowing through a tube.

Euler arrived at his equations by considering fluid behaviour in terms of Newton's second law of motion – which states that the force acting on a body is equal to its mass multiplied by the acceleration it experiences (see page 57). The formulae were absolutely correct, but only applied to the special case of fluids that are 'inviscid', meaning they have zero viscosity – a property quantifying the 'stickiness' of the fluid. For example, treacle has high viscosity while water has relatively low viscosity.

Virtually all fluids have some degree of viscosity, making Euler's fluid equations quite limited in their scope. Figuring out how to model viscous fluids mathematically was not straightforward. The first person to achieve this feat was the French engineer Claude Navier, in 1822. And yet even Navier's equations were not a complete generalization. While they could handle viscosity, they only applied to fluids that are 'incompressible' – meaning that the fluid's density cannot change. In other words, the equations

LEFT The swirls on the left of this image are Von Karman cloud vortices caused by clouds flowing over Alexander Selkirk Island, Chile. Navier-Stokes equations apply to liquids and gases, and govern the movement of everything from a stirred drink to global weather systems.

only apply to fluids that cannot be squashed. Put your finger over the end of a bicycle pump and there's a small amount of springiness in the pump action as you compress the air inside – air is compressible. Almost all other fluids are, too.

Finding a way to represent the compressibility in the mathematics of fluid flow was finally made by the Irish mathematician George Stokes in 1845. Stokes figured out how to mix compressibility into Navier's equations to give a new set of formulae. Known as the Navier-Stokes equations, they are the defining equations of fluid dynamics, describing the flow of air over vehicles, liquids through pipes and determining the behaviour of climatic and meteorological phenomena, such as oceanic and atmospheric currents.

Partial differentials

The Navier-Stokes equations are 'partial differential equations' that link the velocity of the fluid flow and fluid density. Differential equations are a consequence of calculus (see page 53), a method introduced by Isaac Newton and Gottfried Leibniz in the 17th century for computing the rates of change of mathematical quantities. Ordinary differential equations deal with rates of change with respect to one single variable, such as time. Partial differential equations, on the other hand, include terms giving the rate of change with respect to many variables. This is the case with Navier-Stokes, which model how velocity and density alter with respect to time, as well as the three directions of space.

'Stokes figured out how to mix compressibility into Navier's equations to give a new set of formulae. Known as the Navier-Stokes equations, they are the defining equations of fluid dynamics.'

They embody Newton's laws of motion, as well as conservation laws for energy and momentum. In addition, the Navier-Stokes equations are usually supplemented by an extra formula called the continuity equation. This is a statement of conservation of mass – that the total mass of the fluid must remain constant.

Solving the Navier-Stokes equations reveals how the velocity and density change from point to point within the fluid as well as with time. However, obtaining solutions to the equations via traditional pen-and-paper mathematics is impossible for all but the simplest cases because the equations themselves are so complicated. The main reason for their complexity is that they are non-linear – involving the fluid velocity squared – which makes them particularly intractable.

Computer power

Progress has been made in grinding out numerical solutions to the equations. The solutions generally break down into two types, classified according to the qualitative behaviour of the fluid flow. When the fluid moves smoothly, the

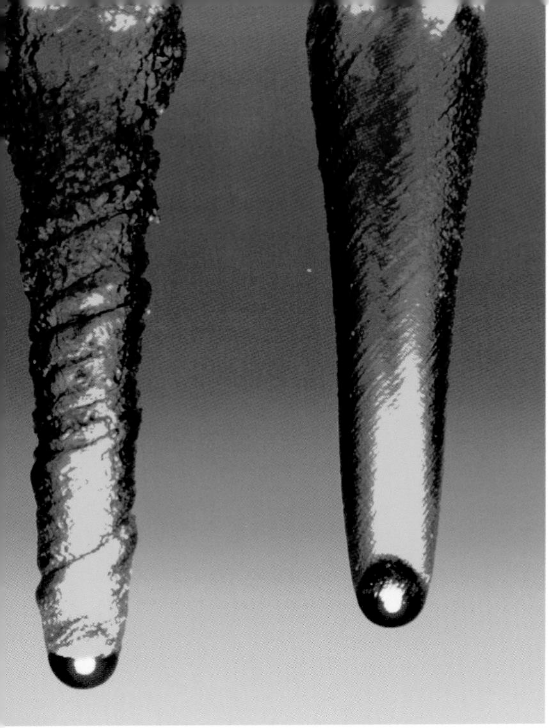

LEFT These three spinning balls are dropped into a fluid to illustrate the effect of viscosity – the friction between a liquid and an object passing through it. The rate of spin has been increased from left to right. Viscosity makes the water stick to the balls, and thus get dragged around by their spin.

flow is referred to as 'laminar'. This means that layers of fluid, such as air over an aerofoil, slip past one another smoothly. On the other hand, the opposite kind of fluid flow is known as 'turbulent'. Turbulent flow is a manifestation of chaos theory: the extreme sensitivity of some physical systems to their initial conditions. In these cases, fluid flow is seemingly random and disordered – as anyone who has been on a plane that has flown through turbulent air can testify.

Usually, the flow of fluid around a solid obstacle is laminar, whereas the flow through free space is prone to turbulence. The region of laminar flow around an obstacle is known as a 'boundary layer'. The speed of the fluid flow in this zone is less than elsewhere, since the fluid sticks to the surface of the obstacle. Minimizing the thickness of the boundary layer reduces drag and is thus a key consideration in aerodynamics. This is also the reason why golf balls have dimples – compared to a purely spherical ball a dimpled ball has a thinner boundary layer, experiences less drag and thus travels further.

Curve ball

Boundary layer theory is also the reason why it is possible to kick a curve ball in soccer. Kicking the ball to one side so that it spins sets up a boundary layer of air that sticks to the spinning ball, slowing the passage of air on one side and speeding it up on the other. In the same way that the Bernoulli principle explains how an aircraft wing generates lift – by setting up a higher airspeed above the wing than below (see page 30) – the differing airspeeds on either side of the ball create a force that makes it curve in flight.

Despite the power of studies of the Navier-Stokes equations carried out on some of the world's most powerful supercomputers, the transition from laminar flow to turbulence is still poorly understood. Turbulence is of fundamental importance in many aspects of aerospace and maritime design. And so, to incentivize the world's top mathematicians and physicists, the Clay Mathematics Institute, in Cambridge, Massachusetts, has listed understanding turbulence from the Navier-Stokes equations as one of its Millennium Prize Problems – with a cash prize of $1,000,000 for anyone who manages to solve it. At the time of writing, the prize is still unclaimed.

29 Greenhouse effect

DEFINITION THE TRAPPING OF HEAT FROM THE SUN BY CERTAIN GASES IN THE EARTH'S ATMOSPHERE

DISCOVERY FIRST PROPOSED BY THE FRENCH SCIENTIST AND MATHEMATICIAN JOSEPH FOURIER IN 1824

KEY BREAKTHROUGH FOURIER CALCULATED THE EARTH'S TEMPERATURE GIVEN ITS DISTANCE FROM THE SUN

IMPORTANCE CLIMATE CHANGE, CAUSED BY MAN-MADE GREENHOUSE GASES, IS PROBABLY THE GREATEST THREAT TO CIVILIZATION

Melting ice caps, rising sea levels, food shortages, and wars for land, fresh water and natural resources – those are some of the grim predictions for the future of civilization on Earth that have been put forward by climate scientists. These forecasts all stem from an idea mooted in 1824 by Joseph Fourier, that the Earth's atmosphere is somehow trapping sunlight to make the planet warmer than it should be.

The greenhouse effect that is warming Earth is very different to the mechanism by which a garden greenhouse warms the air inside it. A greenhouse works because heat comes in as visible radiation from the Sun, which passes freely through the glass, tries to escape as rising warm air, and then gets trapped.

In the case of the Earth, sunlight comes in as radiation at visible light wavelengths. And, like the glass in the greenhouse, the atmosphere is largely transparent to this kind of radiation so it passes straight through and transfers its energy to the ground below. But that's where the analogy ends. The sunlight causes the ground to warm up and emit what's called thermal radiation – mainly in the infrared region, at longer wavelengths than visible light. The atmosphere is not so transparent at these wavelengths and so some of the thermal radiation gets absorbed. Over time, the atmosphere re-emits this radiation evenly in all directions – and some of it is inevitably directed back downwards, making the ground warmer than it would otherwise be.

LEFT Droughts are a consequence of the greenhouse effect because infrared radiation from the Sun is trapped by gases, such as carbon dioxide and methane, causing the planet to heat up.

This is the atmospheric greenhouse effect. It is caused by gases in the atmosphere that absorb infrared radiation – gases such as carbon dioxide, methane and ozone (a molecule made up of three oxygen atoms).

The effect was first hypothesized by the French scientist and mathematician Joseph Fourier in 1824. Given the temperature of the Sun and the Earth's distance away from it, he calculated what the temperature of the Earth's surface should be – finding it to be much lower than is actually observed. To explain the discrepancy, he wondered whether the Earth's atmosphere could somehow be insulating the planet. But he had no idea how.

Infrared absorber

That piece of the puzzle was slotted into place by Swedish chemist Svante Arrhenius, in 1896. He had read about Fourier's temperature calculation, and the shortfall in its result, and hypothesized (correctly) that this was due to the presence of carbon dioxide (CO_2) in the atmosphere. Experimental results had shown that carbon dioxide was an absorber of infrared radiation.

Arrhenius also offered the first prediction about what this meant for the planet. At the time, the industrial revolution was in full swing – coal was being burned as a power source at a faster rate than ever before, churning massive volumes of carbon dioxide into the sky. Arrhenius spotted that this would bolster Fourier's greenhouse effect, causing the planet to warm up.

However, his prognosis was a far cry from the doom and gloom of climate science today. It was cause for celebration. The realization that carbon dioxide could make the world warmer came right after the discovery that the planet had passed through a number of ice ages, during which large areas of its surface had frozen over. Nobody worried about global warming – instead, here was a mechanism that could potentially save us all from the ravages of extreme 'global cooling'. Not only that but warming up the planet, they reasoned, could be great news for agriculture – turning the barren frozen wastes of Russia and Canada into rich farmland.

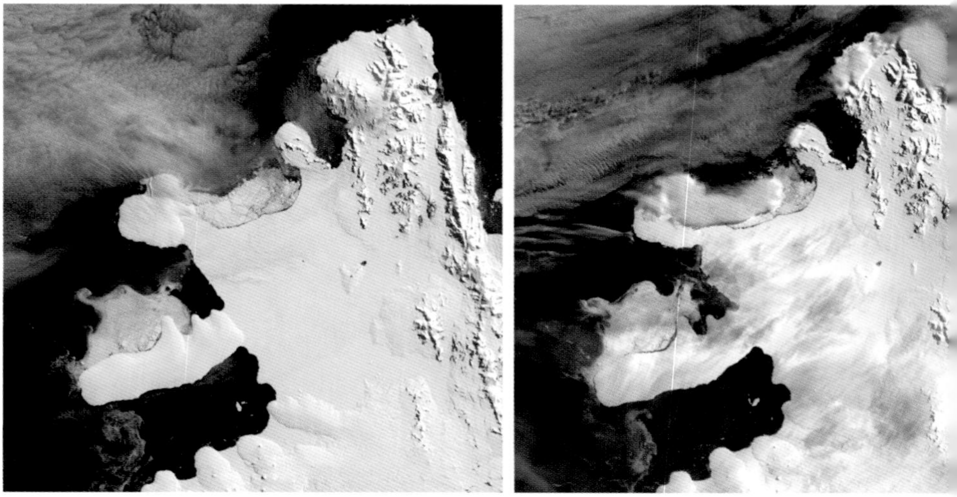

RIGHT Global warming melts the ice caps, raises sea levels and causes flooding. Pictured is the break-up of the Wilkins Ice Shelf in Antarctica in 2008. The first image (left) shows the solid ice shelf. The blue area of the second image (centre) shows the melting edge. The third image (right) shows the melting ice drifting away.

Warming world

So where was the evidence that the world was warming up? This was provided in 1938 by English engineer Guy Callendar. He gathered data on the Earth's average temperature between 1890 and 1935 – finding that it had risen by half a degree Celsius. More evidence came in 1958 when Charles Keeling at the Scripps Institute of Oceanography began a programme of carbon dioxide measurements on Mauna Loa, Hawaii. He showed that between 1958 and 2005, carbon dioxide rose from 315 parts per million (ppm) to 380ppm – compared with 280ppm in the pre-industrial 19th century.

'Over the next hundred years, melting of the ice caps caused by the greenhouse effect could raise sea levels by enough to place some of the world's major cities in danger of severe flooding.'

By the 1960s and 1970s, there could be little doubt that the world was warming. Gradually, the near-term consequences became clear and eclipsed the long-term hazard presented by ice ages. If the climate scientists were right, then over the next hundred years melting ice caps could raise sea levels by enough to place some of the world's major cities in danger of flooding – including Hong Kong, London and New York.

Sceptics' response

Despite the work of Keeling and others, the link with human carbon emissions seemed harder to prove. Climate sceptics argued that the temperature rises were caused by natural variations in the Sun's output. The validity of theoretical climate models was also questioned: for example, the role of clouds is still not fully understood – to some degree they warm the planet (water vapour itself is a greenhouse gas) yet they also help to keep it cool, forming a white blanket that reflects the Sun's warming rays. Nevertheless, the scientific consensus today is that global warming is anthropogenic (created by human activity). In 2007, the Intergovernmental Panel on Climate Change (IPCC) – a body appointed to investigate the phenomenon – published its fourth report and concluded: 'Most of the observed increase in globally averaged temperatures since the mid-20th century is very likely (>90 percent) due to the observed increase in anthropogenic (human) greenhouse gas concentrations'.

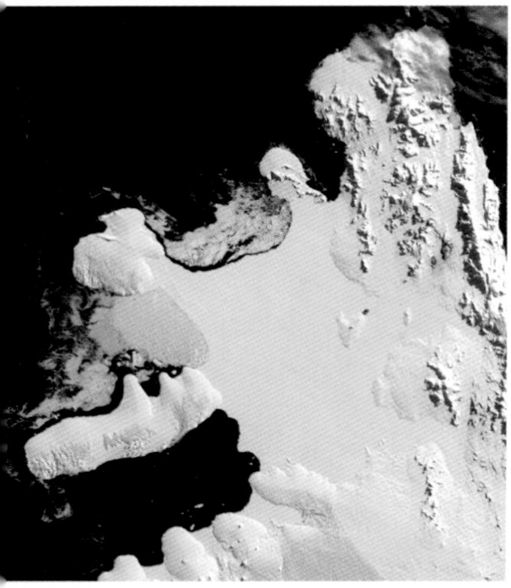

The future is now very much in the hands of international governments to curb carbon dioxide emissions before the Earth's climate passes a critical 'tipping point', a threshold from beyond which it may be very difficult, or even impossible, to return. Worryingly, some scientists think this may have already happened.

30 Ice ages

DEFINITION PERIODS OF EARTH'S PRE-HISTORY WHEN ICY GLACIERS COVERED MUCH OF THE PLANET'S SURFACE

DISCOVERY THE EXISTENCE OF ICE AGES WAS FIRST PROPOSED BY THE SWISS-AMERICAN GLACIOLOGIST LOUIS AGASSIZ IN 1837

KEY BREAKTHROUGH AGASSIZ DISCOVERED GEOLOGICAL EVIDENCE THAT SWITZERLAND WAS ONCE COVERED BY AN ICE SHEET

IMPORTANCE IT MAY BE THAT EARTH IS IN AN ICE AGE RIGHT NOW – AND WE ARE SIMPLY ENJOYING A WARM 'INTERGLACIAL' PHASE

In 1837, scientists studying the geology of Switzerland realized that many of the rock formations they observed could be explained if the whole country had once been buried beneath a colossal ice sheet. The data gathered by geologists there was the first evidence that the Earth sporadically undergoes periods of glaciation known as ice ages.

Our planet is known to have passed through phases, sometimes lasting hundreds of millions of years, during which large portions of its surface froze over. These are the ice ages. The freezing was severe – not just a frosty morning in winter, but the complete smothering of the land beneath thick glaciers to form an ice sheet, reminiscent of the Earth's polar regions today.

In one extreme scenario, called 'Snowball Earth', glaciers from the North and South Poles may even have met at the equator, totally encasing the planet in ice. This is suspected to have happened during one of the longest ice ages, between 850 and 630 million years ago (mya).

The Snowball Earth freeze-over was just one of five major ice ages that have wracked the Earth during its 4.5-billion-year history. These are thought to have happened 2.4–2.1 billion years ago, 850–630mya (Snowball Earth), 460–430mya, 360–260mya, and 2.53mya to 10,000 years ago.

We know the dates of the ice ages because they have left their scars on the landscape. Expanding glaciers ooze and flow across the land. They are extremely heavy – a cubic metre of ice weighs 920kg (2028lb) – and their passage leaves gouge marks on the rock surfaces over which they pass.

LEFT This is part of the 12km-long (7.5-mile) Gorner Glacier in the Valais region of Switzerland. The heat of the daytime Sun has melted some of the ice to form a lake on the glacier's surface. During an ice age, much of the Earth's surface would be covered by glaciers.

Some valleys are severely reshaped by the passage of a glacier. The cross-section of a valley between mountains is usually V-shaped. However, the glacier's movement turns this into a U-shape, as its abrasive ice action grinds away at the valley floor. When the ice melts and recedes, boulders and debris that have been displaced by the ice litter the valley floor.

Moraine debris

This debris material can form striking landforms known as 'moraines', where the debris has been piled up into banks by the force of the advancing ice as if some giant bulldozer has ploughed through the terrain. Another landform characteristic of glacial action is known as a 'drumlin' – a long, tear-drop-shaped hill with its long axis parallel to the direction of the moving glacier and its blunt end pointing upstream.

Occasionally, geologists find boulders in odd locations or in places where their rock type is different to that of the surrounding terrain. This, too, is evidence of glacial action – the theory being that the boulder has been scooped up and carried along by the moving ice and then dropped once the climate has warmed and the glacier has retreated. These rocks are sometimes known as 'glacial erratics'.

'In one extreme scenario, called "Snowball Earth", glaciers from the North and South Poles may even have met at the equator, totally encasing the planet in ice.'

Dating terrain that has been re-sculpted in this way gives geologists an idea of the timing of ice ages. Other clues come from the chemical record of a region, which carries an imprint of temperature variations. And the fossils present at particular time periods offer another pointer – indicating, for example, whether species have migrated in search of warmer climes.

Ice time

The person who put the pieces of the puzzle together was the Swiss-American glaciologist, Louis Agassiz. He came forward with the notion that during its geological history, the Earth has been prone to freezing over. In the 1830s, he and a number of friends – including German botanist Karl Schimper – had been examining what would turn out to be glacial erratics around the Jura Mountains in Switzerland. What he found convinced him that during Earth's long geological history, Switzerland had looked more like Greenland – that is, a flat sheet of ice. For a time, he lived in a hut atop a glacier in the Swiss Alps in order to better understand the movement of the ice.

Agassiz first put forward his ideas in 1837 and, over the subsequent years, developed them into a book entitled *Études sur les Glaciers* (Study on Glaciers), making the case that glacial ice sporadically creeps out from the poles to cover large swathes of the planet. It was his friend Schimper who named these events *Eiszeit*, meaning ice age. Indeed, Schimper's role in the rest of the theory is not clear – some historians have suggested that he deserves credit not just for the name, but for the whole theory. What is certain is that Agassiz

was the first to publish it. At first, the scientific community was not receptive to the theory of ice ages – clinging to the accepted view that the Earth had been steadily cooling since its formation, without any major fluctuations in this trend. However, by the end of the 1870s the overwhelming evidence was leading many geologists to start taking the idea seriously.

ABOVE These boulders, from Spitsbergen (Svalbard), Norway, are known as glacial erratics – rocks scooped up by a moving glacier, then deposited hundreds of miles away when the glacier melted. Glacial erratics are one indicator used by geologists to identify past ice ages.

Milankovitch cycles

The cause of ice ages is not well understood, but is thought to be a combination of many factors, such as volcanic activity, changes in ocean currents which transport warm water around the globe and even variations in the brightness of the Sun. In the early 20th century, Serbian mathematician Milutin Milankovitch formulated a theory whereby periodic wobbles in the Earth's orbit and rotational axis can subtly alter the amount of sunlight that the planet receives, leading to eras of heating and cooling.

The temperature variations produced by these so-called 'Milankovitch cycles' are now believed insufficient to lift the planet in and out of an ice age. They may, however, contribute to what are called 'interglacials' – periods of relative warmth that punctuate the extreme cold of an overarching ice age. Indeed, it may well be that the last ice age did not end 10,000 years ago after all, but that we are simply basking in the cosy glow of an interglacial. If that is the case then the Earth could be heading back into the freezer. No need to panic though – it will not happen for tens of thousands of years. Short-term global warming is a much bigger problem.

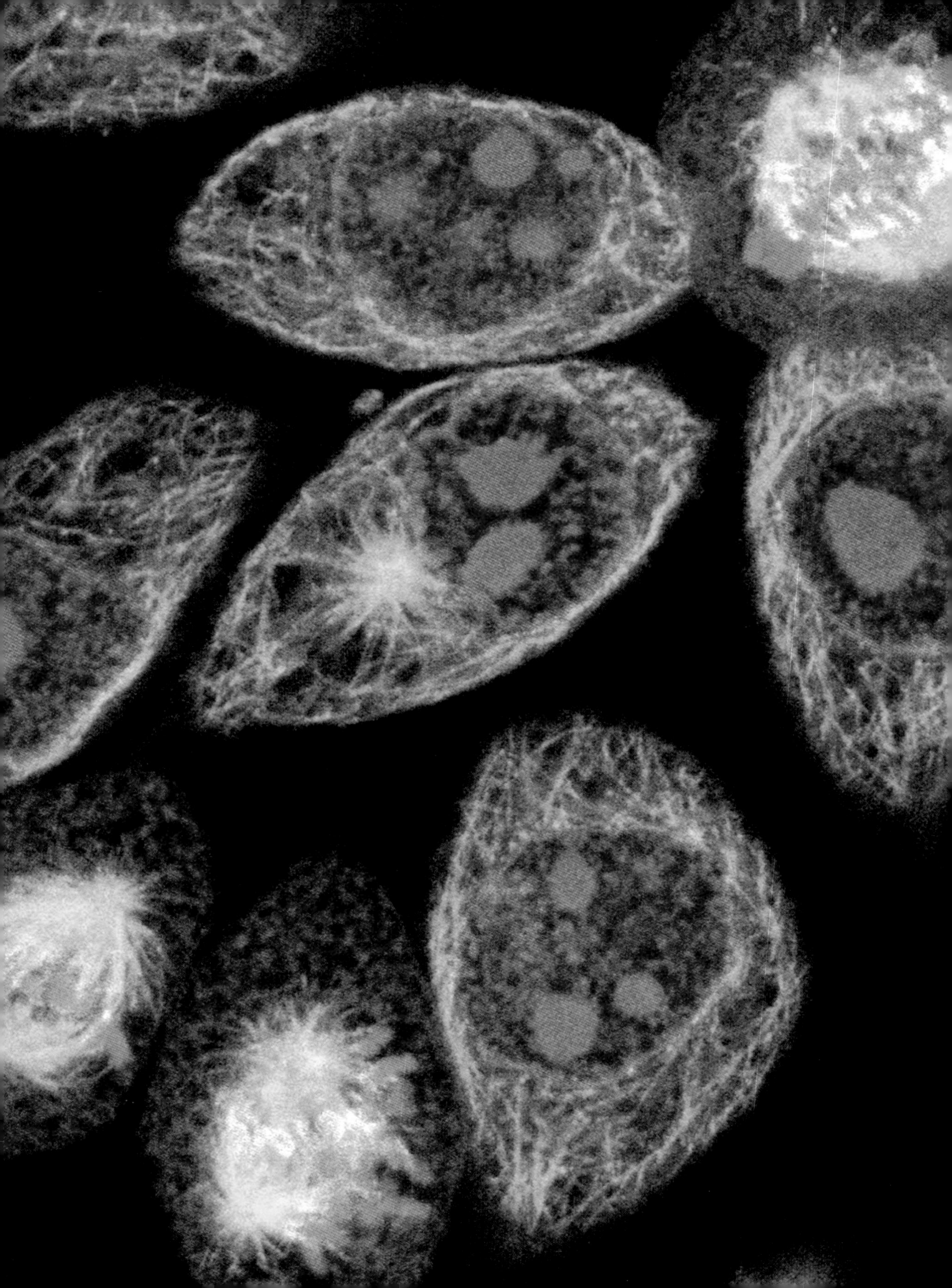

31 Cell theory

DEFINITION THE IDEA THAT SELF-REPLICATING CELLS ARE THE FUNDAMENTAL BUILDING BLOCKS OF LIFE ON EARTH

DISCOVERY REALIZED FOR PLANTS IN 1838 BY MATTHIAS SCHLEIDEN; FOR ANIMALS, IT WAS DISCOVERED IN 1839 BY THEODOR SCHWANN

KEY BREAKTHROUGH IMPROVEMENTS IN MICROSCOPE TECHNOLOGY REVEALED THE CELL STRUCTURE OF LIVING TISSUE

IMPORTANCE THE CELL THEORY GIVES THE FUNDAMENTAL PICTURE BY WHICH ALL LIFE PROCESSES IN BIOLOGY CAN BE UNDERSTOOD

From humble bacteria to the largest animals, all forms of life have something very simple in common – they are made of cells. These tiny building blocks are the fundamental units of life: you have tens of millions of them in one fingernail. This fact was only realized in the 19th century, as sharper and more powerful microscopes revealed the amazing spectacle of cells at work.

In the time of Aristotle, in the fourth century BC, scientists believed that the essence of life lay in the doctrine of 'vitalism' – a pseudoscientific notion that life forces exist above and beyond the fathomable processes of chemistry and biology. That all began to change in the 17th century when researchers got a hint that living things are, in fact, made from biochemical building blocks.

Progress came from the observations of English scientist Robert Hooke. In 1665, he examined slices of dried cork under an early, primitive microscope. Under the microscope's magnification, Hooke saw the structure of the cork was divided into tiny compartments, which he named 'cells'. But he failed to grasp the true significance of what he saw. The cells Hooke saw were dead and what he observed were the dry cell walls without their internal structure.

Multicellular view

The first observations of living cells were made by Dutch biologist Antony van Leeuwenhoek in 1674. He studied *Spirogyra* algae, blood cells, bacteria and microbes: all multicellular life forms that are too small to be seen with the naked eye. But establishing the theory underpinning these observations would not follow until much later, when two German researchers put forward a formal theory of cells. The first breakthrough came in 1838, when

LEFT Cells undergoing 'mitosis', where the cell nucleus splits in half. Mitosis is the first step in 'cell division', where each cell splits to form two new ones. Since all life is made of cells, cell division is the linchpin of biology – allowing new life to be created from old.

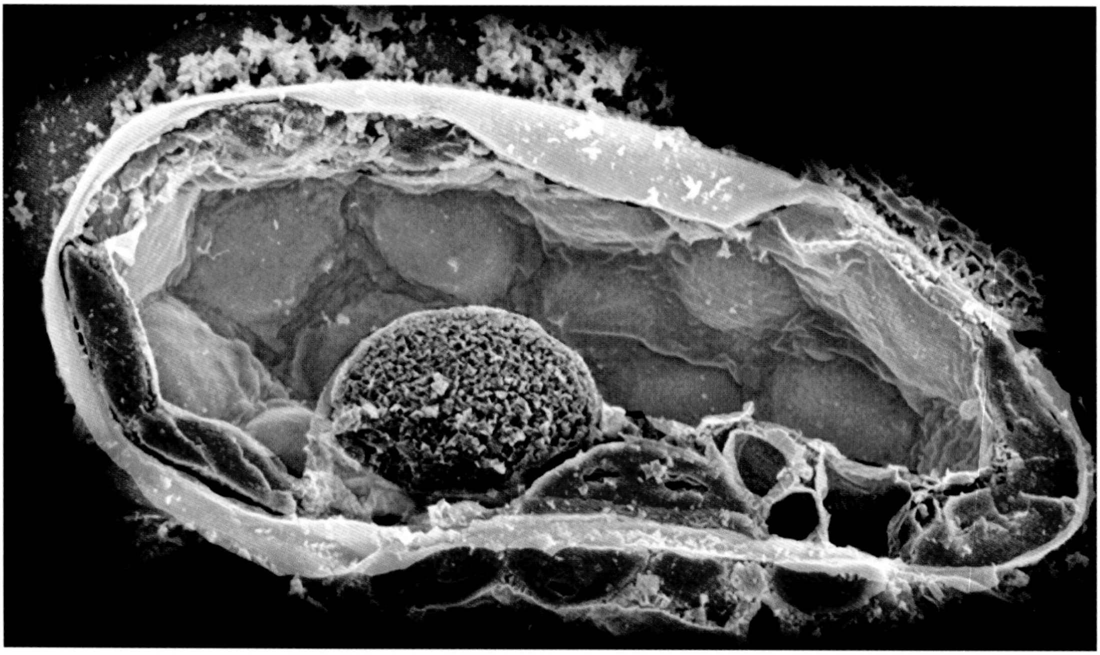

ABOVE Despite their tiny size, cells are incredibly complex structures. This plant cell (imaged using a scanning electron microscope) highlights the outer cell wall, made of tough cellulose fibre. The brown regions inside are chloroplasts, where photosynthesis takes place. In the centre is the nucleus, where all the genetic information resides.

botanist Matthias Jakob Schleiden postulated that the cell is the basis of all plant life. A year later, in 1839, the physiologist Theodor Schwann extended the theory to animals – finding cellular structure in a range of zoological tissue samples. Together, Schleiden and Schwann had done for biology what atomic theory had done 60 years before for physics and chemistry.

Soft cell

Some science historians argue that there was a third player in the story – German biologist Rudolf Virchow. In 1858, he published evidence that cells grow from other cells through a process of cell division. However, other commentators suggest that the Polish-German physiologist Robert Remak may have reached this conclusion first. Either way, the idea was correct – with corroborating evidence coming from Louis Pasteur's research on germs.

The cell theory was an incredible achievement – and much of it is still valid today. For example, we still believe the cell to be the basis of life, that all life forms are made of one or more cells and that cells are made from other cells by cell division. There have been recent additions that take account of developments in biochemistry: for example, we now know that cells contain hereditary DNA information stored inside their nuclei.

So, what else do we know about cells? Cells are made of protein. They are not merely structural building blocks but biological machines performing functions necessary to the survival of the organism of which they are part.

That might be energy generation, making new tissue, fighting infection or moderating other biochemical processes. We also know that cells come in two types: eukaryotes and prokaryotes. Eukaryotes are what make up almost all multicellular organisms – such as plants and animals – and a small number of unicellular (single-celled) creatures. A typical animal cell measures around one hundredth of a millimetre across and weighs about a billionth of a gram. The human body is home to approximately 100 trillion (10^{14}) of them.

DNA storage

Eukaryote cells are distinct because they have a central nucleus, enclosed within a membrane, from where the functions of the rest of the cell are controlled, and where the reproductive DNA is stored. Outside the nucleus is an area referred to collectively as the 'cytoplasm'. It consists of various subunits – called 'organelles' – each performing different functions within the cell. For example, ribosomes are organelles in which new protein is manufactured according to the genetic code written on the DNA in the nucleus. Lengths of nuclear DNA code are transcribed onto a related type of molecule known as 'messenger RNA' and then ferried out of the nucleus to a ribosome site where the sequence is translated and used as a blueprint to bolt together different types of amino acids – the building blocks of proteins. Other prominent organelles are mitochondria, where energy is generated, and lysosomes, which clear up the debris from other cellular processes.

Prokaryotes, on the other hand, are the much simpler of the two cell types. They have no nuclei, just a bundle of loose DNA at their centre. They also have fewer organelles and are smaller than eukaryotes – measuring one ten-thousandth of a millimetre. Prokaryotes form unicellular life forms – for example, bacteria.

'A typical animal cell measures around one hundredth of a millimetre across and weighs about a billionth of a gram. The human body is home to approximately 100 trillion (10^{14}) of them.'

Cell division

Central to the cell theory is the ability of cells to replicate, by dividing. Eukaryotes achieve this through a two-step process. The first step is mitosis, where DNA in the cell nucleus unzips lengthwise and new chemical bases then attach to each of the two strands to form two new strands. In the second step, cytokinesis, the nucleus divides in two with one strand of the new DNA ending up in each half. Finally, the rest of the cell's cytoplasm divides in two, with one new nucleus in each piece.

In prokaryotes, the process is simpler. Called 'binary fission', the first step is rather like mitosis as the DNA at the centre of the cell unzips and then pairs up with new nucleotide bases to duplicate itself. The rest of the cell then expands and breaks in half, and each half takes one copy of the DNA. In this way, every single living cell on Earth came from another cell before it, forming a 3.5-billion-year chain of cellular heritage stretching right back to the origin of life – when the very first cell came together from base chemicals.

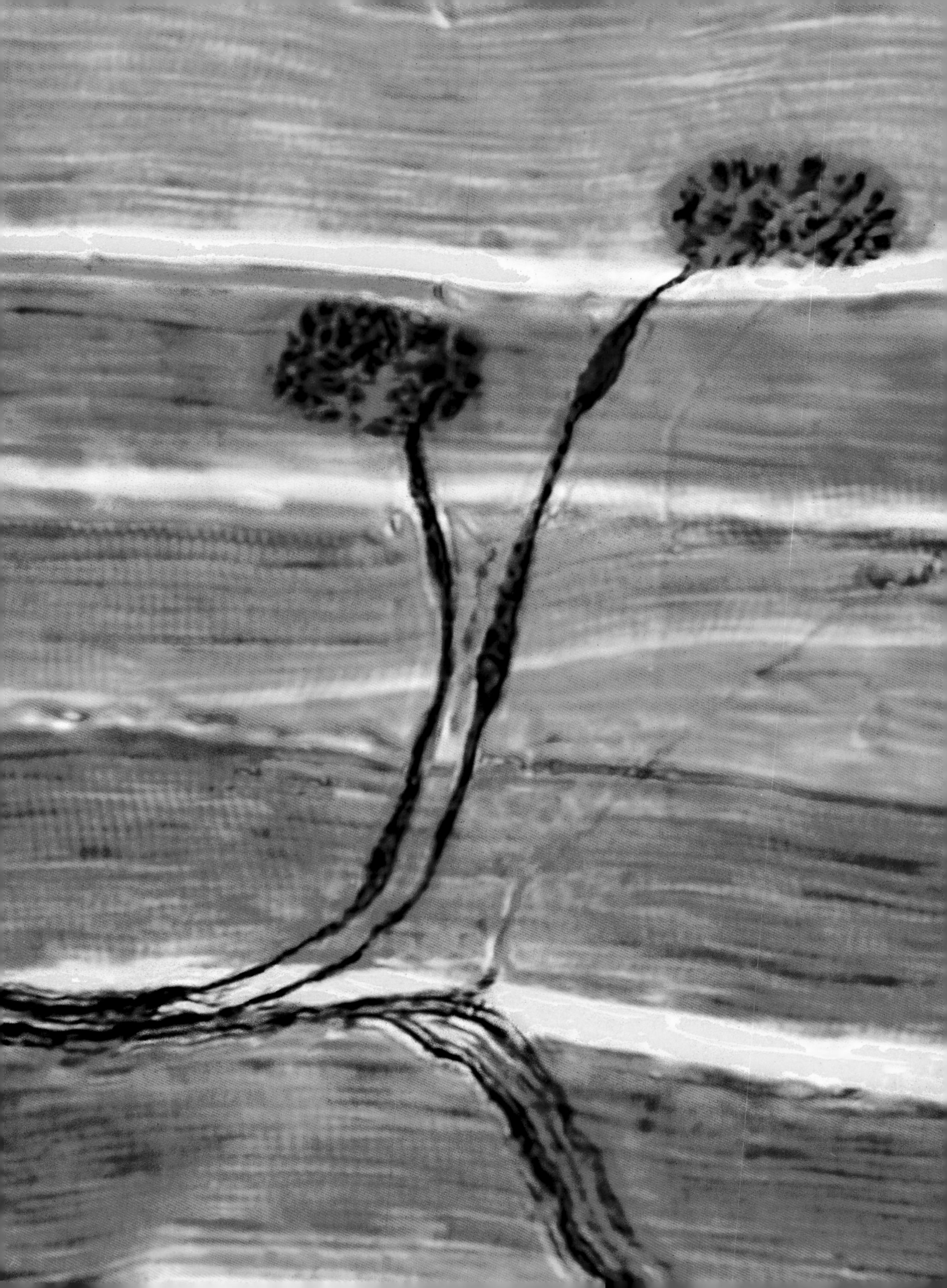

32 Anaesthetics

DEFINITION DRUGS TO DEADEN THE PAIN OF SURGERY, AS WELL AS INDUCE MUSCLE RELAXATION AND UNCONSCIOUSNESS

DISCOVERY SURGERY UNDER ANAESTHESIA WAS FIRST DEMONSTRATED PUBLICLY IN 1842 BY CRAWFORD LONG

KEY BREAKTHROUGH LONG NOTICED THAT RECREATIONAL USERS OF DIETHYL ETHER WERE OBLIVIOUS TO THE PAIN CAUSED BY INJURY

IMPORTANCE WITHOUT THE URGENCY CAUSED BY PATIENT PAIN, SURGEONS COULD ATTEMPT MORE COMPLEX PROCEDURES

Before the days of anaesthetics, surgeons had one priority – to be as quick as possible to limit the suffering of their patients. Happily, those grisly days are long gone. In the 1840s, drugs were found that could not only relieve pain but also induce a temporary state of unconsciousness so that patients did not need to be awake during operations. Suddenly, doctors could attempt longer, more complex operations – it was the beginning of modern surgery.

Before the arrival of effective anaesthetics, surgery was an awful task and an awful experience. Minimizing patient suffering meant that speed was of the essence, and operations were confined to those that could be carried out quickly. Limb amputations were one of the most common – the pre-anaesthetic speed record for sawing off a leg was reportedly 15 seconds, held by the French surgeon Dominique Larrey (who was also Napoleon's top battlefield medic).

Early anaesthetics included alcohol, primitive opiates, herbal preparations and even knocking the patient unconscious with a blow to the head. There is a long history of attempts at surgical procedures under such impromptu anaesthetic measures, stretching back to the Ancient Sumerians in the fifth millennium BC, as well as the ancient cultures of Egypt, Greece and Rome.

LEFT A motor neuron cell, which is responsible for voluntary muscular movement. The neuron is shown embedded in a skeletal muscle. Brain signals initiate movement by passing through the nervous system and down the axons (black lines) of the neuron. Anaesthetics paralyze the motor neurons in the patient during surgery.

Herbal potion and laughing gas

Important medical developments did not start until the 1800s. In Japan, a surgeon named Hanaoka Seishu concocted an anaesthetic herbal potion called Tsusensan that paralyzed muscles, deadened the nerves and left patients unconscious for hours. In 1804, he used Tsusensan to anaesthetize a 60-year-old woman suffering with breast cancer while he performed a mastectomy. He had effectively given the woman a general anaesthetic – even if his methods were somewhat less than scientific.

At around the same time, English chemist Humphry Davy discovered the sedative – and euphoria-inducing – properties of the gas nitrous oxide, also known as 'laughing gas'. It was an instant hit as a recreational drug, but it would be over 40 years before it was used clinically (and successfully so: nitrous oxide is the 'gas' in 'gas and air', still offered to women in labour).

Diethyl ether

Another drug that was exploited recreationally at this time was 'ether' – or diethyl ether, to use its full name. It was a liquid, the fumes of which could be inhaled to induce light-headedness and lift the mood. One of the people who tried it was an American student named Crawford Long, who was studying at the Pennsylvania University School of Medicine. He noticed that ether revellers who sustained minor injuries while cavorting under the influence of the drug only felt pain later on, when it wore off.

When Long graduated and became a practising physician, his dabblings with ether led him to try it as an anaesthetic. His first procedure was carried out on 30 March 1842, when he used diethyl ether to anaesthetize a man so that a tumour could be cut from his neck. It was a success. The pioneering procedure was the first scientific application of a general anaesthetic for a clinical operation. Long later used ether to knock out patients during amputations and to ease labour pains.

'The pre-anaesthetic speed record for sawing off a leg was reportedly 15 seconds, held by French surgeon Dominique Larrey.'

In fact, Long's only real mistake was in not publishing his work in any scientific journals. Four years later, in 1846, the American dentist William Morton administered diethyl ether before performing a tooth extraction. Shortly after, he used the same drug in the removal of a neck tumour – in a public demonstration held at Massachusetts General Hospital. The same year, American doctor Oliver Wendell Holmes coined the name 'anaesthesia', meaning 'without sensation', to describe the process. Owing to the public nature of Morton's experiment, he is often cited as the first to use a general anaesthetic in a clinical environment – even though Long got there first. Long ultimately published his own work in 1849, but by then it was too late.

Royal appointment

Morton's demonstration was followed in 1847 by the first anaesthetic use of chloroform, by the Scottish obstetrician James Simpson, who also prescribed it to ease the pain of childbirth. Its use was initially shunned by some religious figures, who argued that giving birth was meant to be painful. However, their protestations were muted when, in 1853, Queen Victoria asked for chloroform during the delivery of her son, Prince Leopold.

All the anaesthetics used so far were general anaesthetics – subduing the pain response from the whole body, relaxing muscles and rendering the patient unconscious so that there was no memory of the surgery. General anaesthetics today are normally administered in two stages: an 'induction' drug – usually injected – puts the patient under, then a second drug is inhaled through a mask to maintain the anaesthetic until the patient is brought round.

ABOVE The root of the mandrake plant (*Atropa mandragora*) can induce a narcotic effect if eaten. It grows in western Asia and the Mediterranean and was used for centuries by apothecaries to concoct anaesthetics for surgery.

Local anaesthesia

In 1884, a new anaesthetic was discovered that numbed localized areas without causing unconsciousness. It was thus called 'local anaesthetic'. The first variety used was cocaine, but – as you might imagine – this proved extremely addictive. It was replaced in the 20th century by the less-habit-forming drug procaine. Local anaesthetics block the sodium channels inside nerve cells, called axons, by which physical stimuli are transmitted through the nervous system to the brain, where the pain signal is registered.

Amazingly, scientists are still unsure how exactly general anaesthetics work. They, too, block pain channels through the nervous system, but their action extends over the whole body. For anyone destined for the operating theatre, though, this small gap in our knowledge of general anaesthetics is a small price to pay compared to the alternative – the terrifying prospect of going under the knife without one.

33 Second law of thermodynamics

DEFINITION A LAW STATING THAT THE AMOUNT OF ENERGY AVAILABLE IN THE UNIVERSE TO DO USEFUL WORK IS DIMINISHING

DISCOVERY FIRST STATED IN 1850 BY PHYSICIST RUDOLF CLAUSIUS

KEY BREAKTHROUGH CLAUSIUS REALIZED THAT THE SECOND LAW OF THERMODYNAMICS IS A NATURAL CONSEQUENCE OF THE INEVITABLE FLOW OF HEAT FROM HOT BODIES TO COLD ONES

IMPORTANCE KEY TO THE SCIENCE OF HEAT, THE LAW IS ALSO USED IN COSMOLOGY, INFORMATION THEORY AND BLACK-HOLE PHYSICS

The second law of thermodynamics is the physicist's way of telling us that nothing lasts forever. It says that slowly but surely the amount of energy available to carry out useful tasks is getting smaller. The law governs phenomena ranging from how your fridge works to the physics of black holes in space. Some cosmologists have wondered whether it could even bring about the end of the universe.

The second law of thermodynamics is a law of physics that effectively limits the efficiency of an engine – or any other process that converts heat (the energy locked away in a system because of its temperature) into physical movement ('work'). For example, a steam engine that turns hot steam into the motion of the train, which it is pulling, cannot do so with 100 percent efficiency – a certain amount of heat energy is lost to the surroundings.

The first person to realize this, and speculate as to why it should be, was the French physicist Sadi Carnot in 1824. Following the construction of the first steam engines in the late 18th century, Carnot had been investigating ways to make these devices more efficient – at the time, they were only converting about three percent of the heat generated by the fuel into useful work.

Carnot correctly deduced that heat engines require a difference in temperature in order to function – there need to be separate hot and cold parts. For example, in a steam engine it is the expansion of high-pressure hot steam against low-pressure cold air that does the work. If all parts of the engine were at the same temperature then this would not happen. Carnot's discovery was later framed by a mathematical equation, relating the cold

LEFT A thermographic image of the interior of a refrigerator, showing the temperature of different objects according to the wavelength of the infrared radiation they emit. Red is the hottest and mauve is the coldest. The flow of heat from one place to another is governed by a field of physics called thermodynamics.

temperature T_c and the hot temperature T_h to give the fraction of the heat available to useful work as $(T_h - T_c)/T_h$. So if T_h is double T_c then 50 percent of the heat energy can be put to use – no more.

In 1850, German physicist Rudolf Clausius took this idea further. He enumerated the energy that is not available to do anything useful by introducing a new thermodynamic quantity – which he later named 'entropy' (Greek for 'conversion'). It can be thought of as the degree of 'disorder' in a thermodynamic system. A state with low entropy is highly ordered. In terms of Carnot's formula, it has well-partitioned areas of hot and cold, meaning that a lot of usable energy is available. On the other hand, a state with high entropy is very disordered, with no neat partition between hot and cold – meaning the system is at a more or less uniform temperature and consequently there is very little energy available to do work.

Clausius reasoned that entropy must always increase. Take two objects, one very hot and one very cold, and place them side by side. They form a low-entropy system. Heat flows from the hot object to the cold one – not vice versa – ultimately leading both objects to settle at the same temperature – at which point they become a high-entropy system.

Maxwell's demon

Office workers may already be familiar with the idea of entropy. Tidy your desk on Monday and, as the days tick by, it becomes messier, with your papers becoming strewn across the table. But the reverse never happens – start the week with a messy desk and you will never find that it has tidied itself by Friday. In the terms of the second law of thermodynamics, a tidy desk has low entropy while a messy desk has high entropy and, as the law states, entropy always increases. (Office managers may add that someone with a low-entropy tidy desk does more work.)

'In the terms of the second law of thermodynamics, a tidy desk has low entropy while a messy desk has high entropy – and, as the law states, entropy always increases.'

It is always possible to violate the second law of thermodynamics locally. For example, your refrigerator pumps heat from the cold interior and dumps it in your warm kitchen. At first sight, this seems to create a decrease in entropy because the fridge has created a partitioned hot zone and cold zone. However, the fridge has to work in order to achieve this. After all, it will not operate unless it is plugged in. And doing the work generates a rise in entropy elsewhere, which compensates for the localized drop. The entropy of the whole universe is always getting bigger.

Scottish physicist James Clerk Maxwell used this same approach to challenge the second law of thermodynamics. In a thought experiment, he supposed that a little creature – Maxwell's demon – could sit between a hot and a cold object, allowing hot atoms pass from cold to hot, but not vice versa. But his argument breaks down for the same reason.

Hot science

Clausius's law is the second of four laws of thermodynamics. The first law is a statement of energy conservation – that heat flowing into a body is equal to the increase in the body's internal heat energy plus the work it does. The third law says that at zero temperature (as defined on the Kelvin scale, where 0°C = 273.15K) then entropy is also zero. The fourth law? Well, there is no fourth law. Instead, it is called the zeroth law because, even though it was discovered later, it is considered more fundamental than the others. The zeroth law states that if a body *A* is in thermal equilibrium with body *B* (meaning no heat can flow between the two) and body *B* is in equilibrium with body *C*, then bodies *A* and *C* are also in equilibrium.

ABOVE This time-lapse image shows the wheels of a steam train in motion. Steam engines were an early application of thermodynamics – extracting useful energy from the heat generated by burning fuel. A later development was the internal combustion engine, by which most cars are still powered.

The second law of thermodynamics explains the failures of early attempts to build perpetual-motion machines – devices that their inventors claimed could generate unlimited amounts of useful work. It also turned out to be useful in the development of statistical mechanics, a method for calculating the bulk properties of materials based on the physics of their atoms. And ever-rising entropy is one scenario that cosmologists have considered for the ultimate fate of our universe. Mathematical analogies for thermodynamic entropy have even found their way into information theory and the physics describing black holes, where the surface area of a black hole seems to continually grow, just like entropy in the second law.

34 Foucault's pendulum

DEFINITION AN EXPERIMENTAL DEMONSTRATION OF THE EARTH'S ROTATION, USING A LARGE SWINGING PENDULUM

DISCOVERY FIRST PERFORMED IN PUBLIC IN THE PARIS OBSERVATORY BY FRENCH PHYSICIST LÉON FOUCAULT IN 1851

KEY BREAKTHROUGH FOUCAULT SAW THAT THE PLANE OF A SWINGING PENDULUM APPEARS TO ROTATE AS THE EARTH TURNS UNDER IT

IMPORTANCE THE ROTATION OF THE EARTH IS CRUCIAL IN UNDERSTANDING THE PLANET'S WEATHER SYSTEMS

Our planet rotates – it is a motion that creates day and night and makes the stars seem to wheel overhead. This fact is now common knowledge, yet at one time it was not known at all. The Earth's rotation was proven in a landmark experiment by Léon Foucault: he pointed out that if the Earth rotates, then the plane of a pendulum should also rotate with time.

Go outside on a clear night and watch the stars turning overhead. The effect is especially noticeable if you look through a telescope, with stars drifting steadily from one side of the telescope's field of view to the other. There are two possible explanations – either that the heavens are turning around the Earth or planet Earth itself is rotating.

The Ancient Greeks got the answer to the puzzle momentously wrong – opting for the former explanation. They believed that the Earth resides at the centre of the universe, surrounded by a concentric series of revolving spheres, which carried the Sun, the Moon and the planets on their orbits. The outermost sphere carried with it distant stars and turned once per day.

Cracks began to appear in the Greek model during the early 17th century when Galileo's telescopic observations of our solar system provided the first solid evidence that the planets revolve around the Sun and not the Earth. Galileo's observations supported the heliocentric view of the solar system that had been put forward by Polish astronomer Nicolaus Copernicus – in which the distant stars are fixed in space. And if that was true then there was only one other possibility to explain the observed motion of the stars at night – the Earth must turn on its own axis.

LEFT A time-lapse image of a night 'star trail', formed as the Earth rotates beneath the fixed starry background of the night sky. Earth's rotation is clearly demonstrated by Foucault's pendulum, in which the plane of a swinging pendulum is seen to rotate over time as the Earth turns beneath it.

The first public experiment to demonstrate this idea beyond doubt was carried out in 1851 by French physicist Léon Foucault. Born in Paris in 1819, he was a sickly boy and was home-tutored for most of his childhood. In his spare time he became adept at building complex toys, constructing a working steam engine and a telegraph. His mother, impressed by his manual dexterity in building these devices, was convinced that he would become a surgeon in later life. He duly entered medical school, but during his first clinical session he fainted at the sight of blood.

Foucault's aversion to blood forced him to drop out of medical school. Instead, he became assistant to the French doctor Alfred Donné, helping the doctor with his work on microscopy – studying and photographing biological tissue on the smallest scales. After three years working with Donné, Foucault struck out and concentrated on his own scientific research. He made contributions to the theory of light, measuring its speed and providing evidence to support the wave theory of light over the opposing corpuscular view.

But in 1851, an idea grabbed Foucault that took him in a new direction. He became interested in the rotation of the Earth, and realized that if he could design a support for a pendulum that allowed it to swing in any direction (not confined to a particular plane of oscillation) then he could carry out an experiment to verify the phenomenon. If the Earth really is rotating then the plane of oscillation of a freely swinging pendulum, as measured by a stationary observer on Earth, should rotate too.

'If the Earth really is rotating then the plane of oscillation of a freely swinging pendulum, as measured by a stationary observer on Earth, should rotate too.'

How it works

Imagine a pendulum positioned at the North or South Pole. The pendulum bob swings back and forth while the planet turns under it, so that in 24 hours – one complete rotation of the Earth – the plane of the pendulum appears to have rotated through 360° and come back to where it started. At the equator, the effect vanishes – a pendulum swinging here will not rotate at all. At intermediate latitudes, the effect is more complicated. Here, Foucault was able to show that, in general, a pendulum located at a latitude q on the planet's surface (where q is an angle varying from 0° at the equator up to 90° at the poles) will turn at a rate of $15 \times \sin q$ degrees per hour (where sin is the trigonometric function).

At the latitude of Paris (48.8°N), this gave an hourly turning rate of 11.3°. Foucault set up a prototype pendulum in the basement of his house, using the multidirectional mounting that he had invented. The results seemed to agree with his predictions. Next he set up a larger version of the device as a public exhibition in the Meridian Room of the Paris Observatory. Soon after, he set up his largest realization of the idea, hanging from the dome of the Panthéon – using a 28-kg (62-lb) pendulum bob suspended from a wire measuring 67 metres (73 yards).

ABOVE The rotation of the Earth, as demonstrated by Foucault's pendulum, is also responsible for creating hurricanes. Instead of the plane of a pendulum rotating, cloud systems are made to turn relative to the planet's surface – forming destructive high-speed vortices.

The longer wire minimized the movement at the pendulum's mount, meaning less energy was lost to friction. (Even so, those demonstration pendulums in science museums today utilize electric drive systems to compensate for losses.) The results from the Panthéon experiment exactly matched the predictions of Foucault's theory – proving that the Earth really does rotate.

Weather on the turn

The effect that makes pendulums turn is also responsible for the powerful energy within hurricanes. The process is known as the Coriolis effect, after the French mathematician who first predicted it for general rotating systems in the 1830s. The Coriolis effect states that moving objects in the southern hemisphere veer to the left and those in the northern hemisphere veer right.

Hurricanes begin as regions of low pressure. Surrounding air masses flow in to the low-pressure core to fill the vacuum and, as they do, the moving air currents veer either to the left or the right because of the Coriolis effect. Hurricanes release energy at a rate equivalent to detonating a ten-megaton nuclear bomb every 20 minutes. Our understanding of their behaviour follows from the fact that the Earth turns, as Foucault proved 150 years ago.

35 Natural selection

DEFINITION THE THEORY OF NATURAL SELECTION STATES THAT NEW SPECIES EMERGE AS ORGANISMS ADAPT TO THEIR ENVIRONMENT

DISCOVERY FIRST FORMULATED BY CHARLES DARWIN IN 1839 BUT NOT PUBLISHED UNTIL 1859, IN HIS BOOK *THE ORIGIN OF SPECIES*

KEY BREAKTHROUGH DARWIN FOUND THAT BIRDS ON ISOLATED ISLANDS HAVE BEAKS OPTIMIZED TO LOCAL FOOD SOURCES

IMPORTANCE THE THEORY OF NATURAL SELECTION WAS THE STARTING POINT FOR THE FIELD OF EVOLUTIONARY BIOLOGY

Charles Darwin's theory of natural selection revealed where species in the natural world come from. He proposed that existing species adapt to the environment in which they live, and that these adaptations – accumulated over the course of many generations – give rise to new organisms, which are so different that they constitute whole new species.

Charles Robert Darwin joined the crew of HMS *Beagle* in December 1831. The ship was about to embark upon a survey mission, studying coastlines and taking depth readings in the waters around South America – information that could then be used to draft accurate nautical charts.

The *Beagle*'s captain, Robert FitzRoy, having served on the ship during a previous survey mission, had realized that the expedition would benefit greatly from having a geologist on board. Darwin, who had studied natural history at the universities of Edinburgh and Cambridge – and who was keen to travel before settling into a career – got the job.

But the trip was longer than Darwin may have anticipated. The *Beagle*'s voyage lasted from 27 December 1831 to 2 October 1836. During this time Darwin became an accomplished natural historian, geologist and fossil collector. Many of the fossils he discovered seemed to show creatures that were similar, yet slightly different, to modern-day species. While still at sea, he wrote of his scepticism about the idea that species are fixed and unchanging. Upon his return to England, Darwin developed this idea. In particular, one set of observations he had made – of finches found on the Galápagos Islands, off the western coast of South America – proved especially interesting.

LEFT Pictured here are two forms of the peppered moth. The top one is the natural variety – its markings give it camouflage from predators when it rests on lichen-covered tree bark. During the Industrial Revolution in England, trees became coated with soot. As a result, the darker variety of moth emerged – an example of evolution by natural selection.

HMS *Beagle* had visited the Galápagos in 1835, and Darwin had put ashore at four of the islands, where he had surveyed the geology and wildlife – and collected samples of the native species. Darwin noted that the finches found on each island were virtually identical, except for the shape of their beaks. Some had short, thickset beaks; others had long, narrow beaks; and others had beak shapes somewhere in between.

It soon became clear to him that the shape of each bird's beak was optimized to its available food source. The finches ate cacti. But whereas the birds with the long, pointed beaks used them to pierce the fruit to get at the sweet flesh inside, those with large, broader beaks fed on the tougher base of the cactus plants – which were also rich in insects. When the finches were examined in 1837 by English ornithologist John Gould, he confirmed that each finch was indeed a separate species – not just a variety of the same species.

Transmutation of species

The beak variations led Darwin to investigate an idea called 'transmutation of species', put forward in 1809 by the French naturalist Jean-Baptiste Lamarck, and which stated that existing species are able to transform into new species. But how do they do it? In 1838, Darwin hit on the idea of organisms passing on their survival traits to their young. Darwin had become interested in the idea of traits being passed between generations after speaking with animal breeders. The idea of species survival being a contest for food and other resources had occurred to him after reading the works of the Swiss botanist Augustin-Pyramus de Candolle.

'Random mutations creep into each generation, and when the mutations confer a survival advantage then the offspring are more likely to reproduce and pass the new survival trait onto their own young.'

This train of thought led Darwin, in 1839, directly to his theory of evolution by natural selection. In it, he argued that species hand down their survival traits to their offspring. But random mutations also creep into each generation, and when the mutations confer a survival advantage then the offspring are more likely to reproduce and pass the new survival trait onto their own young. And when enough of these beneficial mutations accumulate, the result is a new species.

Fade to black

A palpable example of evolution by natural selection took place during the Industrial Revolution in England. The peppered moth originally had pale wings, dappled with dark markings. This enabled it to blend in against light-coloured lichens growing on trees, thus camouflaging it against predators. But during the 19th century, the pollution that accompanied the Industrial Revolution left much of its habitat blackened with soot. The pale peppered moth was then an easy target for birds. In response, a variant of the moth emerged that had darker wings. Mutations between generations naturally made some moths darker, and in the new climate these moths were better adapted to survive and pass the dark traits down to their offspring. Before long, all the peppered moths living in England's industrial quarters had wings that were coloured inky black.

Darwin was a little slow to cast his new theory into print. He now had a day job working as a geologist, forcing him to pursue his theory of evolution in his spare time. However, during the mid-1850s he became aware that the English biologist Alfred Russel Wallace was also working away on a similar theory. Darwin's competitive desire to be credited with being the first to publish on the subject galvanized him into fresh action, and in 1859 his 509-page *magnum opus* was published by the publishers John Murray. It was titled *On the Origin of Species by Means of Natural Selection, or the Preservation of Favoured Races in the Struggle for Life.*

Darwin received two-thirds of the profits from the book, amounting to £180 from the first edition – equivalent to almost £14,000 in 2010 UK sterling. Many further editions were printed, earning Darwin the equivalent of £225,000 by the time of his death in 1882. In the fifth edition, published in 1869, he coined a phrase that encapsulates the ethos of the theory: 'survival of the fittest'. There was considerable resistance to natural selection from religious figures and this opposition is still echoed today by some 'educators'.

ABOVE Galápagos tortoises mating. This species was studied by Darwin during his visit to the Galápagos Islands and helped him to formulate the theory of evolution by natural selection. The tortoises have a very long life expectancy: one kept in captivity, named Harriet, died at Australia Zoo in 2006, aged 175. She is believed to have been collected by Darwin himself during his expedition.

Dawkins and the selfish gene

The reaction to Darwin's theory from the scientific community was initially mixed. In time, however, new discoveries supported his findings and ruled out rival ideas. These discoveries included American botanist Gregor Mendel's work on inherited traits, and the revelation that traits are carried on 'genes' written into a molecule called DNA.

By the mid-20th century, Darwin's theory was universally accepted by scientists. It is the basis for modern evolutionary biology, with its emphasis on genes as the unit of inheritance and the competition between them as the driving force behind evolution. Evolutionary biology has been developed further by many biologists with Richard Dawkins's 1976 book, *The Selfish Gene*, becoming synonymous with Darwin's theory. The idea of inheritance has also influenced scientific understanding of the spread of knowledge, now described as the transfer of gene-like nuggets of data called 'memes'.

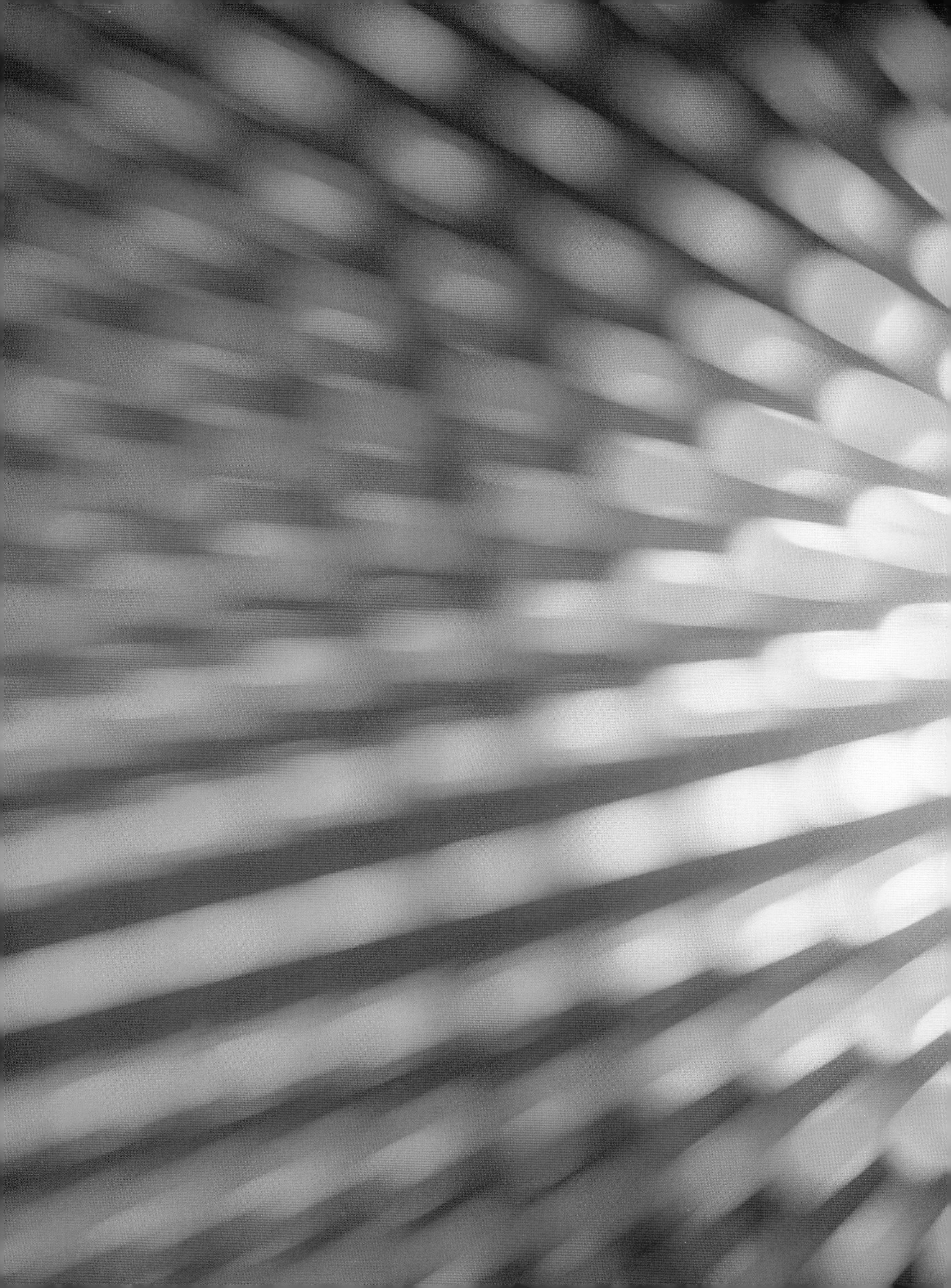

36 Spectroscopy

DEFINITION SPLITTING LIGHT INTO ITS CONSTITUENT WAVELENGTHS AND MEASURING THE BRIGHTNESS AT EACH WAVELENGTH

DISCOVERY ROBERT BUNSEN AND GUSTAV KIRCHHOFF ESTABLISHED SPECTROSCOPY AS A SCIENCE IN 1860

KEY BREAKTHROUGH BUNSEN AND KIRCHHOFF FOUND EACH ELEMENT SHOWS A UNIQUE VARIATION OF BRIGHTNESS WITH WAVELENGTH

IMPORTANCE SPECTROSCOPY REVOLUTIONIZED ALL ASPECTS OF ASTRONOMY, FROM PLANETARY SCIENCE TO COSMOLOGY

There is more to light than meets the eye. In the middle of the 19th century, two German scientists discovered that breaking up light into its component colours – and then looking at the brightness of each colour – could tell them the chemical make-up of the light's source. It was a discovery that would transform astronomy and yield further applications in terrestrial sciences, such as medicine, forensics and archaeology.

Spectroscopy is a vital field of experimental physics and chemistry that figures out what substances are made of by analyzing the light that they have emitted, which has passed through or has bounced off them. It works by breaking up the light into a spectrum – in other words, separating the light into different wavelengths.

Different wavelengths for visible light correspond to different colours, producing the familiar rainbow – red, orange, yellow, green, blue, indigo and violet. Red is the long wavelength end of the visible spectrum, with a wavelength of about 650 billionths of a metre (or nanometres, nm), while violet makes up the short wavelength end, at around 430nm.

Visible light is just one kind of electromagnetic wave. Other sorts exist with wavelengths longer than 650nm and shorter than 430nm. For example, infrared radiation has a wavelength longer than visible red light, ranging from 650nm up to around 300,000nm. At the other end of the scale, ultraviolet is radiation with a shorter wavelength than violet, ranging from about 430nm down to just 10nm. It is still possible to perform spectroscopy at these wavelengths.

LEFT The combination of a cross-screen filter and two polarizers allows visible light to be split into spectral colours. Spectroscopy originated as a study of visible light, but the discipline expanded to include all kinds of electromagnetic radiation and even quantum particles.

At visible wavelengths, light can be split into its constituent colours by shining it sideways through a glass prism. As the light passes through the glass it is refracted (bent) because its speed in glass is less than it is in air. Different wavelengths refract by different amounts, resulting in a spectrum when the light emerges on the other side of the prism.

Diffraction grating

Another way to split light is with a diffraction grating. This is a piece of transparent material or a polished reflecting surface that is scored with multiple lines (tens of thousands per cm for visible light). Light passing through, or reflected off, the grating undergoes diffraction, where it is bent by interference effects at an angle dependent upon its wavelength – thus fanning out the light into a spectrum. You can see the effect on a compact disc – the spiral track into which the data is etched forms a diffraction grating and produces a visible spectrum. Diffraction gratings are preferable to prisms because they absorb less light and can be used at wavelengths outside the visible spectrum, such as infrared and ultraviolet.

> **'By the early 19th century, scientists had noticed that when light was broken up into a spectrum in this way, the brightness of each colour was not always the same.'**

By the 19th century, scientists had noticed that when light was broken up into a spectrum in this way, the brightness of each colour was not always the same. Sometimes there would be distinct dark lines at some wavelengths; other times, sharp bright lines would be present. In the late 1850s, two German scientists joined forces to figure out why: the chemist Robert Bunsen and physicist Gustav Kirchhoff.

Bunsen had just invented the piece of lab equipment that still bears his name today – a gas burner that produces a very hot, clean flame, perfect for scientific work. Bunsen and Kirchhoff saw bright lines in the spectrum of metals that were heated on Bunsen's burner. The position of the lines was anything but random: each metal they studied – such as sodium, lithium and potassium – displayed a characteristic set of lines at definite wavelengths, and from which the type of metal being burned could be reliably identified.

In 1860, they used the technique to discover two new chemical elements. Bunsen saw spectral lines in a sample of mineral water that matched nothing they had seen so far. Upon further investigation, they isolated caesium as a new chemical element. The following year, they also identified rubidium.

Energy levels

It is not just hot, glowing substances that exhibit bright spectral lines. A cold substance illuminated by light also absorbs some of that light at specific wavelengths, forming a pattern of dark lines in the light's spectrum. Kirchhoff found that these dark 'absorption lines' seen in light that had passed through a cold substance were at the same wavelengths as the bright 'emission lines' produced when the same substance was heated.

Despite their systematic studies and groundbreaking experiments, Bunsen and Kirchhoff could not explain the patterns of absorption and emission. Indeed, neither would anyone else for the next 50 years. The explanation finally came from Niels Bohr in 1913.

Bohr's new model of the atom, using the field of quantum theory, had electrons orbiting the atom's nucleus. However, the laws of quantum theory only allowed the particles to do this on a small number of discrete, well-defined orbits. Each orbit has a characteristic energy so that when a photon (particle of light) with an energy that exactly matches the gap between two orbits strikes the atom, it is absorbed and its energy is used to raise the electron to a higher energy level. A photon's energy is linked to its wavelength, and the effect makes the atom selectively absorb light at a well-defined wavelength, producing a dark line in the light's spectrum. Since all elements have a different quantum structure, their electron energy levels and spectral lines vary too.

Astronomical applications

Spectroscopy is one of the most fruitful areas of experimental science. It is a primary tool in all forms of chemical analysis, from medicine to forensics. But perhaps its biggest contribution is in astronomy. Kirchhoff used it to study the light from the Sun, revealing a multitude of spectral lines that proved the Sun is made up of known chemical elements. And soon after, studies of the solar spectrum led to the discovery of the element helium.

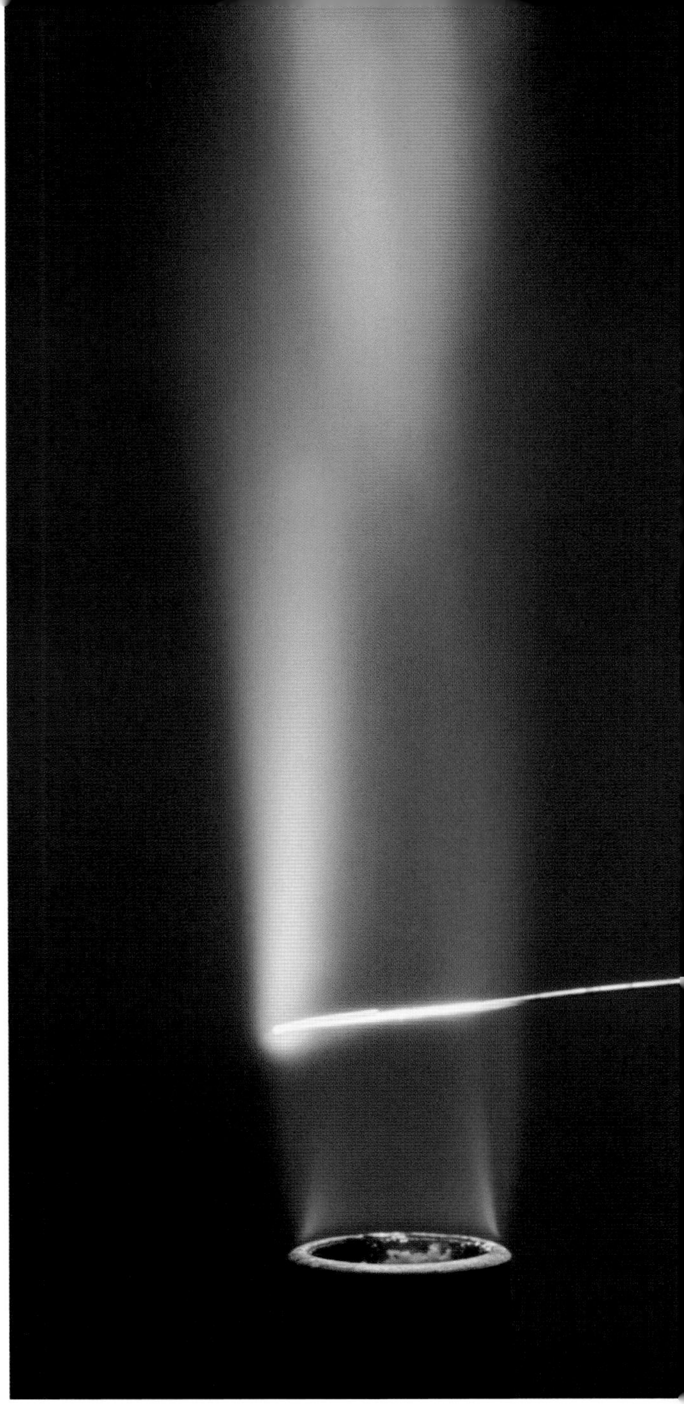

ABOVE Different chemical elements burn with different colours, as revealed in this flame test for the element rubidium.

At around the same time, English astronomer William Huggins used spectroscopy to distinguish between galaxies and gaseous nebulae. He also used the Doppler shift in the spectral lines of moving stars (equivalent to the shift in the wavelength of a moving ambulance siren as it passes by) to infer their speed. Spectroscopy allowed astronomers to discover the expansion of the universe, and now it is used to analyze the atmospheric composition of extrasolar planets – where we might just find alien life.

37 Maxwell's equations

DEFINITION MATHEMATICAL EQUATIONS LINKING ELECTRIC AND MAGNETIC FIELDS, ELECTRIC CHARGE AND ELECTRIC CURRENT

DISCOVERY FORMULATED IN 1861 BY SCOTTISH PHYSICIST JAMES CLERK MAXWELL WORKING AT KING'S COLLEGE, LONDON

KEY BREAKTHROUGH EXPANSION OF ELECTROMAGNETIC THEORY AND DISTILLATION INTO FOUR EQUATIONS

IMPORTANCE UNIFIED ELECTRICITY AND MAGNETISM, AND PREDICTED RADIO WAVES THAT LED TO WIRELESS COMMUNICATION

Mobile phones, television, radio – even wireless internet connections – were all made possible by the theory of electromagnetism put forward in 1861 by a Scottish physicist called James Clerk Maxwell. But his discovery was not just about technology. Maxwell's mathematical equations were, and still are, a guiding light in the advancement of theoretical physics.

Following Alessandro Volta's invention of the battery in the 19th century, the theory of electricity developed rapidly, leading to the development of electrical circuits, electric light, telegraphy and the telephone. But scientists were also discovering more about electricity's intimate link with magnetism. Magnetism was a force with which the Ancient Greeks were already familiar, and the Chinese used the magnetic stone magnetite to make early compasses.

In 1820, the Danish physicist Hans Christian Oersted noticed the presence of a magnetic field around a wire carrying an electric current. As he moved a compass around the wire he saw the needle change direction, revealing that the magnetic field formed a loop around the current. His findings stirred great interest in the physics world, and French physicist André-Marie Ampère subsequently worked out a formula giving the force of attraction between two parallel, current-carrying wires, caused by their magnetic fields.

English scientist Michael Faraday used this principle to build an electric motor, using magnets and current-carrying coils to produce motion. Faraday later discovered that the converse is also true – that a wire moving in a magnetic field generates an electric current by an effect he called 'induction'. Faraday used this principle to build the dynamo – the first electricity generator.

LEFT Kirlian plasma globe, a kind of executive toy made with a glass mantle and containing a central wand that is held at a high voltage. When the outside of the globe is touched, the voltage of the glass changes, causing the electric current to arc spectacularly from the wand. The process is described by Maxwell's theory of electromagnetism.

These results hinted at a connection between the laws of magnetism and electricity. And it was James Clerk Maxwell who dug down deep enough to find the link. Born in 1831 in Edinburgh, Maxwell had an illustrious academic career, finally becoming Chair of Natural Philosophy at King's College, London in 1860.

It was at King's College that Maxwell carried out his seminal work to unify electricity and magnetism. He was familiar with the idea Faraday had put forward that electricity and magnetism were linked through 'lines of force' – inspired by the lines that could be traced out around a magnet using a compass (or iron filings). In 1861, Maxwell used Faraday's theory to distill his knowledge of electricity and magnetism down to a linked set of four mathematical equations. The equations dictated the behaviour of electric and magnetic fields in response to distributions of electric charge and electric current. They became known as Maxwell's equations, and the resulting theory was called 'electromagnetism'. He wrote up his findings as a four-part paper entitled 'On Physical Lines of Force', published in 1861–1862.

Vector calculus

Maxwell's equations were a neat formulation of a complex physical theory, largely due to his use of a field of mathematics called 'vector calculus'. Indeed, much of the field was developed specifically with electromagnetism in mind. Faraday had realized that – unlike other physical quantities, such as mass and electric charge – his lines of force could not be specified by a single number. They had a direction associated with them in three-dimensional space. And quantifying this directionality meant that the lines had to have not one but three component values – one for each dimension of space. These quantities are described as vectors.

'These results hinted at a deep connection between the laws of magnetism and electricity. And it was James Clerk Maxwell who dug down deep enough to find the link.'

The velocity of a ship sailing on the two-dimensional surface of the Earth is another example of a vector. In general terms, the ship's velocity is a vector made up of two components – one part in the east–west direction and the other in the north–south direction. Together, these two components give the speed of the ship and its direction. Vector calculus was developed further in the 1880s by the mathematicians Josiah Willard Gibbs in America and Oliver Heaviside in England.

Oscillating fields

One of Maxwell's four equations is a generalized form of Ampère's law for the force of attraction between two current-carrying wires. However, Maxwell realized that the form of Ampère's equation was not quite complete. So he added a correction, which essentially meant that it is not just electric current that creates a magnetic field, but that a time-varying electric field has the same effect. This meant that electric fields give rise to magnetic fields and magnetic fields give rise to electric fields. Maxwell showed that this led to the phenomenon of self-sustaining electromagnetic waves that could travel through space. When Maxwell computed the speed of the waves – which turned out to be a simple combination of the magnetic and electric properties of empty space – it was extremely close to existing measurements of the speed of light. In a paper published in 1864, Maxwell concluded that light is a wave of oscillating electric and magnetic fields.

Radio waves

German physicist Heinrich Hertz provided the experimental proof in 1888 when he transmitted and received long-wavelength electromagnetic (EM) waves using devices built according to Maxwell's theory. These were radio waves. Other kinds of EM waves are also known, such as infrared, ultraviolet and X-rays – they are all electromagnetic waves, but with different wavelengths. Radio waves have wavelengths measured in the hundreds of metres, for infrared the wavelength is in the hundredths-of-a-millimetre range, while the wavelength of visible light is just a few ten-thousandths of a millimetre.

Maxwell's breakthrough was a milestone in the development of theoretical physics, and has been compared to Newton's *Principia* and the work of Albert Einstein. Indeed, Einstein's special theory of relativity and, later, quantum field theory were motivated by electromagnetism. As well as providing the theory that underpins wireless communication, Maxwell highlighted the vital concept of 'fields', which form the heartland of modern physics research.

BELOW Electric motor from a computer DVD drive. Coils of wire are wound around pieces of iron. When current flows through the wire, the iron becomes magnetized and propels the DVD turntable around. The operation of electric motors and electric generators are both described by Maxwell's equations.

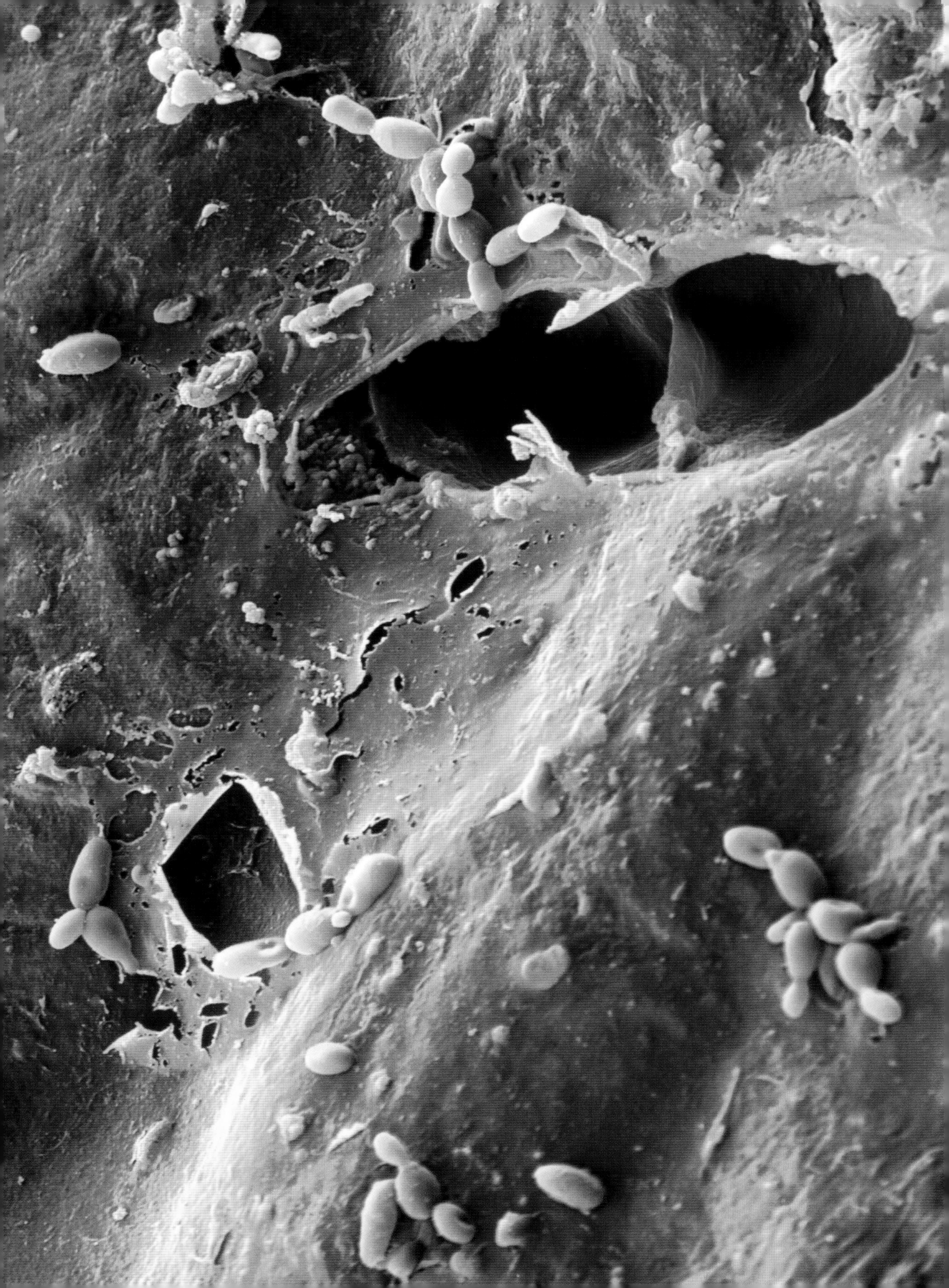

38 Germ theory

DEFINITION A THEORY STATING THAT DISEASES ARE CAUSED BY THE ACTION OF AIRBORNE MICRO-ORGANISMS

DISCOVERY PROVED BEYOND REASONABLE DOUBT BY THE FRENCH MICROBIOLOGIST LOUIS PASTEUR IN 1864

KEY BREAKTHROUGH PASTEUR PREVENTED FOOD SPOILING USING CONTAINERS FITTED WITH FILTERS TO BLOCK AIRBORNE PARTICLES

IMPORTANCE LED TO THE PASTEURIZATION HEAT-TREATMENT PROCESS AND LATER TO THE DEVELOPMENT OF ANTISEPTICS

Today it is second nature – if someone has a virus you do not go near them, for fear that you'll catch it and get sick yourself. But the idea that diseases are caused by germs – tiny micro-organisms – was only established in the second half of the 19th century. This so-called 'germ theory' led to antiseptics and improved hospital hygiene – not to mention the process of pasteurization to kill harmful bacteria in food.

Life springs randomly from inanimate matter – that was the basic gist of the theory of 'spontaneous generation' which scientists and natural philosophers, right up until the 19th century, believed was the cause of disease – and the reason why food that has been left out for too long goes mouldy.

The demise of the spontaneous generation idea first began in the 17th century when the invention of the compound (multi-lens) microscope brought the first glimmer in the understanding of biological processes. Dutch scientist Antony van Leeuwenhoek made the first observations of living biological cells and micro-organisms – life forms consisting of a few cells.

LEFT This electron microscope image reveals the surface of a rotting apple. The apple has become infected with a strain of single-celled yeast fungi (blue) and mould (green). The idea that diseases are the result of infection by micro-organisms such as these is the basis of germ theory.

By the 1830s, the vital role of cells was becoming clearer to scientists. German biologists Matthias Schleiden and Theodor Schwann encapsulated the view in their cell theory, proposed in 1838 and 1839. There was also a growing realization that cells can grow and multiply – set out formally by German biologist Rudolf Virchow. If diseases and infections were caused by cellular life forms, then cell division offered an enticing mechanism to explain how they spread. Some of the first evidence was discovered by Hungarian physician Ignaz Semmelweiss in 1847. While working in the maternity clinic at Vienna General Hospital, he noticed something odd about the mortality rates of mothers giving birth there. Mothers whose babies were delivered by doctors were more likely to subsequently die of fever than those delivered by midwives. It dawned on Semmelweiss that doctors and students regularly performed dissections of cadavers – while midwives did not. He wondered whether the women were becoming contaminated with 'cadaverous particles'.

To find out, Semmelweiss instructed doctors and students to wash their hands in chlorinated lime (weak bleach) before attending the maternity ward. After one month the mortality dropped from 18 percent to two percent.

Bad water

Just a few years later, a cholera outbreak in London was traced to water from a pump on Broad Street. Once the local authorities removed the handle from the pump, cases of the disease declined. It was later discovered that the well feeding the pump was contaminated by effluent from a nearby cesspit.

But it was Louis Pasteur who sounded the death knell for the spontaneous generation theory. In a series of experiments conducted between 1860 and 1864, he proved that disease is caused by micro-organisms. He did this by studying the development of moulds in a 'nutrient broth' – a cocktail of meat extract, protein and sugar that encourages the tiny organisms to grow.

Keep out!

Pasteur found that when samples of broth were left exposed to the air, they would quickly sprout new growths of cells feeding on the nutrients. However, when the samples were placed in sealed jars no growth took place. Crucially, Pasteur investigated what would happen if the seal was replaced by a filter that let in the air but blocked any particles carried by it. Again, nothing grew. The implication was clear: infections are caused by micro-organisms. These can be carried in the air and inhaled, passed through physical contact with infected objects and people (as in Semmelweiss's maternity ward) or ingested through contaminated food and water (as with cholera). In 1864, the French Academy of Sciences accepted Pasteur's conclusions and the germ theory of disease became an established principle.

'It was the French microbiologist Louis Pasteur who sounded the death knell for the spontaneous generation theory. In a series of experiments conducted between 1860 and 1864, he proved that disease is caused by micro-organisms.'

Bug busters

That disease is caused by organisms is good news, since the organisms themselves can be killed. Attention then turned to tracking down and fighting disease-causing microbes. Much of the research in this area was carried out by Prussian physician Robert Koch. He found the bacteria responsible for anthrax (*Bacillus anthracis*, discovered in 1877) and cholera (*Vibrio cholerae*, found in 1883), and he developed a system to isolate the microbes causing any particular illness – framed as a set of axioms, known as Koch's postulates.

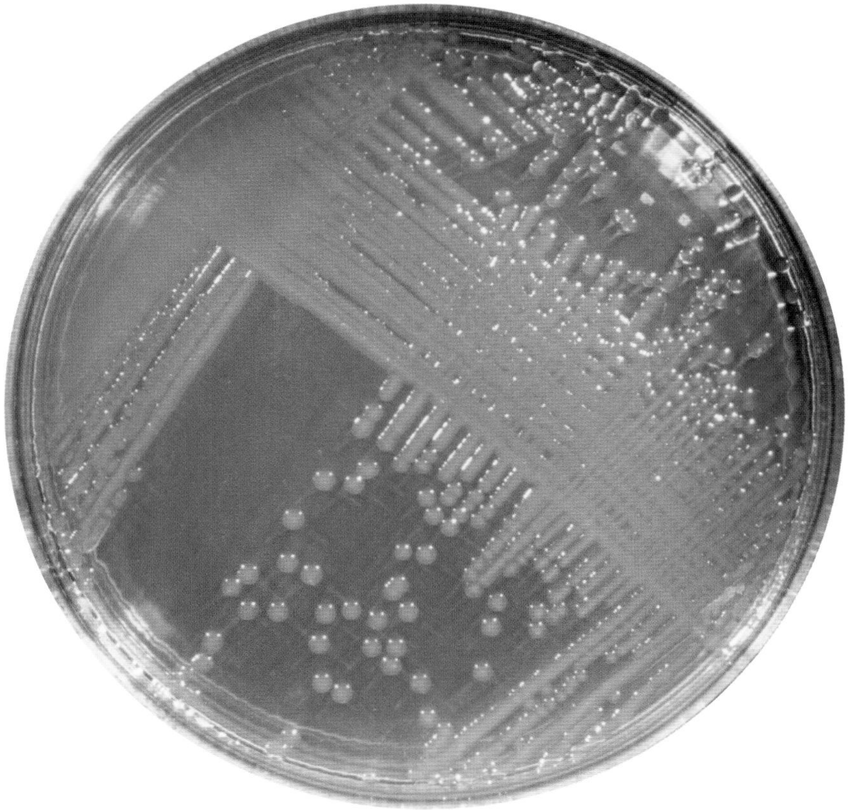

LEFT Cholera bacteria thriving in a Petri dish. The bacteria are lethal to humans, triggering the release of a toxin in the digestive tract that causes severe diarrhoea and dehydration. Cholera is spread by poor sanitation, usually when effluent mixes with drinking water.

Pasteur himself famously pioneered a technique of heat treatment that is often applied to foods to destroy harmful bacteria, thus increasing the food's shelf life. Called pasteurization, it is most commonly applied to milk. Boiling the milk is not an option, as this can cause it to curdle – however, heating it to 72°C for 16 seconds, Pasteur showed, is sufficient to kill all the harmful bacteria and give the unopened milk a shelf life of up to 16 days.

Scientists have since developed ultra-high temperature (UHT) treatment methods too, where milk is exposed to temperatures of over 100°C for a fraction of a second – long enough to kill the bacteria, yet quick enough for the milk not to curdle. UHT milk can be kept unopened for months, even though its taste and nutritional value are impacted by the heat treatment. And it is not just milk that can be pasteurized – canned food, fruit juice and bread are among the comestibles that can be preserved by heat treatment.

The development of the germ theory galvanized the medical establishment to combat infection from harmful bacteria. That meant during surgery – through the work of Joseph Lister on antiseptics – and through measures to improve overall cleanliness and levels of hygiene in hospitals. It also led to the use of antibiotic medications – drugs that fight infection by killing the bacteria responsible.

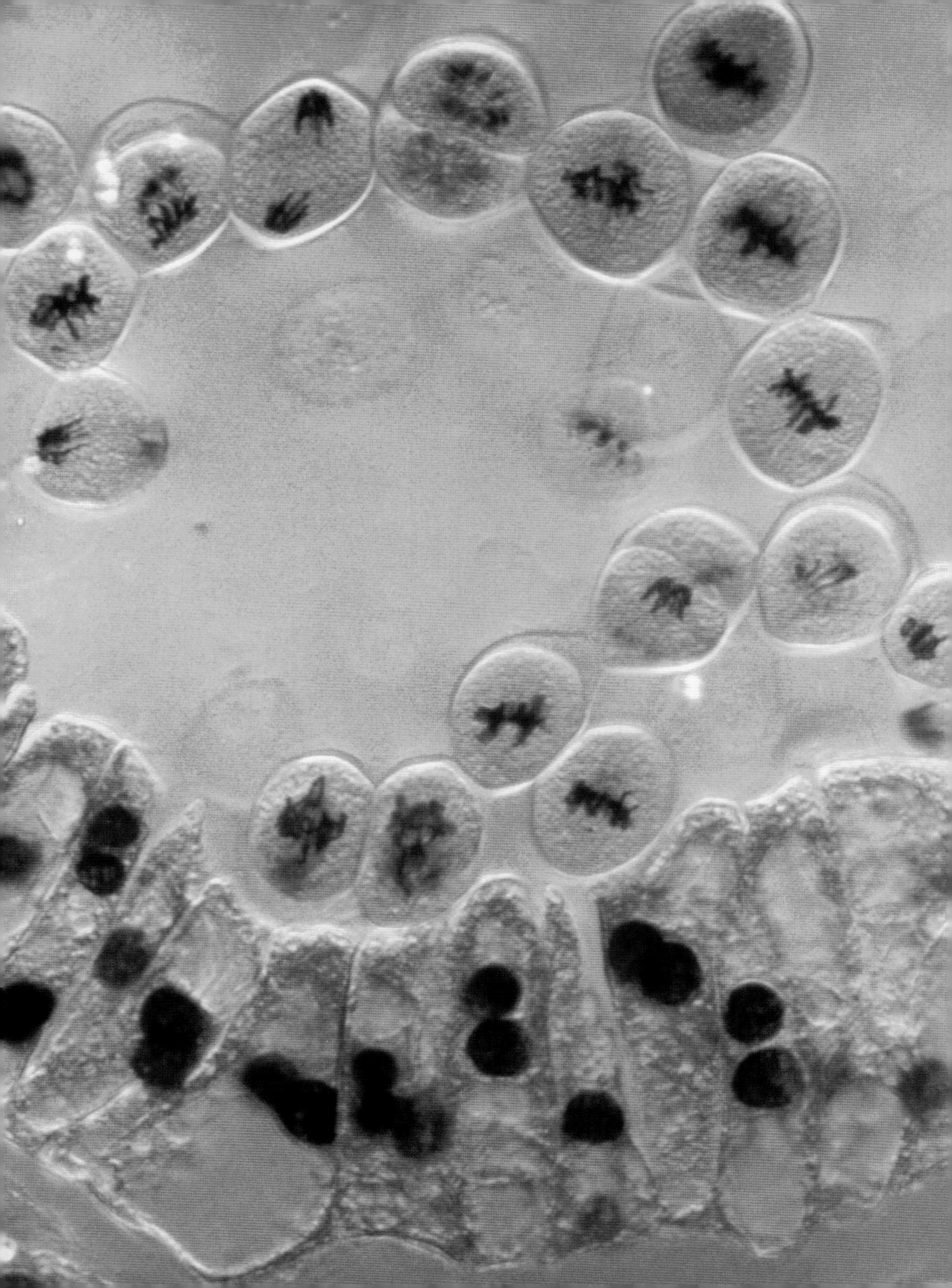

39 Mendel's laws of inheritance

DEFINITION BIOLOGICAL LAWS THAT GOVERN HOW CHARACTERISTICS ARE HANDED DOWN FROM PARENT ORGANISMS TO OFFSPRING

DISCOVERY FORMULATED BY GREGOR MENDEL, AN AUGUSTINIAN MONK, AND PUBLISHED IN 1866

KEY BREAKTHROUGH FINDING THAT BIOLOGICAL DATA PASSES FROM ONE GENERATION TO THE NEXT IN DISCRETE CHUNKS

IMPORTANCE MENDEL'S DISCRETE CHUNKS WERE LATER GIVEN A NEW NAME: GENES. HIS WORK IS THE FOUNDATION OF GENETICS

From eye colour to the likelihood of developing certain illnesses, we inherit our traits and characteristics from our parents. The laws governing how these traits are passed on were first formulated by Austrian monk Gregor Mendel. His work provided the missing piece in Darwin's theory of evolution, and changed the face of biology.

Gregor Mendel's new experiment would tax his monastic patience to the limit. Mendel, a monk at the Augustinian Abbey of St. Thomas in Brno, in the modern-day Czech Republic, was about to spend seven years of his life studying peas. Yet his dedication would lead him to make what has been described as the single most important contribution to biological science.

Between 1856 and 1863, Mendel patiently grew and cross-pollinated some 29,000 pea plants in the monastery's garden. In one experiment, he crossed a smooth, yellow variety of pea with one that was green and wrinkly. He expected a new, slightly wrinkly strain of pea, with a colour somewhere between green and yellow. But Mendel was in for a surprise. The peas in the new generation were either just as smooth or just as wrinkly as their parents, and either green or yellow – there was no middle ground. More surprising still, in some cases the traits had become jumbled up, making some peas yellow and wrinkly while others were green and smooth.

It was as if the traits were being passed from one generation of pea to the next in indivisible, discrete chunks. Mendel referred to the chunks of biological information that are responsible for the inheritance of traits as 'factors'. Today, we know them by a different name – genes.

LEFT Cells undergoing 'meiosis', whereby the nucleus divides as a precursor to cell division. The new cell nuclei contain one set of chromosomes, rather than the usual two. Meiosis is used to create reproductive sperm and egg cells that combine during conception to form an embryo that carries genetic data from both its mother and father.

ABOVE Four female generations of a family. The great-granddaughter shares roughly half of her genetic make-up with her mother, a quarter with the grandmother, and an eighth with the great-grandmother. Traits encoded by genes are passed down according to Mendel's laws of inheritance.

By studying many generations of his pea plants, Mendel could chart how the genes for colour and wrinkliness were handed from plants to their offspring. And he was able to formulate two laws describing the process.

Law of segregation

The first law is the 'law of segregation'. Mendel realized that each pea plant has two copies of the gene for any particular trait – one from each parent. So, a plant with a green parent and a yellow parent might have two copies of the colour gene – one green and one yellow. The law of segregation states that when this plant reproduces, only one of these copies can go into each of its reproductive cells – and so be passed to its offspring.

But if a pea plant can have both the genes for green and yellow peas, then what determines their actual colour? Mendel hypothesized that some genes are 'dominant', while others are 'recessive'. For peas, the yellow gene is dominant while the green gene is recessive. That means that if a pea plant has one of each then the peas it produces will still be yellow. Even though the presence of a recessive gene may have no bearing on the appearance of a plant, the gene is still there and can still be passed to future generations – where it can manifest itself should it happen to combine with another recessive copy. For example, two pea plants each with one yellow and one green colour gene will be yellow. But they can still have green offspring – if they both happen to pass on the recessive green gene.

What determines this? That is where Mendel's second law comes in. Also known as the 'inheritance law', it states that the particular copy of a gene that goes into a plant's reproductive cells during their creation – a process known as meiosis – is determined entirely at random. So a plant with two different copies of a particular gene – say, 'yellow' and 'green' – will randomly generate some ovules and pollen with the green version and some with the yellow version.

Magic numbers

Mendel confirmed the laws by cross-breeding yellow and green pea varieties. Because of the second law, the genes for these traits were mixed randomly in the next generation of plants, leading to the combinations yellow-yellow, yellow-green, green-yellow and green-green in equal proportions. However, because of the recessive-dominant nature of the genes, the first three combinations all produce yellow pea plants while only the last combination makes green ones. That is, the ratio of yellow to green plants should be 3:1 – and that's exactly what Mendel observed.

But, of course, Mendel's peas were more interesting than that. They were not just yellow or green; they were either smooth or wrinkly, too. Mendel found that the gene for smoothness is dominant over the gene for wrinkliness. In other words, the ratio of plants with smooth peas to those with wrinkly peas was also 3:1. And, by multiplying these shape and colour ratios together, the theory correctly predicted the ratio of yellow-smooth to yellow-wrinkly to green-smooth to green-wrinkly to 9:3:3:1.

In fact, Mendel started out with 'pure' strains of peas – each of his generation-zero parent plants had only yellow and smooth genes, or green and wrinkly genes. That meant that the first-generation plants were all yellow and smooth. It was only with the second generation that the fun began as the recessive green and wrinkly genes – still hidden in the first generation – could combine and produce the magic 9:3:3:1 ratio.

'Mendel referred to the chunks of biological information that are responsible for the inheritance of traits as "factors". Today, we know them by a different name – genes.'

Father of genetics

Peas are, in fact, a special case, where traits are determined by a single gene. In honour of Mendel, these are known as Mendelian traits. In humans, characteristics such as dimples or a widow's peak hairline – as well as diseases such as sickle-cell anaemia and cystic fibrosis – are also examples of Mendelian traits. In general, however, genetic inheritance is more complex, with traits determined by complex interactions between genes. Nevertheless, Mendel had revealed how parents pass on characteristics to their children. It was the missing piece in Darwin's theory of evolution, providing the mechanism by which natural mutations of a species take place. Although Mendel published his laws of inheritance in 1866, they were not embraced until the 20th century, by which time Mendel was long dead.

40 Antiseptics

DEFINITION CHEMICALS USED IN MEDICINE TO KILL MICRO-ORGANISMS THAT CAN LEAD TO INFECTION

DISCOVERY THE USE OF ANTISEPTIC METHODS DURING SURGERY WAS ESTABLISHED BY ENGLISH SURGEON JOSEPH LISTER IN 1867

KEY BREAKTHROUGH LISTER REALIZED THAT LOUIS PASTEUR'S GERM THEORY COULD EXPLAIN SURGICAL INFECTIONS

IMPORTANCE TRANSFORMED WORKING METHODS IN HOSPITALS, ESPECIALLY IN OPERATING THEATRES, SAVING COUNTLESS LIVES

Anyone who's ever squirted antiseptic cream on a cut or gargled with TCP (trichlorophenol) to shift a sore throat can thank English surgeon Joseph Lister. In 1867, Lister pioneered the use of antiseptics in medicine – using Louis Pasteur's germ theory of disease to link micro-organisms with the infection of surgical incisions. His work revolutionized hospital hygiene and led to the use of antiseptics as a preventative measure beyond the operating theatre – for example, with alcohol-based hand gels.

The development of antiseptics followed directly from the germ theory put forward by French microbiologist Louis Pasteur in 1864. Prior to this time, surgery was a filthy business with wounds frequently becoming infected, leading to a condition known as 'sepsis'. In amputations alone, the number of patients dying from sepsis was in excess of 40 percent. And this only got worse as the discovery of anaesthetics in 1846 removed the need for surgeons to be as quick as possible, meaning wounds were left open for longer.

LEFT Fibres of a wound dressing infected with the superbug bacterium MRSA (methicillin-resistant *Staphylococcus aureus*), shown as red blobs. MRSA can lead to infections, pneumonia and blood poisoning. Use of antiseptic dressings and hand gels minimize its spread, but MRSA is resistant to most antibiotics.

Chemical agents

There is evidence showing that the use of chemicals to prevent putrefaction goes back thousands of years. The Ancient Egyptians embalmed their dead – the technique of mummification – using chemicals such as 'natron' to extract moisture from corpses, which then prevented them from decomposing. Ancient Greek and Roman physicians laced wound dressings with wine and vinegar to ward off infection. Meanwhile, many 'bog bodies' have been recovered by archaeologists from peat-cutting sites in a surprisingly good state of preservation, suggesting that natural chemicals within the peat halt the process of decay.

Accordingly, some doctors had already realized that infection rates were a symptom of poor hygiene. For example, in 1750, the Scottish military physician John Pringle published results of experiments showing the infection-fighting properties of acids and alcohol in response to diseases spread though unclean conditions. His research was based on six years of active military service, and additional observations on the spread of typhus in prisons. Despite this, medicine retained woefully poor levels of hygiene.

But Joseph Lister was about to change all that. Lister was born on 5 April 1827 to a Quaker family living in Essex. He graduated from the Royal College of Surgeons in 1854, having studied medicine at the University of London, and took up a position as a surgical assistant at Edinburgh Royal Infirmary. In 1859, he moved to Glasgow, being appointed as a professor of surgery at the university and in 1861 becoming a surgeon at Glasgow Royal Infirmary.

'Having heard that carbolic acid had been used to lessen the odour of sewage in the English city of Carlisle, Lister began using the chemical to treat surgical instruments, wound sites and dressings.'

Carbolic cleaners

It was while working in Glasgow that Lister heard about Louis Pasteur's germ theory, published in 1864. This theory explained disease as being caused by micro-organisms. Lister immediately realized that the same mechanism could also explain post-operative sepsis. Having heard that carbolic acid had been used to lessen the foul odour of sewage in the English city of Carlisle, Lister began using the same chemical to treat surgical instruments, wound sites and dressings. He even employed someone during surgery to spray a mixture of carbolic acid and steam into the air in operating theatres during surgery.

Like Ignaz Semmelweiss (see page 157), Lister also required that surgeons in the hospital wash their hands – using a weak carbolic acid solution – and that they should always wear clean gloves and gowns from one procedure to the next. In addition, he gradually replaced wooden-handled surgical instruments (which could absorb microbes) with all-metal designs.

Spreading the word

Lister's methods were very successful, dramatically reducing incidence of fatal infections, such as gangrene. He wrote up his findings as a series of papers entitled 'Antiseptic Principle of the Practice of Surgery' that were published in the *British Medical Journal* in 1867. Lister had turned the use of antiseptics into a science – a science that was now there in black and white for the rest of the medical community to read.

Antiseptics work by depriving micro-organisms of the environment they require in order to survive. The reason why the ancient human bodies preserved in peat bogs have sustained so little decay is the special combination of low temperature, high soil acidity and complete lack of oxygen – which is a harsh physical environment for micro-organisms to live in. And this is how clinical antiseptics work, too. They create an environment that is highly toxic to harmful microbes, and yet relatively non-toxic to human tissue.

Alcohol-based antiseptics (of the sort present in the hand gels used on all hospital wards) work by causing proteins in the micro-organisms to coagulate, disrupting the healthy operation of the cell. Surprisingly, a solution of 70 percent alcohol and water is a more effective antiseptic than 100 percent alcohol, because the extra water makes the solution better able to penetrate to the interior of germ cells where it can do more damage.

Antibiotics and disinfectants

Other chemical compounds that can kill microbes are more subtle and sophisticated in their action. Where the need to protect healthy cells is stringent, antibiotic drugs are sometimes used for fighting internal infections deep inside a patient's body (see 'Penicillin' on page 245). But there also exist fiercer chemicals, such as disinfectants, that are used to blast microbes from inanimate surfaces.

ABOVE Tollund Man, the mummified corpse of a man thought to have lived in the third century BC, recovered in 1950 from a peat bog in Denmark. The body has been preserved in part by the high acidity of the peat, which has killed the micro-organisms normally responsible for decay. This is exactly how some antiseptics work.

By 1900, Lister's methods were in general use around the world. Ultimately, the skin irritation caused by antiseptics has led operating theatres to switch to aseptic instead of antiseptic methods – where the emphasis is on not bringing micro-organisms into the theatre in the first place, rather than killing them once they are in there. Instruments and garments are sterilized in extreme heat prior to surgery, rather than being treated with chemicals in situ.

Nevertheless, Lister's methods revolutionized the understanding of infection in medicine and the need for the highest standards of hygiene in hospitals. Joseph Lister was made Baron of Lyme Regis in honour of his achievements. The bacteria species *Listeria* and the trademarked antiseptic mouthwash Listerine are still named after him today.

41 Periodic table

DEFINITION A TABLE LAYING OUT ALL THE CHEMICAL ELEMENTS ACCORDING TO THEIR PROPERTIES

DISCOVERY FIRST WRITTEN DOWN BY THE RUSSIAN CHEMIST DMITRI MENDELEEV, AT ST PETERSBURG STATE UNIVERSITY, IN 1869

KEY BREAKTHROUGH ARRANGING KNOWN ELEMENTS INTO ROWS AND COLUMNS SO THAT EACH COLUMN HAD SIMILAR PROPERTIES

IMPORTANCE PREDICTED MISSING ELEMENTS AND WAS A GUIDING LIGHT IN DEVELOPING THEORETICAL MODELS OF THE ATOM

Memorized today by students everywhere, the periodic table is regarded by many as the greatest discovery in chemistry. It was discovered in 1869 by Russian chemist Dmitri Mendeleev, and organizes the known chemical elements according to their physical properties. The insights it provided into the order of the chemical world resulted in the prediction of undiscovered elements and steered the course of future theoretical research.

Dmitri Mendeleev was born in 1834, near Tobolsk in western Russia – the youngest of at least 14 children. He was educated at the Tobolsk Gymnasium and later at the Main Pedagogical Institute in St Petersburg after the death of his father forced his family to relocate there. He excelled at science and his first job was as a science teacher in Simferopol in Crimea.

Mendeleev returned to St Petersburg after a couple of years and soon became interested in scientific research. He worked on the capillary action of liquids – the way a fluid will rise up a narrow tube unassisted – and he contributed to the science of spectroscopy, which in 1861 became the subject of his first book. His work earned him a professorship in chemistry at St Petersburg State University in 1863. It was here that he began musing over the chemical elements – and how to organize them according to their properties.

Mendeleev began his task by initially ordering the 62 known chemical elements according to their atomic mass. This is the mass of each atom of the element, given by adding up the total number of positively charged protons and neutrons in its atomic nucleus. Listing the elements from left to right, from lowest to highest mass, Mendeleev noticed that if he arranged his

LEFT Coloured field-ion micrograph of a platinum crystal. The geometric patterns reveal the structure of the crystal. Platinum is a rare, but versatile, chemical element that is used in making jewellery, electronics and automotive catalytic converters. It is just one of the 118 known chemical elements.

sequence into straight rows, each of 18 columns, then the elements that lined up in each column tended to have very similar chemical properties. It was as if the properties were 'periodic', as atomic mass increased. For example, the far-left column of the table contained the elements sodium, potassium and lithium – all of which are solid at room temperature, highly reactive, have low melting points and tarnish when exposed to air. By contrast, the elements in the far-right column are all gaseous at room temperature, and are colourless, odourless and extremely unreactive. The columns within the table were referred to as 'groups', while the rows are known as 'periods'. He published his 'periodic table' in 1869.

Gaps to fill

Mendeleev was not the first to try and arrange chemical elements this way. But he was the first to order them primarily according to their properties, even if this meant leaving gaps in the table. For example, at the time, arsenic (As) was the next heaviest element known beyond zinc (Zn), which occupies group 12 of period 4. However, Mendeleev believed arsenic fitted better with group 15 rather than group 13 and so he left two gaps in the table. He predicted new elements would be discovered to fill the gaps. And over the next two decades, chemists discovered gallium (Ga) and germanium (Ge), which occupy groups 13 and 14.

'Mendeleev predicted that new elements would be discovered to fill the gaps in the table. And they duly were – over the next two decades, chemists discovered the elements gallium (Ga) and germanium (Ge)'

Mendeleev's determination to place elements where he thought they should go, according to their properties, even led him to break the order of increasing atomic mass in some cases – for example by placing tellurium (atomic mass 127.6) to the left of iodine (atomic mass 126.9). This led to the true ordering of the table as it is known today – not in terms of mass, but of atomic number. This is the electrical charge of the atomic nucleus, equal to the number of positively charged proton particles that the nucleus contains. The fact that Mendeleev's table is ordered by atomic number was later proven in 1913 by English physicist Henry Moseley.

Quantum chemistry

At around the same time, the development of quantum theory was providing the necessary physics by which to understand the structure of the periodic table. Chemical reactions are all about interactions between the electrons orbiting in the outermost regions of atoms. Quantum theory revealed that electrons are organized into a number of levels, each accommodating a fixed number of electrons. Since the overall electric charge of an atom is zero, the total number of negatively charged electrons equals the number of positive protons in the nucleus – which in turn equals the atomic number. As this increases from element to element, each electron level steadily fills up – until it is full and the process repeats. It soon became clear that each of Mendeleev's periodic groups contained elements with the same number of

1 H HYDROGEN																	2 He HELIUM
3 Li LITHIUM	4 Be BERYLIUM											5 B BORON	6 C CARBON	7 N NITROGEN	8 O OXYGEN	9 F FLUORINE	10 Ne NEON
11 Na SODIUM	12 Mg MAGNESIUM											13 Al ALUMINIUM	14 Si SILICON	15 P PHOSPHORUS	16 S SULFUR	17 Cl CHLORINE	18 Ar ARGON
19 K POTASSIUM	20 Ca CALCIUM	21 Sc SCANDIUM	22 Ti TITANIUM	23 V VANADIUM	24 Cr CHROMIUM	25 Mn MANGANESE	26 Fe IRON	27 Co COBALT	28 Ni NICKEL	29 Cu COPPER	30 Zn ZINC	31 Ga GALLIUM	32 Ge GERMANIUM	33 As ARSENIC	34 Se SELENIUM	35 Br BROMINE	36 Kr KRYPTON
37 Rb RUBIDIUM	38 Sr STRONTIUM	39 Y YTTRIUM	40 Zr ZIRCONIUM	41 Nb NIOBIUM	42 Mo MOLYBDENUM	43 Tc TECHNETIUM	44 Ru RUTHENIUM	45 Rh RHODIUM	46 Pd PALLADIUM	47 Ag SILVER	48 Cd CADMIUM	49 In INDIUM	50 Sn TIN	51 Sb ANTIMONY	52 Te TELLURIUM	53 I IODINE	54 Xe XENON
55 Cs CAESIUM	56 Ba BARIUM	57-71 LANTHANOIDS	72 Hf HAFNIUM	73 Ta TANTALUM	74 W TUNGSTEN	75 Re RHENIUM	76 Os OSMIUM	77 Ir IRIDIUM	78 Pt PLATINUM	79 Au GOLD	80 Hg MERCURY	81 Tl THALLIUM	82 Pb LEAD	83 Bi BISMUTH	84 Po POLONIUM	85 At ASTATINE	86 Rn RADON
87 Fr FRANCIUM	88 Ra RADIUM	89-103 ACTINOIDS	104 Rf RUTHERFORDIUM	105 Db DUBNIUM	106 Sg SEABORGIUM	107 Bh BOHRIUM	108 Hs HASSIUM	109 Mt MEITNERIUM	110 Ds DARMSTADTIUM	111 Rg ROENTGENIUM	112 Cp COPERNICIUM	113 Uut UNUNTRIUM	114 Uuq UNUNQUADIUM	115 Uup UNUNPENTIUM	116 Uuh UNUNHEXIUM	117 Uus UNUNSEPTIUM	118 Uuo UNUNOCTIUM

57 La LANTHANUM	58 Ce CERIUM	59 Pr PRAESODYMIUM	60 Nd NEODYMIUM	61 Pm PROMETHIUM	62 Sm SAMARIUM	63 Eu EUROPIUM	64 Gd GADOLINIUM	65 Tb TERBIUM	66 Dy DYSPROSIUM	67 Ho HOLMIUM	68 Er ERBIUM	69 Tm THULIUM	70 Yb YTTERBIUM	71 Lu LUTETIUM
89 Ac ACTINIUM	90 Th THULIUM	91 Pa PROTACTINIUM	92 U URANIUM	93 Np NEPTUNIUM	94 Pu PLUTONIUM	95 Am AMERICIUM	96 Cm CURIUM	97 Bk BERKELIUM	98 Cf CALIFORNIUM	99 Es EINSTEINIUM	100 Fm FERMIUM	101 Md MENDELEVIUM	102 No NOBELIUM	103 Lr LAWRENCIUM

ABOVE The periodic table in its final form. Elements are arranged in order of atomic number – the number of positively charged proton particles in their atomic nuclei. The elements are colour-coded according to type. Some of the two-letter abbreviations are taken from the element's Latin name. For example, gold uses 'Au' from the Latin word *aurum*.

electrons in their outermost level – and this is what accounts for their similar chemical properties. For example, the reactive elements in group 1 all have a single electron in their outermost level, while for all the unreactive elements in group 18 the outermost level is full. Quantum theory also explains the odd gaps in the top three periods, which arise because the innermost energy levels only have room for a small number of electrons and so fill up quickly.

Similarly, the outer layers can accommodate a large number of electrons – and this is the reason for the two pull-out bars at the bottom of the table. These have been clipped out of periods 6 and 7 to neaten the form of the table. The top one, called the 'lanthanide' sequence, should go in between barium (Ba) and hafnium (Hf) in period 6. The bottom one, the 'actinide' sequence, fits between radium (Ra) and rutherfordium (Rf) in period 7.

Modern elements

There are now 118 known elements in the periodic table, 94 of which occur naturally on Earth. The rest are radioactive with half-lives much shorter than the age of the planet and so have decayed away. The only way to see these latter elements is to manufacture them in nuclear reactors and particle accelerators. This was first done in the early 20th century with technetium – which, like gallium and germanium, was another of Mendeleev's predictions.

Mendeleev was nominated for the Nobel Prize in Chemistry in 1906 and 1907, but his disputes with members of the Nobel Committee dashed his chances of winning. Today, the chemical element mendelevium (symbol Md, atomic number 101) still carries his name – a fitting tribute to a man who dedicated his life to improving our understanding of the chemical world.

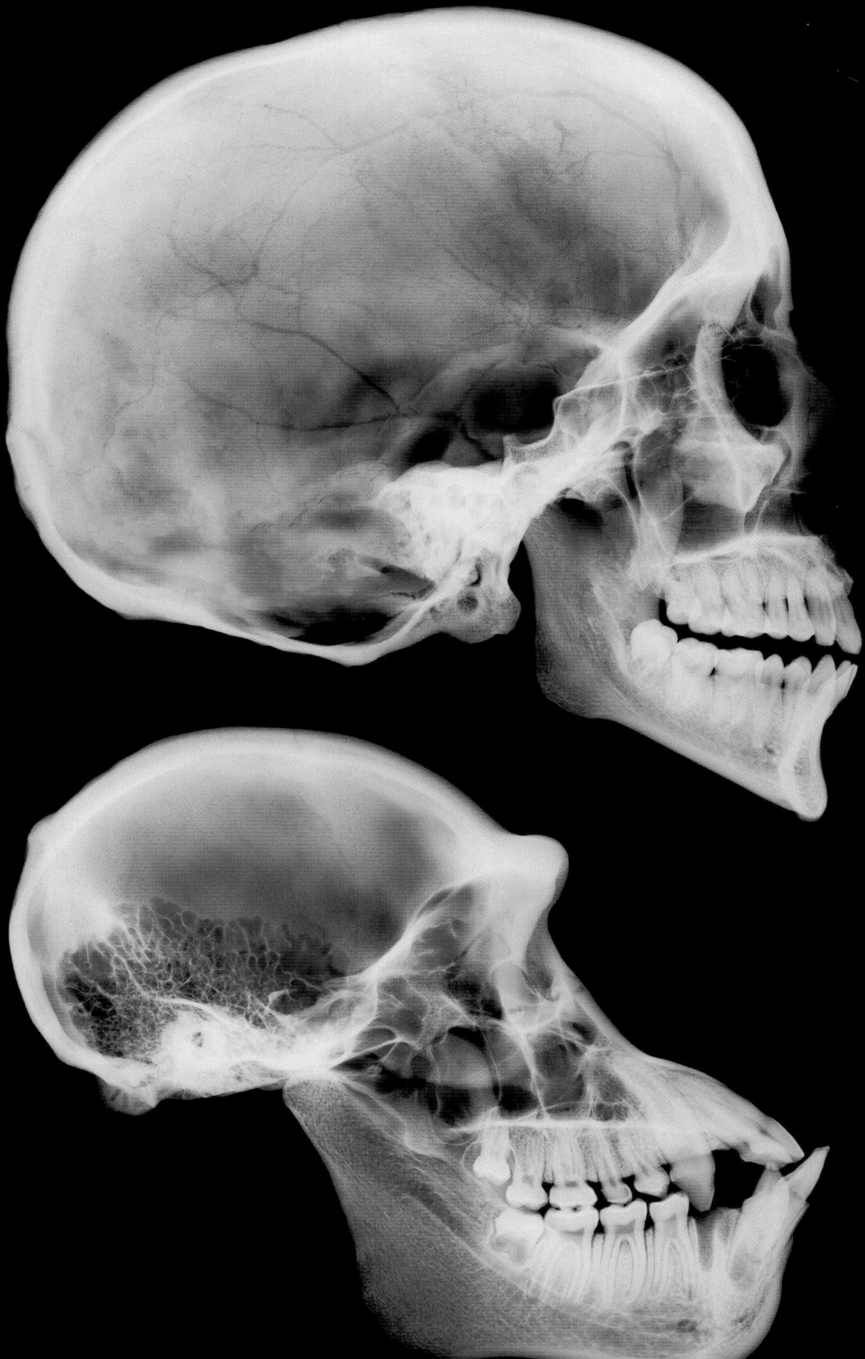

42 Out of Africa

DEFINITION THEORY THAT MODERN HUMANS APPEARED IN AFRICA 200,000 YEARS AGO, FROM WHERE THEY COLONIZED THE WORLD

DISCOVERY ENVISAGED BY CHARLES DARWIN IN 1839 BUT NOT PUBLISHED UNTIL 1871, IN HIS BOOK *THE DESCENT OF MAN*

KEY BREAKTHROUGH REALIZATION THAT HUMANS EVOLVED FROM PRIMITIVE SPECIES – THE OBVIOUS CANDIDATES BEING GREAT APES

IMPORTANCE THE 'OUT OF AFRICA' THEORY MEANS THAT EVERY HUMAN BEING, FROM EVERY CORNER OF THE GLOBE, IS RELATED

Up until the late 19th century, scientists thought that the human species had multiple origins – and that different races each emerged and evolved separately. Charles Darwin argued that humanity had a common point of origin in the African continent, and was determined to put his idea on a scientific footing. Archaeological and genetic research has now confirmed this idea, commonly known as the 'Out of Africa' hypothesis.

In 1859, Charles Darwin's *The Origin of Species* caused a revolution in religious and scientific circles alike, with its suggestion that individual species are not fixed and immutable – but that new ones slowly emerge via a process of evolution. A subtext – not explicitly mentioned in the 1859 book – was that human beings also fell into this category and that we, like all other animals, are descended from an earlier ancestor.

Twelve years later, Darwin stated his ideas on human evolution in no uncertain terms. In 1871, he completed *The Descent of Man, and Selection in Relation to Sex* (usually abbreviated to *The Descent of Man*). Originally published as two 450-page volumes, the book puts across Darwin's case that human beings evolved from the great apes.

The theory had already occurred to Darwin in the late 1830s following his expedition aboard HMS *Beagle*, which had led him to his ground-breaking theory of natural selection. As he consolidated his data on his return, he had sought the opinions of palaeontologists, ornithologists and other scientific experts. Darwin had visited a zoo to view a great ape, and commented how closely it resembled a human.

LEFT These two X-rays show the skull of a modern human (top) and that of a chimpanzee (bottom). The difference in the size of the brain cavity is striking, and is the reason why humans have been the more successful of the two species. Both are thought to have evolved from a common primate ancestor that lived in Africa between five and seven million years ago.

Darwin's observations led him to speculate in *The Descent of Man* that humankind's evolution from the apes had taken place in Africa – the home to both gorillas and chimpanzees – from where the new species had migrated to colonize the entire world. It was a theory that would later become known as Out of Africa.

Mental faculties

Many of Darwin's scientific contemporaries had criticized the inference in *The Origin of Species* that humans are simply a species of animal that has evolved. The principal objection was over the emergence of human mental abilities, which many believed were so far in advance of the apes that the gap between us and them was unbridgeable. In *The Descent of Man*, Darwin addressed his critics by drawing direct parallels with other animals, citing evidence that dogs and monkeys also demonstrate a primitive capacity for higher mental reasoning and emotion – and pointing out that the evolution of the mind has much in common with the evolution of the body.

Darwin also drew upon his personal encounters with primitive native peoples during the *Beagle* expedition to highlight the differences in human mental faculties that exist across the world. His thesis was that different human races are not different species, as most other scientists believed, but simply subspecies – variations on the original species of humans that had set out from Africa – that had subsequently evolved to suit the different living environments they encountered as a result of global migration.

> **'Researchers were able to analyze the mtDNA and Y chromosomes of races around the world, revealing a DNA paper trail that led back to eastern Africa around 200,000 years ago.'**

Indisputable evidence

The evidence for the Out of Africa theory did not accrue until the 1920s, when the first fossilized remains of early humans were discovered. One of the first key finds was the 'Taung skull' in South Africa, unearthed in 1924. Although Neanderthal and *Homo erectus* bones had been recovered during the late 19th century, the skull found at Tuang was much older (dating from two-to-three million years ago) and was clearly intermediate in form between humans and apes.

A slew of similar fossil discoveries across Africa over the following decades supported the theory. But it was the development of genetic techniques to trace the genealogy of early humans that was the real clincher.

Archaeogenetics

Genes – the essential information about our traits and characteristics – are stored within our cells on a biological molecule called DNA. Most DNA gets shuffled and mutated between generations but two areas in particular – mitochondrial DNA (mtDNA) and the Y chromosome – are always passed unchanged from one generation to the next. Two fossil specimens containing the same mtDNA or Y chromosome are, without question, related to one another. Beginning in the late 20th century, researchers were able to analyze the mtDNA and Y chromosomes of multiple races around the world, revealing a DNA paper trail that led back to eastern Africa around 200,000 years ago. This ancient African race had evolved significantly larger brains, giving them the intelligence to conquer all the other early hominid races that they encountered as they swept around the globe, including Neanderthals and *Homo erectus*. This hominid species is now called *Homo sapiens*, meaning 'wise man'. This is the taxonomic name used for modern humans.

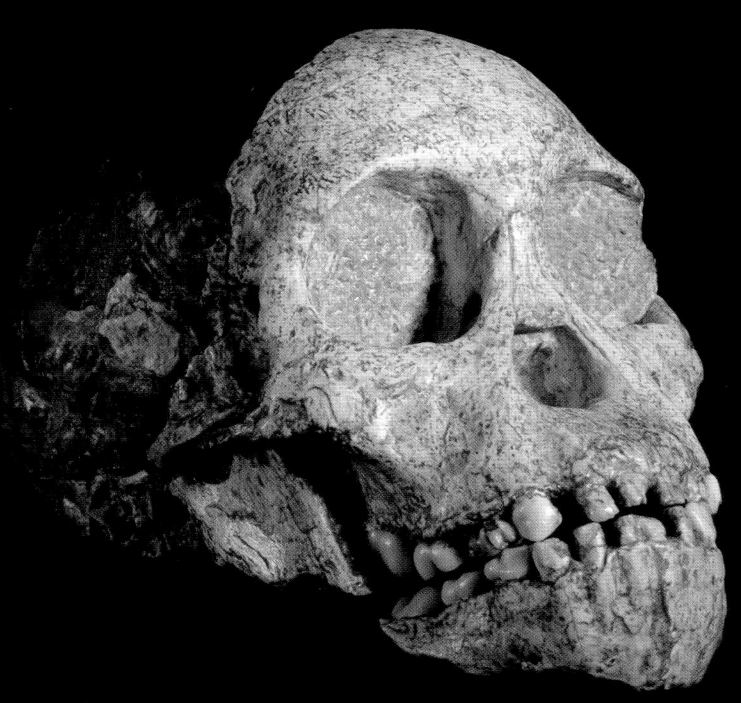

ABOVE This is the 'Taung Child' skull, the fossilized skull of an *Australopithecus africanus*, an early human ancestor believed to have inhabited southern Africa around two-and-a-half million years ago. Tests suggest that the specimen was three years old at the time of death. *A. africanus* had a larger brain than chimpanzees and human-like teeth.

Social Evolution

The Descent of Man also brought some eye-opening consequences for day-to-day human life. 'Social Darwinism' put a 'survival of the fittest' slant on modern life, in which the fittest individuals – those having the skills best suited to succeed in society – would rise up to the positions of power. A sinister by-product of this was the notion that helping 'less than fit' individuals to survive, through welfare measures and social policy, was weakening society and the strength of individual nations. This led to the alarming idea of eugenics, advocated by certain scientists in the 19th and 20th centuries, in which these 'less-fit' people should be discouraged – or forcibly prevented – from breeding.

Some biologists believe that human evolution has now pretty much ended – overtaken by our ability to correct our own weaknesses directly using technology: think of spectacles, sat nav and medicine. Others argue that the same rapid development of technology – and the need for us to keep up with it if we want to continue enjoying its benefits – is forcing our brains to evolve faster than ever before.

We may even be entering an era known as 'transhumanism' – where the modifications we can now make to the human body through scientific disciplines, such as genetics and cybernetics, are empowering us to dictate the course of our future evolution. If this theory is right, transhumanism may soon breed new species that are as far removed from modern humans as we are from our primate ancestors.

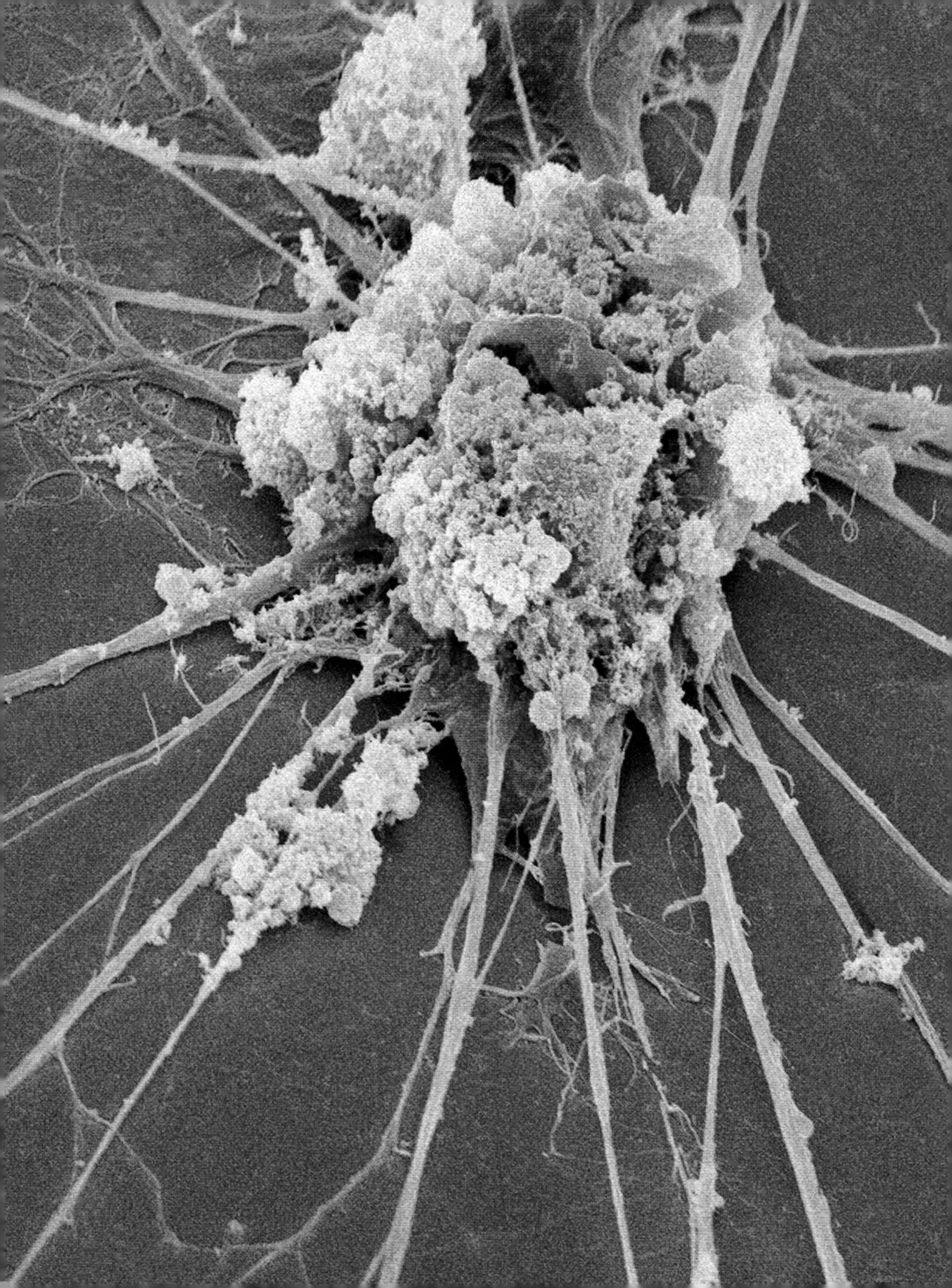

43 Nervous system

DEFINITION THE NETWORK THROUGH WHICH SENSORY SIGNALS ARE RELAYED TO THE BRAIN AND COMMANDS SENT BACK TO MUSCLES

DISCOVERY THE STRUCTURE OF THE NERVOUS SYSTEM WAS FIRST NOTED BY HISTOLOGIST SANTIAGO RAMÓN Y CAJAL IN 1889

KEY BREAKTHROUGH RAMÓN Y CAJAL DEVELOPED DYE TECHNIQUES THAT REVEALED NERVE CELLS UNDER THE MICROSCOPE

IMPORTANCE THE NERVOUS SYSTEM IS THE BODY'S COMMAND CENTRE AND THE SITE OF NEUROLOGICAL DISORDERS

The nervous system is the information superhighway of the human body and, indeed, the bodies of all other animal life forms. Our bodies are criss-crossed with fibres called nerves that transport information to the brain. The brain coordinates the body's response to that information which, in turn, gets transmitted back down the nervous system as electrical impulses that activate all the necessary muscles.

Nerves – and the brain itself – are made of a special kind of cell known as a neuron. The outer membrane of a neuron looks very different to the blobby appearance of ordinary cells, with straggly appendages called dendrites and axons across which electrical and chemical signals can be transmitted. Dendrites are short projections that surround the main body of the cell, and which are used to receive signals. The axon – each neuron has only one – is used for transmitting outgoing signals. The axon of a single cell may stretch from one side of the body to the other.

LEFT A scanning electron microscope image of a nerve cell (neuron) grown in a laboratory culture. The central body of the cell has axon and dendrite fibres extending out from it to send and receive nerve impulses. In this way, signals from the senses are relayed to the brain, while the brain in turn sends signals to control the muscles of the body.

CNS and PNS

The nervous system breaks down into two main divisions – the central nervous system (CNS) and the peripheral nervous system (PNS). The central nervous system includes the brain, the spinal cord and the retinas in your eyes. The peripheral nervous system radiates out from the spinal cord as a network of fibres that provides nervous responses from the arms, legs and the rest of the body. The fibres consist of bundles of axons extending from neuron cells rooted along the length of the spinal cord. In the peripheral nervous system, all signals are transmitted electrically along axons. In the central nervous system, however, this is supplemented by neuron-to-neuron communication through junctions between axons and dendrites, called synapses. A current travels to the end of an axon, where it triggers the release of chemicals called neurotransmitters. The presence of these chemicals is detected by a dendrite in the adjoining neuron, which then alters its behaviour accordingly – the signal has been passed on.

Neurons in the CNS that have longer dendrites than axons are sometimes known as sensory neurons. These transmit information from receptors – for example, touch receptors under your skin, or the light sensors in your retina – to the brain. Likewise, neurons that have longer axons than dendrites are often called motor neurons – these cells are responsible for relaying impulses from the brain to activate the muscles.

Neural circuits

The two kinds of neural cells work together to form neural 'circuits'. A typical circuit might involve sensory neurons picking up a signal, for example cells in your retina detecting a glass of orange juice on the table in front of you. This is followed by processing in the brain that reaches the conclusion that you would like to drink the juice, and the brain's motor neurons sending a signal to muscles in your arm to pick up the glass and raise it to your mouth. However, some neural circuits can be much shorter, and sometimes do not involve the brain at all. For instance, reflex responses, such as touching a hot stove and withdrawing your hand, never go near the brain – they are processed in the spinal cord, thus quickening the response time.

'Reflex responses, such as touching a hot stove and withdrawing your hand, never go near the brain – they are processed in the spinal cord, quickening the response time.'

These are motor responses involving what are known as voluntary muscles – muscles that you must make a conscious decision to move. They form a branch of the peripheral nervous system's motor division called the somatic nervous system. There is also another parallel branch called the autonomous nervous system. This automatically directs motor signals from the brain to subconsciously operate essential life processes, such as breathing and making the heart beat.

Unruly youth

The structure of the nervous system was first elucidated by Spanish physiologist Santiago Ramón y Cajal. He was born in 1852 in the town of Petilla de Aragón. He was a rebellious youth – at the age of 11, he was detained for destroying the town gates with an improvised cannon. While grounds for concern, this incident did at least prove one other facet of his character – he was very intelligent.

Ramón y Cajal graduated from medical school at the University of Saragossa in 1873. After a spell working as a medical officer in the Spanish Army, he returned to the world of academia – holding professorships at the universities of Valencia, Barcelona and Madrid.

Drawing nerves of silver

When Ramón y Cajal was not teaching, he carried out research into histology – the anatomy of living tissue at the cellular level. This is done by studying the material through a microscope and – in his day, at least – making expert drawings of what he saw. His speciality was brain and spinal cord tissue, which he studied using a silver nitrate staining agent. The agent had been invented by Italian physician Camillo Golgi and worked by leaving a black deposit on nerve fibres, which highlighted them against the surrounding tissue.

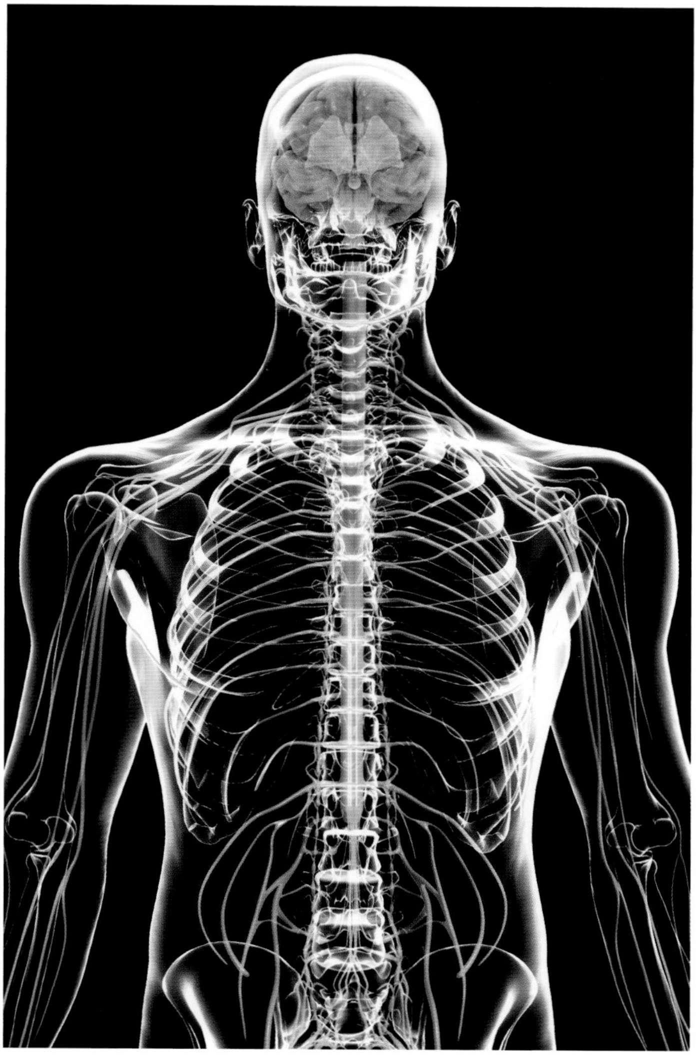

ABOVE Computer-generated image of a simulated X-ray. The image shows the brain and central nervous system (CNS) descending through the spinal cord. Radiating outward from the CNS are the connections that make up the peripheral nervous system (PNS).

Ramón y Cajal refined Golgi's technique, enabling him to see the fine structure of the cells in brain tissue and nerve fibres for the first time. Indeed, prior to this it was not known for sure whether nervous tissue was even made up of distinct cells.

Ramón y Cajal made several breakthrough observations that are evident in his drawings. He established the neuron as the fundamental unit of the nervous system, identified the axonal growth cone that enables axons to develop, showed how nerve cells connect and observed the synaptic junctions between neurons across which electrical or electrochemical signals are passed.

He published four landmark medical texts: *Manual de Histología Normal y Técnica Micrográfica* (Manual of Normal Histology and Micrographic Technique), *Les Nouvelles Idées sur la Fine Anatomie des Centres Nerveux* (New Ideas on the Fine Anatomy of the Nerve Centres), *Textura del Sistema Nervioso del Hombre y de los Vertebrados* (Textbook on the Nervous System of Man and the Vertebrates) and *Die Retina der Wirbelthiere* (The Retina of Vertebrates). In 1906, Ramón y Cajal and Golgi jointly received the Nobel Prize in Physiology or Medicine. He is regarded today as the father of modern neuroscience.

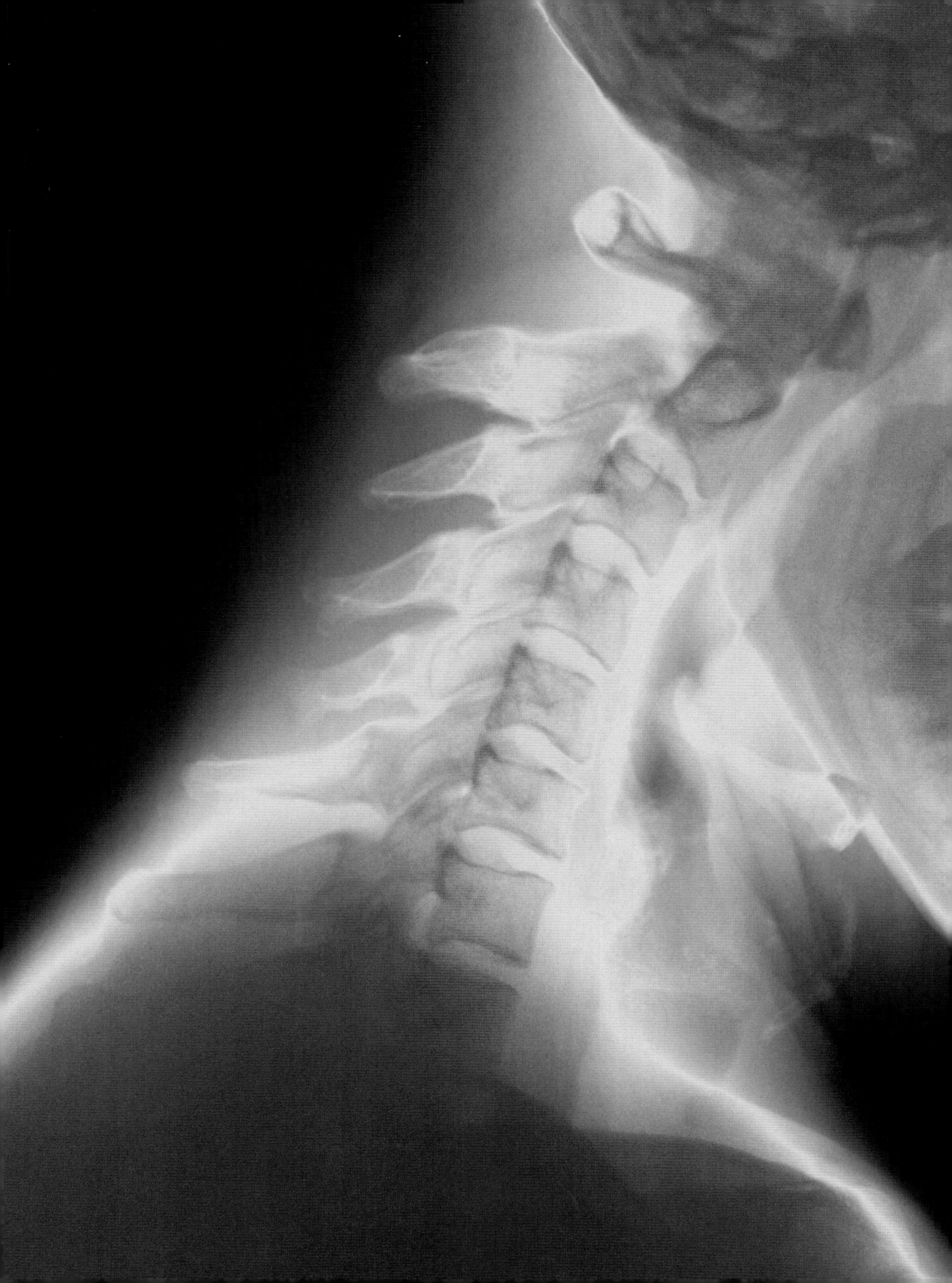

44 X-rays

DEFINITION A HIGH-FREQUENCY FORM OF ELECTROMAGNETIC RADIATION, ABLE TO PENETRATE SOLID OBJECTS

DISCOVERY THE NATURE OF X-RAYS WAS UNRAVELLED BY WILHELM RÖENTGEN AT THE UNIVERSITY OF WÜRZBURG IN 1895

KEY BREAKTHROUGH RÖENTGEN CORRECTLY ASCRIBED ODD EFFECTS SEEN NEAR 'CROOKES TUBES' TO A NEW KIND OF RADIATION

IMPORTANCE X-RAYS ARE AN INDISPENSABLE TOOL IN MEDICINE, AND HAVE APPLICATIONS IN AIRPORT SECURITY AND ASTRONOMY

Some of the most amazing breakthroughs in science have happened largely by accident. The discovery of X-rays is one of them. When mysterious effects on photographic plates were noticed near devices called Crookes tubes, German physicist Wilhelm Röentgen decided to investigate. He found that the effects were caused by a new kind of radiation that can penetrate solid matter. He called the radiation X-rays.

X-rays are a form of electromagnetic radiation. They have very high frequency (up to 10^{19} Hz) and short wavelength (down to around one hundred-millionth of a millimetre). Because the energy of an electromagnetic photon increases in proportion to its frequency (as shown by German physicist Max Planck), X-rays' high frequency gives them extremely high energy – enabling them to pass through all but the densest intervening objects. Scientists got the first whiff of the existence of X-rays in experiments carried out with devices called Crookes tubes during the late 19th century.

A Crookes tube – named after its inventor, English physicist William Crookes – is a long glass tube fitted with a negative electrical terminal (a cathode) at one end and a positive terminal (an anode) in the form of a metal plate sticking up about two-thirds of the way along from the cathode. Most of the air in the tube is sucked out during the fitting process. Scientists found that when a very high voltage was applied to the two electrode terminals – typically tens of thousands of volts – then the glass at the far end of the tube would begin to glow. What is more, the outline of the anode plate would show up as a dark patch in the glowing glass – as if it was blocking some kind of ray being given off by the cathode.

LEFT An X-ray of a broken neck: an injury that could lead to paralysis. The image has been coloured for clarity and shows the break beneath the second vertebra down from the skull. The first two vertebrae are also partially dislocated (subluxation). X-rays have proven to be a powerful diagnostic tool in medicine.

These 'cathode rays' were later found to be high-speed electron particles emitted from the cathode, and attracted by the opposite electric charge of the anode. Electrons that strike the anode are stopped, but those that miss have enough energy to whistle past and strike the glass at the far end – where their energy is converted into light by a process called fluorescence. But something else was going on, too. Researchers investigating the properties of Crookes tubes had found that attempts to photograph the tube in action always went awry, with the photographic plates becoming fogged and overexposed.

Röentgen's detective work

In 1895, German physicist Wilhelm Röentgen decided to explore the mystery. Röentgen was born in 1845, in Lennep, Germany. His family moved to the Netherlands while Röentgen was a boy, but his schooling did not go well there – not only was he was expelled, but also denied admission to any other school in the Netherlands or Germany. Nevertheless, he eventually secured a place at Zurich Polytechnic in 1865 to study engineering. After obtaining a doctorate from the University of Zurich, he held academic positions at the University of Strasbourg and Germany's Academy of Agriculture.

In 1888, Röentgen became professor of physics at the University of Würzburg, which is where – seven years later – he would make the landmark discovery of X-rays. Röentgen knew it probably was not visible light that was responsible for clouding the photographic plates in the other experiments. So to exclude it, he first wrapped his Crookes tube in a light-excluding cardboard sheath. His plan was to then hunt for the mysterious invisible radiation by holding a special screen that he had coated with fluorescent paint close to the shrouded tube. However, when he fired up the tube to test his cardboard shrouding

BELOW 'Backscatter' X-rays are used in security scanning. This example shows a truck stopped on the Mexican–Guatemalan border. It reveals 37 illegal immigrants crammed into containers packed between the cargo of bananas. Backscatter technology detects radiation reflected from the target.

he noticed the screen – placed on the nearby desk – had already begun to glow. He repeated the procedure, finding that the rays lighting up the screen passed straight through books and even thin sheets of metal.

Guinea-pig spouse

During these experiments Röentgen had noticed that the radiation also passed through his body. He asked his wife, Anna, to place her hand in between the tube and a photographic plate. The result was the first medical X-ray, showing the bones of the human hand. Anna was clearly quite troubled by the experience of having viewed her own skeleton – exclaiming, 'I have seen my own death!' Röentgen named the mysterious radiation X-rays, though he only intended this to be a temporary name.

The Crookes tube was later shown to be giving off X-rays because of the collisions between the high-speed electrons and the metal anode. Maxwell's equations (see page 153) showed that an accelerating or decelerating electric charge radiates energy (the same principle upon which a radio transmitter works) – the X-rays were being produced by the massive deceleration of the charged electron particles upon hitting the metal.

'Anna was clearly quite troubled by the experience of having viewed her skeleton – exclaiming, "I have seen my own death!" Röentgen named the mysterious radiation X-rays.'

X-rays soon became an invaluable tool in medicine. They pass through soft body tissues, but are absorbed by denser materials, such as bone, tumours and kidney stones, to reveal breaks and other damage. X-ray imaging can even show up blood vessels if the patient is first injected with a 'contrast agent', which increases the X-ray absorbency of the blood.

Radiation therapy and X-ray astronomy

X-rays do, however, damage tissue – and exposure can lead to cancer (as it did for many early researchers). For this reason, they are used sparingly and X-ray-machine operators are screened thoroughly to prevent regular exposure. This effect on tissue means that X-rays are also used therapeutically to destroy malignant cells in radiation therapy. X-rays have other applications in materials science, where the diffraction pattern (see page 150) created by shining X-rays through a material can reveal its atomic arrangement. They are used in airport security to check the contents of luggage, and in engineering for checking the integrity of structures.

Many of the high-energy processes that take place in space also emit X-rays, and studying these with X-ray telescopes has brought new scientific insights about the universe. Since X-rays are naturally absorbed by the Earth's atmosphere, X-ray telescopes must be stationed in space. Most prominent among these is the Chandra X-ray Observatory, which has returned images of matter falling into black holes, superdense spinning stars (pulsars) and dark matter – invisible material thought to account for a large fraction of the mass of the universe.

45 Radioactivity

DEFINITION THE EMISSION BY SOME CHEMICAL ELEMENTS OF A FLUX OF PARTICLES AND RADIATION AS THEIR ATOMS BREAK APART

DISCOVERY BY FRENCH PHYSICIST HENRI BECQUEREL IN 1896

KEY BREAKTHROUGH BECQUEREL FOUND THAT A PHOTOGRAPHIC PLATE WRAPPED IN BLACK PAPER CAN BE FOGGED BY PLACING A SAMPLE OF URANIUM SALTS ON IT

IMPORTANCE RADIOACTIVITY WAS A STEP TOWARDS ENJOYING THE BENEFITS OF THE ATOM – AND UNDERSTANDING ITS DANGERS

Few of us would want to be without electricity. The bulk of the world's electricity is generated by nuclear power – a vital area of science that developed directly from studies of radioactivity. The steady outflux of radioactive energy from specific elements was first discovered in 1896 by French physicist Henri Becquerel.

Radioactivity is the spontaneous disintegration of the nucleus of an atom. Atoms of some chemical elements are inherently unstable and, now and again, one of them will fly apart. Averaged over many trillions of atoms in a sample, radioactivity produces a steady stream of particle debris emanating from the substance. This effect can make radioactive materials feel hot to the touch. Indeed, much of the Earth's internal heat is thought to be created by radioactive rocks deep underground.

Radioactivity was discovered before there was any clear understanding about the nucleus of the atom. The man responsible for this discovery was a French physicist called Henri Becquerel. He was an engineer by trade, but had become fascinated by the discovery of X-rays in 1896 by German physicist Wilhelm Röentgen. Becquerel wondered whether the rays might be connected to the phenomenon of phosphorescence – the way in which some materials can absorb light and then emit it again slowly as a dim glow that may persist for several hours.

LEFT A scanning electron microscope view of a crystal made up of single uranium atoms. Uranium is a radioactive chemical element – meaning that the nuclei of its atoms are unstable and sporadically break apart, spitting out particles. Uranium is used as the explosive in nuclear fission bombs and as a fuel for nuclear power.

In 1896, he set about an experiment to find out the relationship between X-rays and phosphorescence. Certain substances were known to show a particularly strong phosphorescent response – these were known as 'phosphorescent salts'. Becquerel had read how Röentgen had been led to the discovery of X-rays by noticing that they fogged sealed photographic plates. He tried to reproduce this effect by wrapping a photographic plate in black paper and then placing phosphorescent salts on top of it. But of all the phosphorescent salts Becquerel tried, none of them left a mark on the photographic plate – until he tried a kind of salt called uranium. This left a noticeable dark fog across the plate. But there was a problem.

BELOW One of the drawbacks of nuclear power is its harmful radioactive waste. Here, a container of low-level radioactive waste is processed by a technician. The waste is diluted first to reduce its potency, allowing the technician to work behind a barrier made of transparent glass, rather than lead.

Invisible radiation

Becquerel found that the effect took place even if the uranium was kept in the dark, so that phosphorescence could not possibly be the cause. He concluded that some kind of invisible radiation from the uranium had passed through the paper to fog the photographic plate. The Polish-born, French physicist Marie Curie came up with a name for it: radioactivity, because the material gave out energy like a radio transmitter.

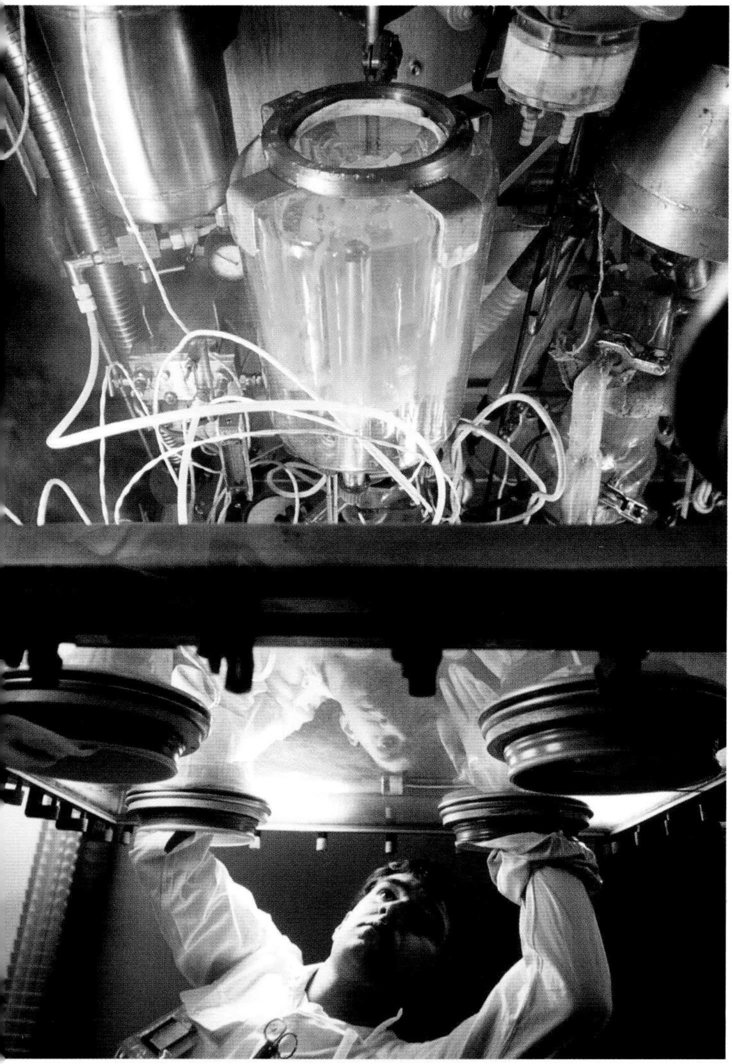

Becquerel and several other researchers investigated the nature of this radiation in more detail. They used various barriers to try to block it, to find out how energetic it was, and they applied electric and magnetic fields to it to determine its electric charge, mass and speed.

They found that radiation actually comes in three forms – which were called alpha, beta and gamma. Alpha radiation was found to have a short range of just a few centimetres, and can be stopped by a piece of paper. It is made up of heavy particles and has a strong positive electrical charge – alpha particles are, in fact, helium nuclei, each containing two neutrons and two protons. Beta radiation is also made of particles, though negatively charged and much lighter than alpha (it is now known to be made of fast-moving electrons). It has a range of tens of centimetres and blocking it requires a sheet of aluminium.

The final type, gamma radiation, has no charge and no mass. It is a kind of very high-energy electromagnetic radiation (higher energy, even, than X-rays). There is no limit to the range of gamma radiation, and blocking it usually requires thick lead shielding. The reason why the nuclei of some chemical elements are stable while others decay comes down to complex considerations in nuclear physics. Generally speaking, heavier atomic nuclei are harder to hold together than smaller ones – which is why most heavy atomic nuclei, such as uranium and plutonium, are radioactive. Some small atomic nuclei have the same quality – for example, the element technetium is radioactive despite being lighter than silver.

Radioactivity is the result of random quantum processes taking place within the nucleus of the atom. And so it is impossible to say with certainty when any particular atomic nucleus of a radioactive element will break apart. Instead, physicists tend to look at a large number of nuclei – and treat their behaviour statistically. The main quantity they use is the half-life of the particular radioactive element, which is the time taken for a selected sample to decay by a factor of a half. This is applied in radiocarbon dating – where the decay of a naturally occurring, radioactive form of carbon (with a known half-life of 5730 years) is used to date ancient samples under examination.

'Becquerel concluded that some kind of invisible radiation from the uranium had passed through the paper to fog the photographic plate. The French physicist Marie Curie came up with a name for it: radioactivity.'

Transmutation

In the early 20th century, New Zealand-born physicist Ernest Rutherford deduced that radioactive decay actually turns one kind of element into another – a process known as transmutation. For example, in the process used for radiocarbon dating, a neutron in the nucleus of a carbon atom converts into a proton plus an electron. The electron is emitted as a beta particle, leaving an extra positive charge on the remaining nucleus. The particular species of a chemical element is determined by the charge on the nucleus, and so the carbon has been transmuted into a new element – in this case, nitrogen.

Radioactive materials are hazardous to humans and other life forms. This power to harm was demonstrated dramatically in 1986 when the Ukrainian nuclear power plant at Chernobyl suddenly exploded, scattering a radioactive smoke cloud over a wide area – poisoning the land and causing radiation sickness among living organisms. The radioactive fallout from a nuclear bomb has much the same effect.

Nevertheless, the discovery of radioactivity has brought colossal advances in our understanding of the subatomic world. And this knowledge, in turn, has also given us benefits in medicine, Earth science, and of course in generating nuclear electricity. Henri Becquerel won the 1903 Nobel Prize in Physics for the discovery. Meanwhile, today, his name is still used as a unit of radioactivity – with one Becquerel (Bq) being equal to the radioactive decay of one atomic nucleus per second.

46 Vitamins

DEFINITION ORGANIC CHEMICALS REQUIRED BY THE BODY IN SMALL QUANTITIES AS NUTRIENTS, AND INGESTED AS PART OF THE DIET

DISCOVERY SUGGESTED IN 1897 BY CHRISTIAAN EIJKMAN; CONFIRMED IN 1906 BY FREDERICK HOPKINS

KEY BREAKTHROUGH BOTH MEN FOUND THAT ANIMALS BECAME ILL WHEN DEPRIVED OF SPECIFIC FOODS

IMPORTANCE VITAMINS ARE ESSENTIAL TO THE HEALTHY FUNCTIONING OF VIRTUALLY EVERY PART OF THE BODY

Whether we eat healthily or keep a bottle of supplements in the cupboard, most people in the western world now enjoy a diet that is rich in vitamins. These essential nutrients were discovered around the turn of the 20th century by Dutch physician Christian Eijkman and English biochemist Frederick Hopkins, working independently.

Throughout human history, people have appreciated the importance of a balanced diet as the key to a long and healthy life. Ancient civilizations knew that a failure to eat certain foods could lead to illness – and, in turn, feeding those same foods to patients with those illnesses could cure them.

The Ancient Egyptians, for example, realized that liver was a remedy for night blindness – an inability to see in low levels of light – and that people who ate liver regularly rarely contracted this illness in the first place. Meanwhile in the mid-18th century, Scottish surgeon James Lind discovered that the disease scurvy, common amongst sailors in the Royal Navy, could be prevented by supplementing the sailors' diets with citrus fruit (which is the origin of the slang term, 'Limey').

LEFT Vitamins are micronutrient chemicals required by the body for healthy functioning. Pictured is a high-magnification microscope image of folic acid – part of the B group of vitamins. It is essential for making proteins and red blood cells. It is prescribed to women in early pregnancy to reduce the risk of spinal and brain defects in the embryo.

Today, we know that night blindness is caused by a deficiency in vitamin A, something in which liver is rich. We also know that scurvy is caused by a deficiency in vitamin C – which can be found aplenty in citrus fruit. Vitamins are an essential component of our diets. The bulk of the food we eat comprises the 'macronutrients', which are protein, fat and carbohydrate. But we also need minute quantities of other 'micronutrient' chemicals – in order to ensure the healthy functioning of our cells, help stave off disease and assist other biochemical processes. Some of these chemicals can be synthesized naturally in the body, but others cannot and must therefore be ingested as part of the diet – and these are the vitamins.

Beriberi rice diet

Vitamins were first discovered in 1897 by the Dutch physician Christian Eijkman. Born in 1858, Eijkman had studied medicine and gone on to become a medical officer in the Dutch army. In this capacity, he was assigned to the Dutch East Indies (modern-day Indonesia) to investigate a disease that had been wracking the region. Called beriberi, it was a disease of the peripheral nervous system causing fatigue, vomiting, pain, paralysis and, in some cases, even death. But its cause was unknown.

The Dutch government set up a medical laboratory in Batavia (now Jakarta) for Eijkman to investigate the disease. He kept a number of chickens in the laboratory for experiments. These birds were fed on rice that was left over from meals prepared for the local Dutch military staff. But when a new cook refused to allow his food to be given to the birds, Eijkman was forced to purchase his own rice for them. Shortly after doing this, however, he was alarmed – yet fascinated – to discover that the birds had developed beriberi.

'The bulk of the food we eat comprises the major, or "macronutrients", which are protein, fat and carbohydrate. But we also need minute quantities of other chemicals – in order to ensure the healthy functioning of our cells.'

The new food was polished rice, in which the outer skin of the rice had been stripped away by a milling process. Previously, the birds had been eating unpolished rice – with the skins intact. The new rice had not spoiled, and Eijkman could find no obvious contagious agent in his new food source that was causing the disease. He concluded that there must be some kind of dietary nutrient in the skins of the rice, capable of warding off beriberi, that was being lost in the polishing process. And indeed when, in 1897, Eijkman fed them unpolished rice again, the disease soon disappeared. He was assisted in this work by Dutch physician Adolphe Vorderman.

Accessory factors

Word spread of Eijkman's discovery. The English biochemist Frederick Hopkins, on learning of the work, carried out his own research, beginning in 1901. Hopkins discovered tryptophan, an amino acid (chemicals used for building proteins), which he found could not be manufactured by the body.

Hopkins had also read about the experiments of Russian surgeon Nikolai Lunin, who had fed one group of rats the individual primary nutrients found in milk – protein, carbohydrate and fat – while another group received whole milk itself. The first group died while the second group thrived. Clearly, Hopkins concluded, there was more to milk than its primary nutrients. He began to suspect that there may be many chemical nutrients, which he called 'accessory factors', that we must ingest in order to remain healthy.

LEFT Electron microscope images of different types of rice. From left to right: Italian risotto rice, long-grain white rice, wild rice, partly boiled rice and milk rice. Comparative studies of the nutritional value of different rice varieties led to the discovery of vitamins in the late 19th century.

He published his research in 1906. In 1912, Polish biochemist Casimir Funk isolated the compound in Eijkman's unpolished rice, the absence of which was causing beriberi. It was a substance called thiamine – one of a family of compounds known as amines.

Funk generalized his discovery, suggesting that all of Hopkins's accessory factors were amines, which he dubbed 'vital amines' – or vitamines. Although this turned out to be untrue, the name had already stuck. All scientists could do was try to de-emphasize the amine component of the name and so they simply removed the 'e', leaving the word as we have it today – 'vitamins'.

To the letter

Thiamine is now known as vitamin B_1. It is one of nine water-soluble, and four fat-soluble, chemicals that are essential to life. They are labelled A, B_1, B_2, B_3, B_5, B_6, B_7, B_9, B_{12}, C, D, E and K. The gap in the letters is the result of vitamins being reclassified (vitamin H is now known as B_7), or discarded as not vitamins after all (the chemical originally designated as vitamin L_1, for example, is now known to be non-essential). Vitamins are part of a larger family of micronutrients that also includes minerals and acids.

It was always necessary to obtain these nutrients from a varied diet, although individual requirements vary. But in the 1930s, the first vitamin supplements became available, enabling everyone to maintain their recommended daily amount (RDA). Many of us today take a multivitamin a day as a top-up, if nothing else. The simple preventative measure of ensuring people get their vitamins has had a hugely uplifting effect on health and wellbeing the world over. And for their discovery Eijkman and Hopkins shared the 1929 Nobel Prize in Physiology or Medicine.

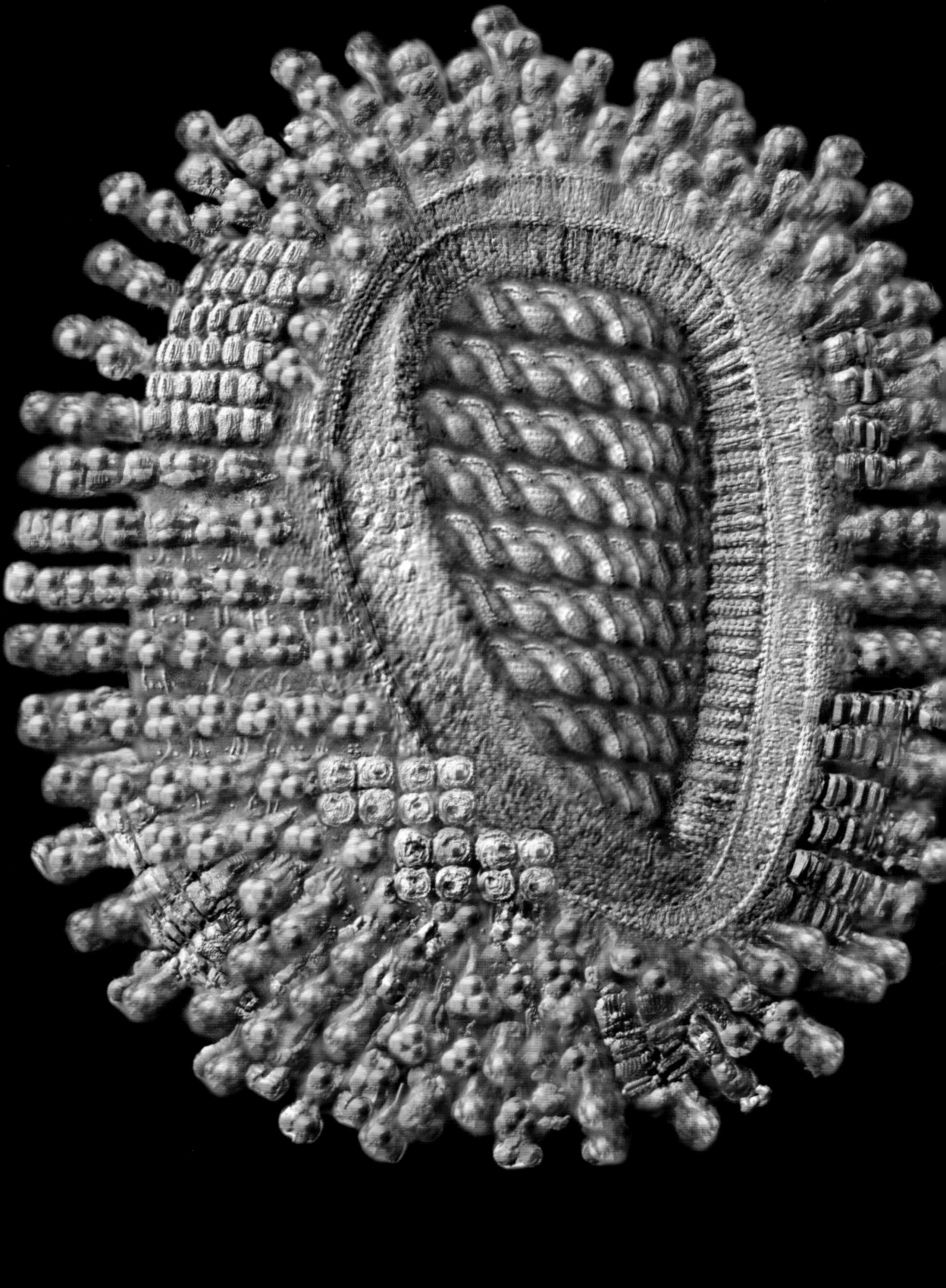

47 Viruses

DEFINITION DISEASE-CAUSING BIOLOGICAL PARTICLES THAT ARE MUCH SMALLER THAN BACTERIA AND OTHER MICROBES

DISCOVERY FIRST HYPOTHESIZED BY DUTCH BIOLOGIST MARTINUS BEIJERINCK IN 1898

KEY BREAKTHROUGH PASSED AN INFECTIOUS SOLUTION THROUGH A BACTERIA-BLOCKING FILTER, FINDING IT REMAINED INFECTIOUS

IMPORTANCE VIRUSES ARE RESPONSIBLE FOR FLU PANDEMICS – AND OTHER DEADLY ILLNESSES, INCLUDING AIDS AND CANCER

For most people, catching the flu is a routine hazard of winter and something that – at worst – will mean a couple of weeks off work. But in 1918, a flu outbreak killed 20–100 million people worldwide. It was a shocking demonstration of the deadly threat posed by viruses – infectious agents tens of times smaller than a bacterium.

Viruses are disease-causing agents that are much smaller than bacteria, the infectious particles that were identified in the 1860s by French microbiologist Louis Pasteur in the formulation of his germ theory of infection. Whereas a bacterial cell is typically a thousandth of a millimetre across, viruses can be 50 times smaller than this.

A virus cell is too small to see through a standard microscope, as Louis Pasteur himself discovered. While he was able to track down the bacteria for diseases such as cholera, he could not find the agent that causes rabies – an infectious disease of the central nervous system, now known to be transmitted by virus particles. In the years that followed, other cases emerged of diseases caused by mystery pathogens. For instance, Russian biologist Dmitry Ivanovsky tried to isolate the cause of a disease affecting tobacco plants by passing juice from infected leaves through a device called a Chamberland filter (named after Pasteur's assistant Charles Chamberland). The filter had holes smaller than any known bacteria – and yet the leaf juice remained infectious.

It was Dutch microbiologist Martinus Beijerinck who, in 1898, took the logical step and inferred that the disease afflicting the tobacco plants was being caused by a new kind of infectious agent. Beijerinck used the word

LEFT Artwork showing the structure of an influenza (flu) virus particle. In the centre is a coil of ribonucleic acid (green) that carries the virus's genetic information. This is surrounded by a protein shell (pink) and a membrane (blue and yellow). The exterior of the membrane is studded with more proteins, which help the virus to infect target cells.

'virus' to describe it – from a Latin word for poison that had been in general use since the 14th century. Because of their ability to pass through the finest filters, Beijerinck believed viruses to be liquid in nature. Later research, however, showed them to be tiny particles. This was demonstrated explicitly in 1931 when German scientists obtained the first images of minuscule virus particles using a device called an electron microscope.

Fine details revealed

Ordinary microscopes, which illuminate their targets with light, were unable to detect virus particles. However, the smallest details visible through a microscope decrease in proportion to the wavelength of the light source used. So if you are using infrared light (with a wavelength of 800 billionths of a metre) and can see tiny microorganisms of some size, call it x. Now switch to violet light (with wavelength of 400 billionths of a metre) and you will see down to microorganisms half that size, i.e. $x/2$. Since the wavelength of electrons is 100,000 times smaller than that of light (see 'Wave-particle duality' on page 237), they were able resolve the finer-scale detail needed to view viruses and, later, catch them in the act of attacking cells.

> **'Even though viruses are not alive they must still reproduce in order to spread – a task that they accomplish in the sneakiest way: by hijacking the reproductive machinery of healthy cells that they infect.'**

These images, and further research, revealed that viruses are extraordinarily simple entities – so simple, that they do not even deserve the status of being a life form. Most consist of a blob of genetic material – in the form of either DNA or RNA – surrounded by a 'coat' of proteins. Some viruses, such as herpes simplex, are also surrounded by a fatty outer envelope.

Even though viruses are not alive they must still reproduce in order to spread – a task that they accomplish in the sneakiest way: by hijacking the reproductive machinery of cells they infect. The virus breaks the outer membrane of the cell and then injects its genetic material. Protein-manufacturing sites inside the cell unknowingly process this foreign genetic code and use the information to rattle out copies of the virus. The new viruses burst out of the cell, destroying it, and then spread to infect new cells – where the process repeats.

This continuous – and exponentially increasing – destruction of cells is extremely bad news for the life form playing host to the virus. Human beings were to discover this the hard way in 1918, when an outbreak of the Spanish influenza virus spread around the world – becoming a so-called 'pandemic' (from the Greek word *pan*, meaning 'all', and *demos*, meaning 'people') – killing between 20 and 100 million. About 500 million people, nearly one-third of the global population at the time, became infected.

The type of flu responsible for the 1918 outbreak was very similar to the strain that caused the 2009 swine flu outbreak – H1N1. The letters refer to the composition of different proteins that make up the virus's outer coat.

These come in two main types – haemagglutinin (H) and neuraminidase (N) – with the numbers indicating particular subvarieties. Another strain that made headlines in 2006 was H5N1 bird flu. Although an ongoing global risk, the H1N1 swine flu pandemic in 2009 was less devastating than the 1918 pandemic – killing approximately 18,000 worldwide. Given that swine flu was transported almost instantaneously by air travel, the limited death toll is a great testament to advances in both viral treatment and prevention.

Vaccines

Conventional measures used to treat viruses are two-pronged. Firstly, there are vaccines: preventative measures that train the body to recognize and attack the protein coat or outer envelope of a particular virus. The body's immune system will attack the virus anyway, but speed is essential in defeating a virus. Once inside the body, the virus spreads exponentially fast – and so training the immune system to act quickly can mean the difference between stopping the infection spreading or succumbing to it. Vaccines expose the subject to virus particles that have either been killed or inactivated by removing their genetic material.

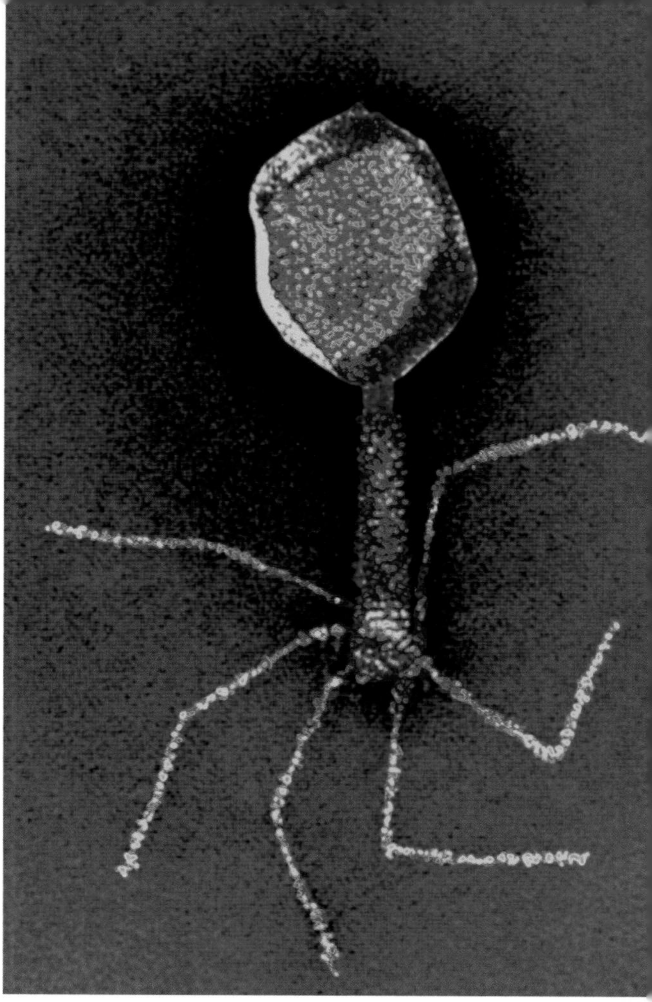

ABOVE It looks like something from a science fiction movie, but this is an electron micrograph of a 'bacteriophage' virus – a virus that preys upon bacterial cells, in this case *E. coli.* The head of the virus (green) contains the genetic material (shown red), which is injected into the host cell by the narrow collar, rather like a syringe. The legs at the base of the collar are used by the virus to grasp hold of target cells.

Antivirals

The second line of defence, for use on people who are already infected, is antiviral drugs. These disrupt the chemical processes on which the virus relies in order to replicate, slowing its spread and giving the body's immune system time to overcome it before the symptoms worsen. Antiviral drugs aim to block the action of the neuraminidase enzyme on which the virus relies to release copies of itself from infected cells. However, the effectiveness of antivirals has been questioned for more serious cases of influenza.

Not all viruses are bad. In 1915, English biologist Frederick Twort discovered a strain that preys solely upon harmful bacterial cells – called a 'bacteriophage' virus. Meanwhile, viruses are often used in gene therapy as a 'vector' to deliver new genetic material into a patient's cells. Despite this, viral infection is still responsible for many of the deadliest diseases on the planet, some of which – including AIDS and cancer – have resisted all attempts to formulate a vaccine. The search for treatments for viral conditions remains a priority – but one that would have been impossible were it not for the discovery of viruses a little over 100 years ago.

48 Rocket science

DEFINITION MATHEMATICAL EQUATIONS GOVERNING THE MAXIMUM SPEED AND PAYLOAD OF A ROCKET

DISCOVERY FIRST FORMULATED BY RUSSIAN SCIENTIST KONSTANTIN TSIOLKOVSKY IN 1903

KEY BREAKTHROUGH TSIOLKOVSKY ANALYZED THE BEHAVIOUR OF A ROCKET IN FLIGHT USING NEWTON'S LAWS OF MOTION

IMPORTANCE ROCKETRY BECAME AN EXACT SCIENCE ALLOWING ROCKETS TO BE USED IN SPACE EXPLORATION

Rockets have been around for a very long time. They were first built in the ninth century by the Chinese, who combined charcoal, sulphur and saltpetre (potassium nitrate) to make gunpowder. Their rockets were used in battle, but were little more than fireworks. Modern rockets are engines that convert energy into motion using jets of high-speed gas to push the rocket forwards.

Rocket artillery was welcomed into the British Royal Arsenal in 1805, where inventor William Congreve was instrumental in developing them as weapons. He built rockets consisting of iron cylinders stabilized with long sticks – much like a modern firework (albeit a lot bigger). Congreve's largest rocket weighed 14.5kg (32lb) and had a range of around 3km (1.9 miles).

Each of Congreve's rockets packed an explosive warhead (although these often went off in flight) and were also inaccurate. English rocketeer William Hale went some way towards solving the latter problem in the 19th century by angling the direction of the rocket engine's nozzle to make the rocket spin, in the same way that a rifle bullet spins to prevent it veering off course.

But the big developments came in the early 20th century, thanks to the work of Russian schoolmaster Konstantin Tsiolkovsky. He turned rocketry into a science governed by hard-and-fast rules and mathematical equations, allowing the behaviour of a rocket to be accurately predicted. It was Tsiolkovsky's research that transformed rockets from crude weapons of warfare into the instruments that would take human beings into space, allowing them to travel beyond the confines of Earth's atmosphere and to the Moon – as well as carrying robotic space probes far across our solar system.

LEFT The exhaust plume from a Russian Soyuz-Fregat rocket as it clears the launch gantry at the Baikonur Cosmodrome, in Kazakhstan, May 2007. As Newton showed centuries earlier in his laws of motion, the downward force of the exhaust sets up an equal and opposite force that pushes the rocket up towards space.

Tsiolkovsky's greatest contribution was his rocket equation, published in 1903. The equation gives the increase in speed that a rocket can produce in terms of its initial (fully fuelled) mass, its dry mass (with all fuel burned), and the speed of the exhaust gases produced by the engine. Scientists call this total boost in speed 'delta-v' – 'v' for velocity, and 'delta' the mathematical shorthand for 'a change in'.

Action and reaction

Tsiolkovsky arrived at his equation from the conservation of momentum. This is a physical property giving the impetus to a moving object, and is given by simply multiplying an object's mass by its speed. Conservation of momentum states that the total momentum of a physical system must always stay the same, and is one reason why a rifle kicks back against your shoulder when you fire it. The bullet travels forwards at high speed, carrying a good deal of momentum with it. The body of the gun then kicks back towards you with an equal but opposite momentum.

The bullet gets its momentum from a low mass multiplied by a high velocity, while the recoiling rifle has high mass and a relatively low velocity – multiplying together to give the same momentum but in the opposite direction. Similarly, the exhaust gas of a rocket has relatively low mass but jets out of the back at very high speed. To conserve momentum, the relatively large mass of the rocket then lurches forwards – steadily accelerating as the fuel is turned into hot exhaust. Tsiolkovsky's equation says that, for example, a rocket with 90 percent of its initial mass taken up by its fuel load can muster a delta-v equal to 2.3 times the speed of the exhaust gases. That means that if the exhaust is moving at a speed of 2500 metres per second (2734 yards per second), then the top speed of the rocket from launch is 5750m/s (6288 y/s).

'It was Tsiolkovsky's research that transformed rockets from crude weapons of warfare into the instruments that would take human beings into space.'

Fast as this may seem, a rocket with this speed is still insufficient to place a spacecraft in orbit around the Earth. (And that makes no allowance for the power of the rocket that is used in overcoming the force of gravity.) Isaac Newton's universal law of gravitation gives the minimum speed that a rocket needs to reach to go into orbit around any planetary body. And for the Earth the figure is 7800m/s (8530 y/s).

Stack 'em up

Tsiolkovsky's genius was in coming up with a solution to this problem before even a single space rocket had flown. What if, he mused, the payload of the rocket – the ten percent of the launch mass that is not fuel – was itself another, smaller rocket? This rocket would already be travelling at 5750m/s (6288 y/s) when its engines were lit up. And when it had finished burning it would have gained an additional 5750m/s (6288 y/s), taking its final speed to 11,500m/s (12,576 y/s) – enough to hit orbit. The idea was known as multi-staging.

LEFT NASA's Saturn V Moon rocket remains the biggest, most powerful space launcher ever constructed. Pictured here are the five F1 engines of the Saturn V's first stage that would propel the 3000-tonne rocket off the pad. The Saturn V made a total of 13 flights during the Apollo Moon programme.

The drawback is that the amount of useful payload that can be carried is reduced. A rocket made up of *n* stages, each of which can lift 0.1 (10 percent) of its total mass as payload, can lift a payload equal to $(0.1)^n$ of the rocket's overall mass. For a two-stage rocket, that is 0.01 (1 percent); for a three-stage rocket, it is 0.001(0.1 percent) – and so on. The solution, Tsiolkovsky showed, is to make your rocket bigger. Crucially, the mathematics revealed that a multi-stage rocket can carry a higher overall payload to orbit than a single-stage rocket of the same launch mass. And that is why multistaging was the strategy adopted by NASA to get to the Moon in the 1960s, with their three-stage Saturn V – the biggest rocket ever flown.

Scientists are now devising engines that maximize the rocket's exhaust speed. These include nuclear rockets and ion engines, which use an electric field to accelerate charged fuel particles to terrific speeds of 29,000m/s (31,714 y/s). These rockets – which are essentially elaborations on Tsiolkovsky's ideas – could power the spacecraft that will take human beings to the stars.

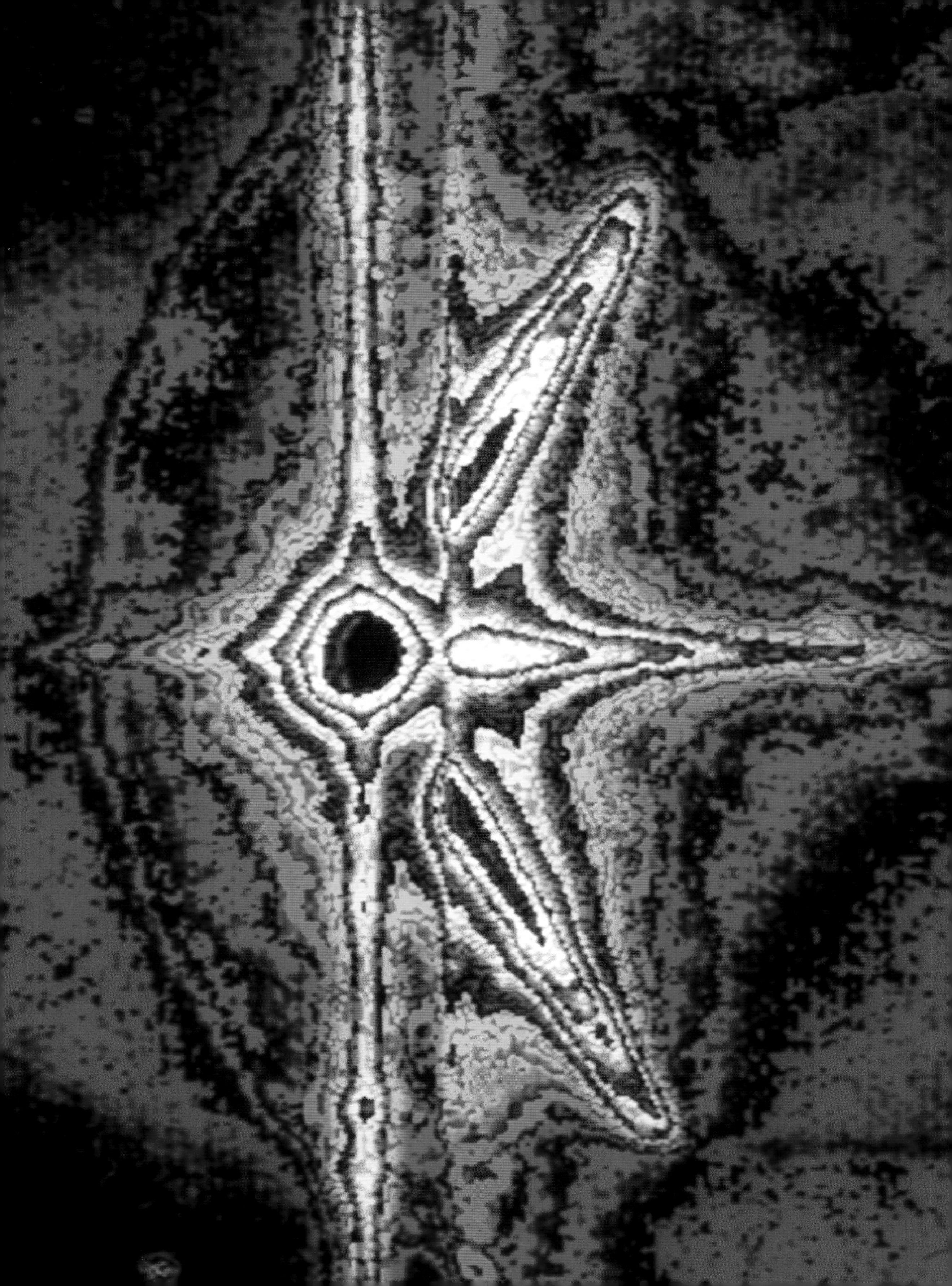

49 Photons

DEFINITION A PHOTON IS A PARTICLE OF LIGHT

DISCOVERY FIRST FRAMED IN A MODERN CONTEXT BY ALBERT EINSTEIN IN 1905

KEY BREAKTHROUGH EINSTEIN SAW THAT COLLISIONS BETWEEN ELECTRONS AND LIGHT PARTICLES COULD EXPLAIN THE PHOTOELECTRIC EFFECT

IMPORTANCE DEMANDING THAT LIGHT COMES IN 'QUANTIZED' PACKETS WAS THE FIRST STEP TOWARDS QUANTUM THEORY

By the 19th century, physicists thought they had explained the nature of light – it is a wave. But in 1905, Albert Einstein constructed a new model, in which light is made of particles instead. He was right, and the light particles became known as photons. This single, bold insight showed physicists the way to understanding the behaviour of matter on the tiniest scales – giving birth to the field of quantum theory.

Physicists and philosophers have long agonized over the nature of light. Early theories adhered to the corpuscular view – that light is a shower of solid particles that rain down upon illuminated objects. In the 17th century, Christiaan Huygens and others suggested that light is instead a wave. In 1803, Thomas Young seemed to prove this was correct, when he carried out an experiment demonstrating light's waviness (see page 105).

But 100 years later, the nature of light was looking uncertain again. Experimental results emerged that could not be explained by treating light as a wave. The wave theory stated that the energy delivered by a light beam is determined by its intensity – the brightness of its source. And yet in a phenomenon called the photoelectric effect, electrons can be knocked out of any given metal by a light beam which is above a certain cut-off frequency (usually in the ultraviolet region of the electromagnetic spectrum). This suggested that light energy increased not with intensity, but with frequency.

LEFT An infrared microscope image showing laser light emitted from a 'photonic lattice' of gallium aluminium arsenide. Red laser light has been fired at the material from the left. Photons (particles of light) are absorbed by the lattice and then re-emitted at infrared wavelengths, as shown by the spikes (right).

During this period, physicists were also trying to explain the characteristics of thermal radiation – the electromagnetic radiation given off by hot bodies. Place a metal poker in a fire and it glows red – this is thermal radiation. As the temperature of the fire increases, the frequency of the thermal radiation emitted increases as well. But no one at this time could come up with a theory that successfully explained the observed relationship between the radiation and the temperature of its source.

Planck's constant

At the end of the 19th century, the German physicist Max Planck came up with the solution. He formulated a law for the intensity of the light at each frequency emitted by a hot body as a function of its temperature. Planck's law explained the observations perfectly – predicting a peak in the graph of intensity versus frequency, the position of which moves to higher frequency as the temperature increases.

But there was a slight oddity with the theory. To make it work correctly, Planck had to suppose that light energy can only be emitted or absorbed in discrete units – given by multiplying the frequency of the light by a tiny number, equal to 6.6×10^{-34} (6.6 divided by a 1 followed by 34 zeroes). This number became known as Planck's constant, denoted by the letter h.

Planck had originally intended this to be simply a quirk in the way light, and other forms of electromagnetic radiation, are emitted and absorbed. But another physicist was about to make a rather more radical interpretation of his mathematics. This physicist's name was Albert Einstein.

Einstein was trying to make sense of the mysterious photoelectric effect, first observed by German physicist Heinrich Hertz in 1887. Hertz had realized that in order for electrons to be knocked out of the metal, the light must somehow collide with them. Einstein's leap of insight was to revert to the view that light really is made up of particles, and then suppose that Planck's formula, of energy = $h \times$ frequency, gives the energy of each particle. Then the photoelectric effect is not mysterious at all. Electrons and light particles simply collide like billiard balls. Only when the energy of an incoming light particle is high enough will it be able to knock an electron from the rigid lattice holding it in place. And, according to Planck's formula, this can only happen when the particle's frequency is high enough – exactly as observed. Einstein published his findings on the photoelectric effect in 1905, and won the 1921 Nobel Prize in Physics for the discovery.

'Einstein's leap of insight was to revert to the old view that light really is made up of particles and then suppose that Planck's formula gives the energy of each particle.'

Hard evidence

Planck hated the idea that light is made up of particles. But he and the rest of the physics community finally accepted it when the particles were detected experimentally in 1923. American physicist Arthur Compton, working at Washington University in St Louis, was able to observe directly particles of light bouncing off solid electrons – in a phenomenon that has

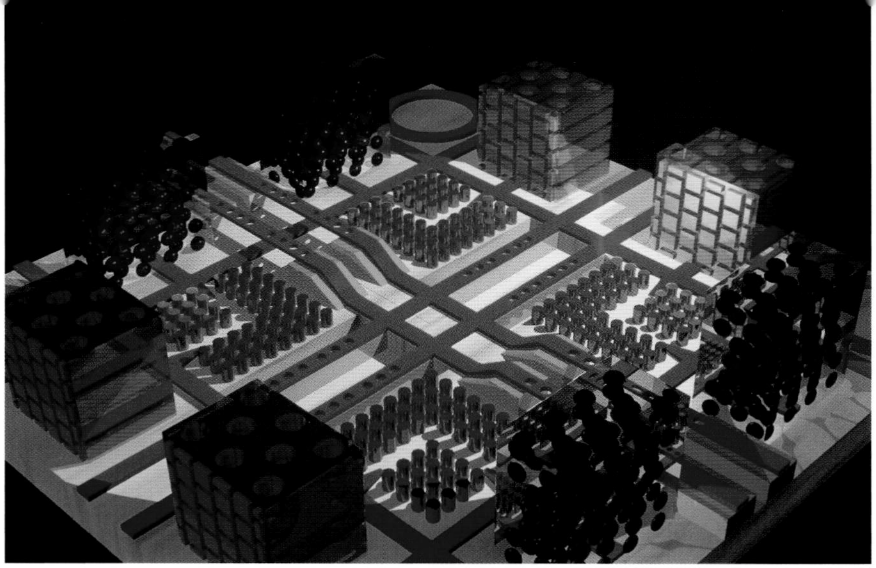

LEFT Artwork showing a variety of 'photonic crystals', which are materials that can trap and manipulate photons of light. These crystals could find applications in optical computers, which are devices that use photons rather than electrons to store and process data, making them faster and more efficient.

since become known as 'Compton scattering'. All that remained was for these light particles to be named. And that was done in 1926 by American chemist Gilbert Lewis, who used the word 'photon' (from the Greek word *phos*, meaning light).

So where did Thomas Young go wrong in his experiment that seemed to prove light was a wave? He did not. Other research was now revealing that light – and, indeed, matter – behave as both particles and waves at the same time. This is a branch of physics called wave-particle duality.

It was all part of the emerging discipline of quantum theory. This was a new approach to the physics of the subatomic world that would prove to be one of the most successful and far-reaching scientific ideas of the 20th century – and which all grew from the initial idea that light and energy come in discrete packets, or 'quanta'. Ironically, Einstein – who had been the key instigator of this revolution – came to loathe the theory and its radical implications for the nature of reality.

Quantum electrodynamics

Photons played a key part in the quantum theory of the electromagnetic field (called 'quantum electrodynamics', see page 303) – not only serving as the particles of electromagnetic radiation, but also as the carrier of the electromagnetic force. Any electromagnetic phenomenon, such as two magnets repelling one another or the attraction between opposite electric charges, is understood through the vital exchange of photon particles.

Photons are now of particular interest to computer scientists developing the next generation of computers. They believe that storing and processing information using particles of light will be vastly more efficient than using the electrons that encode information inside computers today.

50 Special relativity

DEFINITION THE THEORY OF RELATIVE MOTION BETWEEN FAST-MOVING OBJECTS, CONSISTENT WITH JAMES CLERK MAXWELL'S ELECTROMAGNETIC THEORY

DISCOVERY FORMULATED BY ALBERT EINSTEIN IN 1905

KEY BREAKTHROUGH REALIZING THE SPEED OF A LIGHT BEAM IS THE SAME FOR ALL OBSERVERS, IRRESPECTIVE OF HOW THEY MOVE

IMPORTANCE SPECIAL RELATIVITY REWROTE THE LAWS OF MOTION AND YIELDED THE FAMOUS FORMULA $E=MC^2$

In the early years of the 20th century, a simple thought experiment by a young German scientist changed the face of physics forever. Albert Einstein had wondered what it might be like to ride a bicycle alongside a beam of light. The quest for the answer to that question led him to a new theory of relative motion that would make its presence felt from the largest scales of the universe right down to the interactions between tiny subatomic particles.

The special theory of relativity is all about relative motion. For the speeds at which most of us travel, you might think that is a fairly simple concept. For example, if two cars travel towards each other at 50km/h (31mph), the relative speed between the cars is 100km/h (62mph) – that is the speed a passenger in one car will see the other approach. By the same token, if the two cars travel side-by-side at 50km/h (31mph) their relative speed is zero.

At the age of 16, Albert Einstein asked himself whether this would also apply to a beam of light. What if a particularly energetic cyclist pedalled so fast that they could pull alongside a light beam and observe it as if it was stationary?

LEFT A car passing another vehicle moving in the opposite direction at the same speed. Relative to either car, the other appears to be moving at twice its own speed. Einstein showed that if the cars moved at close to the speed of light, then this simple rule of adding velocities no longer applies.

Einstein soon realized this was not so straightforward. In 1861, Scottish physicist James Clerk Maxwell had put forward a theory of electromagnetism. Maxwell's equations provided a beautiful description of all the phenomena of electricity and magnetism. And they provided a new description of light beams. According to Maxwell, light – like all other electromagnetic waves, from radio to X-rays – is made up of an electric and a magnetic field vibrating at right angles to one another. But the theory predicted that the speed of light – 300,000km per second (186,400 miles per second) – is an immovable 'constant' of nature. In other words, it makes no difference how fast you travel, the speed of light is always the same. This bothered Einstein because it was clearly at odds with his notion of a super-fast cyclist reducing the relative speed of a light beam to zero.

Einstein concluded that the only way out of this anomalous situation is if the true laws of relative motion are different to the simple addition and subtraction required to calculate the relative speed between two cars. The correction would be negligible at these low speeds but would become significant as the cyclist approached light speed.

New mathematics

And that was exactly what Einstein found when he crunched through the mathematics of his new theory – re-deriving the laws of relative motion under the restriction that the speed of light remain a constant, denoted in his formulae by the letter c. Rather than simply adding together the speeds, v, the new equations of relative motion included factors of $(1 + v^2/c^2)$ – so that when v was small they boiled down to the laws of everyday experience, but when v approached c everything changed.

'An astronaut jetting into space at 86 percent lightspeed for ten years (as measured in their frame of reference) would come back to find that 20 years had passed during their absence on Earth.'

Einstein's new mathematics – the special theory of relativity – threw up some unusual predictions. Firstly, to keep the speed of light constant Einstein found that the size of moving objects had to get shorter as viewed by a stationary observer. For example, an object moving at just over 86 percent light speed appears to contract to just half its size in its direction of motion. There was an even weirder effect, too. Just as space gets squashed up in a moving frame of reference, time gets stretched out in a bizarre phenomenon called time dilation. Ticks on a clock moving at 86 percent lightspeed take twice as long as ticks on a clock that is not moving – fast-moving objects literally experience less time than slow-moving ones. That means that an astronaut jetting into space at 86 percent lightspeed for ten years (as measured in their frame of reference) would come back to find that 20 years had passed on Earth.

Same place, different time

The absolute order of time and space – a sacrosanct idea since the time of Newton – is completely disrupted by special relativity. The theory even makes it impossible for differently moving observers to agree whether or not two events take place at the same time. For example, someone switching on a light at the centre of a moving railway carriage will see the light from the bulb illuminate opposite ends of the carriage at the exact same moment. Someone standing on a platform watching as the light is switched on, however, sees the light move at the same speed – but from their point of view the back of the carriage is now rushing towards it while the front is moving away. And so the light reaches the back first. Special relativity made physicists consider

space and time – hitherto viewed as distinct concepts – to be different facets of the same thing, which they called 'spacetime'. In this regard, time really is the fourth dimension. And this meant that quantities that previously only had a spatial component now had a temporal component as well.

ABOVE The special theory of relativity makes some bold statements about the nature of space and time. One of these is 'time dilation', which says that for an astronaut rapidly speeding away from Earth, time passes more slowly than it does for those watching him or her leave. The astronaut ages more slowly than those left on Earth.

Energy possession

Applying this idea to energy led to one of the most famous equations in physics. Kinetic energy is energy of motion, which massive objects have by virtue of their movement through space. In the spacetime view, they also have energy by virtue of their movement through time – Einstein showed that all stationary masses possess energy, E, given by the formula $E=mc^2$ (mass times the speed of light squared). Whenever an object loses mass, an amount of energy equal to the mass lost times c^2 must be given off. This applies equally to burning a piece of coal in a fire as it does to splitting a nucleus of uranium in nuclear fission – mass and energy are equivalent.

Einstein's energy analysis also revealed that there is an ultimate speed limit in the universe. He found that the faster an object moves, the more energy it takes to make it go faster. At the speed of light, the energy required to speed up becomes infinite – in other words, special relativity demands that nothing can travel faster than light. The veracity of these astonishing predictions has been confirmed in particle accelerators, where particles are sped up to near lightspeed. These mighty machines rely on special relativity in order to function and they test the theory to the limit on a daily basis.

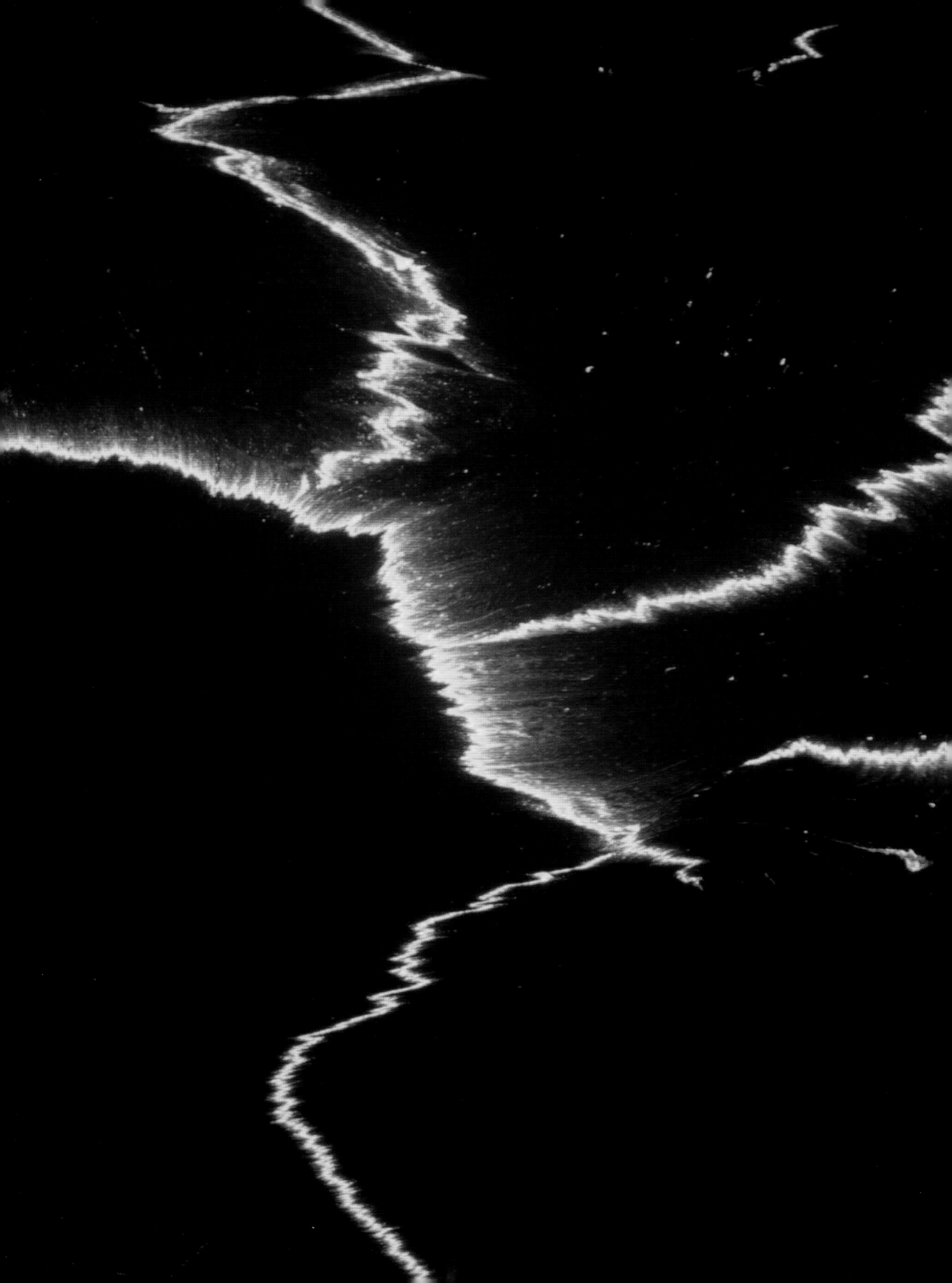

51 Earth's core

DEFINITION THE DENSE HEART OF OUR PLANET IS MADE FROM HEAVY IRON AND NICKEL THAT HAVE SUNK THERE UNDER GRAVITY

DISCOVERY ITS EXISTENCE WAS DEDUCED BY THE IRISH GEOLOGIST RICHARD OLDHAM IN 1906

KEY BREAKTHROUGH OLDHAM FOUND SEISMIC WAVES SLOW DOWN IN THE PLANET'S CENTRE INDICATING A REGION OF HIGHER DENSITY

IMPORTANCE THE SEISMIC WAVE TECHNIQUE OLDHAM PIONEERED HAS SINCE BEEN USED TO MAP THE REST OF THE EARTH'S INTERIOR

What lies at the heart of our planet? At the beginning of the 20th century, nobody knew. Then, an ingenious technique for probing the Earth's innards was put forward that took measurements of the seismic waves from earthquakes and resulted in the discovery of the Earth's core. Subsequent applications revealed the internal structure of the Earth in all its glory.

In the 18th and early 19th centuries, James Hutton and Charles Lyell had propounded the view that the Earth's interior is hot, and that this heat drives geological processes at the planet's surface, including the formation of rock. The process was known as 'plutonism'. But their ground-breaking research did not shed any light on the Earth's interior structure. Most scientists accepted that there must be some kind of interior structure because of the existence of the Earth's magnetic field. The field has a strength of between 30 and 60 microteslas, about 1000 times weaker than a fridge magnet. In the 16th century, physicist William Gilbert had demonstrated that the Earth's field was consistent with it being produced by iron at the planet's centre.

The first hint of physical evidence to support this idea came in 1906 when Irish geologist Richard Oldham discovered that the planet has a core with a markedly different composition to the surrounding rock.

Born in Dublin in 1858, Oldham was educated in geology by his father, Thomas, who was professor of geology at Trinity College, Dublin. He supervised geological surveys of Ireland and India, assisted by son Richard. In the early 1900s, Richard Oldham carried out his own study of an earthquake that had rocked the Indian state of Assam in 1897. A monster

LEFT The cooled, solidified surface of this lava lake has cracked under the stress caused by currents in the hot, liquid lava below. Lava is molten rock from the Earth's interior that has oozed up through weak points in the planet's crust. Not all of the Earth's innards are liquid, though – seismic studies show that the planet's inner core is made of solid nickel and iron.

quake with a Richter magnitude in excess of eight, it caused massive damage and killed an estimated 1500 people. The quake's power also meant that its tremors were felt around the world. And this is what interested Oldham.

He collated data that had been gathered following the quake by seismic monitoring stations across the globe. He looked at two kinds of seismic waves. P-waves, or primary waves, are formed by compression of the rock through which they are passing. A P-wave is rather like a wave on a stretched spring that causes the coils of the spring to bunch together as the wave travels by. In contrast, S-waves, or secondary waves, are more like the waves you get on a guitar string. Unlike P-waves, where the displacement caused as the wave passes is parallel to its direction of motion, in an S-wave the displacement is perpendicular.

'Oldham had discovered the Earth's outer core, a layer of molten iron and nickel that begins at a depth of around 2890km (1796 miles). Its density is typically 11 tonnes per cubic metre, while the temperature is between 4000 and 6000°C.'

Wave motion

P-waves and S-waves travel through rock at different speeds, at a ratio of about 1.7:1 for surface rock densities, with P-waves being the faster of the two. The difference in the arrival times of the two types at a monitoring station can then be used to infer the distance to the earthquake's epicentre. Generally, for earthquakes relatively near the surface, the difference in the arrival times multiplied by eight gives the distance in kilometres. However, if the waves pass through the deep Earth then their speeds and even their trajectories can be altered. This was what Oldham found in the records from the great 1897 Assam quake. But he also noticed something strange. In measurements taken by seismic monitoring stations diametrically opposite the globe from the earthquake's epicentre, the P-waves were slowed down dramatically.

Oldham deduced from this that there must be a region at the Earth's centre where the density suddenly jumps up, becoming much higher than it is in the surrounding rock. The P-waves passing through this region were being slowed down by the high-density material. Oldham had discovered the Earth's outer core, a layer of molten iron and nickel that begins at a depth of around 2890km (1796 miles). Its density is typically 11 tonnes per cubic metre, while the temperature is between 4000 and 6000°C.

Core values

Within the molten region is an inner core, where the pressure of the overlying rock and metal is great enough to squash the iron and nickel back into solid form. The inner core is approximately 2440km (1516 miles) in diameter, the density there approaches 13 tonnes per cubic centimetre and the temperature reaches a blistering 7000°C. The inner core was discovered in 1936 by Danish seismologist Inge Lehmann, who carried out more detailed studies of P-wave and S-wave data to infer the presence of a solid centre to the planet. In the 1950s, she also discovered the so-called 'Lehmann discontinuity' – a dramatic increase in both P-wave and S-wave speed at a depth of around 220km (137 miles) beneath the Earth's crust (its outer layer). Oddly, the Lehmann discontinuity is only seen below oceanic crust – not the thicker crust layer making up the continents. Its nature is still not fully understood.

Mantle in motion

The Lehmann discontinuity occupies a broad layer of rock between the crust and the core that is referred to as the mantle. It is divided into two zones, the upper mantle and lower mantle. The heaving and roiling motion of the viscous semi-molten upper mantle dragging on the crust above is thought to be what causes the movement of tectonic plates (see page 225). In the lower mantle, however, the rock is solid and less mobile. Temperatures in the mantle range from a few hundred degrees at the base of the crust to 4000°C at the top of the core. William Gilbert's hunch about geomagnetism stemming from the Earth's iron core turned out to be right. Earth's geomagnetic field is now thought to be caused by the dynamo effect set up by swirling currents of conducting liquid metal in the outer core.

BELOW This bright ring around the Earth's South Pole is known as the Aurora Australis – the Southern Lights. It is caused as the Earth's magnetic field pulls in electrically charged particles streaming through space from the Sun. The particles collide with gas atoms in the atmosphere, which radiate the collision energy as light. The Earth's magnetic field is created by the dynamo effect of swirling currents of molten metal in the planet's core.

But there is still a mystery to be solved. Every now and again the geomagnetic field mysteriously changes direction – with North and South Poles literally switching places. It happens over timescales of tens of thousands of years, as revealed by studies of volcanic rock. As molten magma has oozed up from the Earth's interior during past eruptions, its atoms and molecules have become aligned with the direction of the Earth's magnetic field at the time. The direction of the field has then become frozen in place as the rock has cooled and solidified. Many such field reversals have taken place during the Earth's history. The most recent one occurred an estimated 780,000 years ago and the oldest one (that we have evidence for) took place around 2.4 billion years ago.

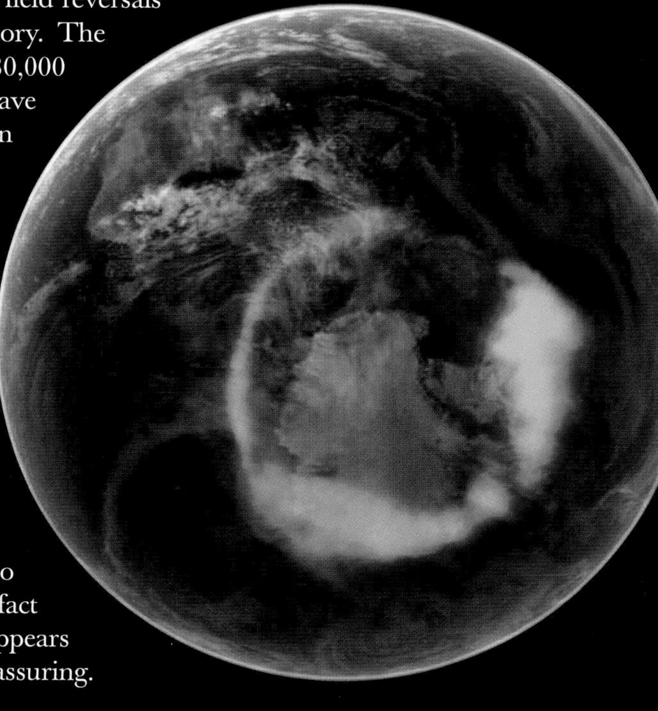

The cause of these field reversals remains unclear, though some researchers have suggested that chaotic flow patterns in the molten metal of the core could be responsible. The Earth's magnetic field protects life on the planet by batting away high-energy radiation particles that arrive from space. There has been some concern that during a reversal, the magnetic field's ability to do this may be compromised, leaving Earth exposed to harmful cosmic radiation. However, the fact that the Earth – and life on its surface – appears to have survived many such reversals is reassuring.

52 Radiometric dating

DEFINITION THE APPLICATION OF RADIOACTIVE DECAY TIMES TO CALCULATE THE AGE OF ROCKS OR ARCHAEOLOGICAL ARTEFACTS

DISCOVERY THE FIRST RADIOMETRIC AGE ESTIMATES WERE PUBLISHED IN 1907 BY BERTRAM BOLTWOOD

KEY BREAKTHROUGH BOLTWOOD'S AGES FOR ROCK LAYERS VASTLY EXCEEDED THE OLDEST ESTIMATES FOR THE AGE OF THE EARTH

IMPORTANCE IT CONVINCED GEOLOGISTS OF THE EARTH'S TRUE AGE, AND FOUND OTHER APPLICATIONS IN ARCHAEOLOGY

In the years after radioactivity was discovered, scientists realized that it offered a perfect geological stop clock – able to accurately gauge the passing of millions of years, impervious to any environmental influences. The first person to implement this idea was American chemist Bertram Boltwood. His results immediately increased estimates for the Earth's age by a factor of 15.

In 1897, the venerated Scottish physicist Lord Kelvin declared the Earth to be between 20 million and 40 million years old. He had deduced these figures by assuming that the Earth formed in a molten state, and then calculated the time for its surface to cool down. Kelvin had made two serious flaws in his calculation. First of all, he neglected to include the effects of 'convection' inside the Earth. This is a method of heat transport summed up by the statement that hot liquids rise. Hot substances expand and become less dense, making them rise – while cooler ones fall. Applied to the Earth, it brought more heat to the surface, meaning that cooling would take longer.

Secondly, and more importantly, Kelvin neglected the effect of radioactivity, discovered just the year before. Radioactivity is an inherent instability in the structure of the atomic nuclei of some chemical elements, which makes them spontaneously break apart over time – spewing out particles and energy. The extra heating effect of this energy was to skew Kelvin's cooling time calculations – as scientists in the 20th century discovered.

LEFT One of the most high-profile applications of radiometric dating was to deduce the age of the Turin Shroud, once believed to be the death shroud worn by Jesus Christ.

New Zealand-born physicist Ernest Rutherford had partitioned radiation given out by radioactive atoms into three types. One of these – called alpha particles – essentially boiled down to nuclei of the element helium, consisting of two proton particles and two neutrons. Rutherford was the first to wonder whether alpha radiation could be used to estimate the age of rocks. He thought it reasonable that alpha particles could be trapped within rock and imagined breaking up a sample to ascertain its precise helium content – this, together with knowledge of the alpha radiation sources in the rock and the rate at which they give off particles, could give an estimate of the rock's age.

Parents and daughter nuclei

Rutherford's idea also turned out to be flawed because helium can, in fact, escape from rock over time. But it was enough to inspire a young Bertram Boltwood at Yale University to carry out his own research. Rather than using helium levels, he looked at the amount of 'parent' and 'daughter' atomic nuclei in the rock. Rutherford had shown that when an atomic nucleus (the parent) decays, the particles it gives off alter its structure enough to change it into a new element (the daughter). For example, radioactive carbon decays by converting a neutron in its nucleus into a proton plus an electron. The electron is emitted as radiation while the proton remains, turning the carbon nucleus into nitrogen.

'When Boltwood worked through the mathematics of the age calculations for his samples, he obtained figures as old as 570 million years – clearly dashing Kelvin's rather conservative estimate of between 20 and 40 million years.'

It was also known that the radioactivity of a given chemical element decays with a characteristic half-life – after which the number of parent nuclei will have decayed to half its original value. After another half-life period has elapsed, the parent number will have fallen to a quarter of its original value. And so on. Boltwood saw that by measuring the number of parent and daughter nuclei, he could work out the amount of radioactive decay and – using the known half-life of the parent – calculate how much time had elapsed since the decay had begun.

Dating uranium decay

Boltwood soon applied his method to the decay of the radioactive elements uranium and radium into lead, and began analyzing rock samples. Just as he had expected, rock samples taken from higher rock strata – laid down more recently – are younger in age and have undergone less radioactive decay, as evidenced by their lower daughter-to-parent ratios. When Boltwood worked through the age calculations for the Earth, he obtained figures as old as 570 million years – dashing Lord Kelvin's rather conservative estimate. He published his findings, in 1907, in the *American Journal of Science.*

Boltwood's calculations were, nevertheless, slightly off-beam. The radioactive processes were not simple decays from a single parent to a single daughter, but rather they were long decay chains where a parent decays first to a radioactive daughter, which in turn decays to a radioactive granddaughter, and so on, with each step of the process having its own characteristic half-life. When Boltwood published his original findings, not all the steps in the decay of uranium and radium were well understood. But once these gaps

in our knowledge were plugged, his figures could be refined. The news for Kelvin was then even worse – correcting Boltwood's calculations actually upped the age of the oldest rocks to 1.3 billion years.

ABOVE The Wolfe Creek meteorite crater on the edge of the Great Sandy Desert, in Western Australia. Geologists have applied radiometric dating techniques to meteorite fragments to determine the age of the Earth. Meteorites formed at the same time as the Earth but, unlike the planets, have changed little since. The studies reveal that the Earth is 4.54 billion years old.

Meteoric evidence

In the 1950s, scientists obtained the best estimate for the age of the Earth by applying Boltwood's radiometric dating techniques to meteorites. Meteorites are chunks of rock and metal that circled the Sun in the same cloud of material from which the Earth and the other planets formed. Unlike the planets, however, these rocks have undergone relatively little heating and mixing (which could contaminate radiometric measurements). Studies of uranium and lead levels in the Canyon Diablo meteorite (recovered from the Barringer Crater in Arizona), carried out in 1956 by geochemist Clair Patterson, revealed the most accurate determination yet – 4.54 billion years, equal to 4540 million years, or over 100 times Kelvin's longest estimate.

Today, radiometric dating is not just applied in geology. It is also a cornerstone of archaeological research, where the radioactive decay of shorter half-life chemicals – in particular, the isotope carbon-14 – is used to measure the age of artefacts that are less than 60,000 years old. In 1988, this technique was applied to the Turin Shroud – which theologians had claimed was the linen burial shroud of Jesus – and placed its origin between 1260 and 1390, much later than the era in which Christ is believed to have lived.

53 The Burgess Shale

DEFINITION A SEDIMENTARY ROCK FORMATION IN CANADA HOLDING THE RICHEST DEPOSITS OF FOSSILS ANYWHERE ON THE PLANET

DISCOVERY THE SITE WAS FOUND IN 1909 BY AMERICAN PALAEONTOLOGIST CHARLES WALCOTT

KEY BREAKTHROUGH HAVING FOUND FOSSILS AT THE BASE OF MOUNT STEPHEN, WALCOTT CLIMBED UPWARD AND FOUND THEIR SOURCE

IMPORTANCE THE SITE DATES FROM THE CAMBRIAN EXPLOSION, THE MOST IMPORTANT ERA IN THE EVOLUTION OF LIFE ON EARTH

The Burgess Shale is a rock formation found on Mount Stephen in Canada. It is a source of fossils from the Cambrian period, 500 million years ago, when life forms made the transition from unicellular microbes to complex plants and animals. The diversity of life at this time was greater than it ever had been or ever would be, and the fossils could tell us why.

Fossils are the remains of life forms – animals, plants and microbes – that over millions of years have become mineralized. The process happens in water, as sediments are laid down and compressed to form sedimentary rock. Inevitably, biological material gets caught up in this process, and leaves its imprint in the resulting rock formation. The result is a fossil.

Fossils can be found in abundance at many locations around the world. And, from a modern perspective, this should not really surprise us. We know from radiometric dating that the Earth is billions of years old. We also know from continental drift (see page 225) that the Earth's surface rock is continually heaving this way and that – making it easy for layers that started life at the bottom of the ocean to end up on the side of a hill today, where their constituent fossils can be discovered by passers-by.

And it is to the benefit of science that this is the case. Fossils are our only window through which to study the species that roamed the surface of the Earth and swam in the planet's oceans millions, and even hundreds of millions, of years ago. Without them, we would have no idea about the evolutionary tree of life on Earth – nor any idea where we, or the other creatures that we share the planet with, ultimately came from.

LEFT This trilobite (*Kootenia burgessensis*) is one of the fossils recovered from the Burgess Shale quarry. With its beetle-like shape, the head and body parts are clearly visible. Trilobites are anthropods that lived 500 million years ago when complex plants and animals first arose on planet Earth. The Burgess Shale provides an excellent record of these and other creatures.

The Burgess Shale fossil quarry is located on a mountain ridge near the town of Field, in Canada's Yoho National Park. It is named after the nearby Mount Burgess, and the Burgess Pass that meanders nearby. Although there are many fossil sites dotted around the world, none of them is as significant – both in terms of quantity and quality of specimens.

'Stone bugs'

The site was discovered by Charles Walcott, of Washington DC's Smithsonian Institute. Walcott first visited the area in 1907, after hearing stories from railway workers of 'stone bugs' found on the slopes of Mount Stephen. He confirmed their accounts, finding buglike trilobite fossils, fossilized plants and other creatures. Two years later, Walcott returned and located the source of the fossils. He reasoned that the fossils he had found first had been dislodged from higher slopes and then slid down. His intuition proved correct and he found the exposed outcrop of fossil-rich sedimentary rock, now known as the Burgess Shale, near the top of the mountain.

'Walcott first visited the area in 1907, after hearing stories from railway workers of "stone bugs" often found lying around the lower slopes of Mount Stephen. He confirmed their accounts, finding a number of bug-like trilobite fossils, as well as other fossilized creatures and plants.'

Walcott immediately began excavating. He returned the next year, and every year after that until 1924 – by which time the 74-year-old Walcott had amassed an incredible 65,000 specimens from the site. The fossils recovered from the Burgess Shale were remarkable. Firstly, and most obvious, was their quality. The rock had clearly formed from very fine sedimentary grains, which had picked out small details in the fossils. Even much of the soft tissue in the creatures and plants was beautifully preserved. Meanwhile, the fossils themselves had survived the ravages of time remarkably intact.

Fossil factory

The site is thought to have been formed by mudslides, which periodically buried the busy throng of life around a submerged reef. That accounts for the large number of fossils found at the site. Their quality is thought to be due in part to the site's geography – situated at the foot of an ancient cliff, which protected the fossils from geological erosion.

Probably more important still was the time period from which most of the fossils hailed – 505 million years ago, in the middle of a period of prehistory known as the Cambrian explosion. Life had already been around on the Earth for some two billion years – but all it had been able to muster were primitive, single-celled organisms. Then, over the space of a few million years, there was an increase in the complexity and diversity of life – all of a sudden, the evolution of life on Earth shifted into high gear.

At this time, all life on Earth existed in the oceans – the migration to land had not yet taken place. This era of frenetic evolutionary activity saw the emergence of the ancestors of all future life on the planet – from house cats

LEFT Scientists examine the layers of sedimentary rock at the Burgess Shale fossil quarry. These strata were formed by an avalanche of fine mud that was deposited on the prehistoric seabed, trapping and preserving the bodies of primitive ocean-dwelling creatures.

to dinosaurs, from platypuses to you and me. Indeed, features common to species today, such as heads, eyes and arms, are seen for the first time in the Burgess Shale fossils. No one knows quite why, or how, the Cambrian explosion happened when it did, although further study at the site may well hold the answer.

Burgess revisited

Walcott himself was quite conservative in his interpretation of the fossils – trying to classify much of what he found according to known taxa. Following his death in 1927, his work on the Shale was largely forgotten and his specimens were locked away in a room at the Smithsonian. It was not until the late 1960s, when the Geological Survey of Canada collected further fossils from the site, that its true potential was discovered.

Research on Walcott's specimens soon resumed, led by trilobite expert Harry Whittington of the University of Cambridge, together with two graduate students, Derek Briggs and Simon Conway Morris. They immediately realized the enormity of Walcott's discovery and, between 1971 and 1985, published a colossal volume of research papers detailing the mass of fossils that Walcott had retrieved. The central thrust of their findings was that many of the life forms simply cannot be classified in terms of modern-day taxa – life 500 million years ago was vastly more diverse than it is now. The implications of this for the theory of evolution are still being debated.

The Burgess Shale was made a World Heritage Site by UNESCO in 1981. It remains today the single most valuable fossil snapshot of the Earth's past that human beings have ever been fortunate enough to lay eyes on.

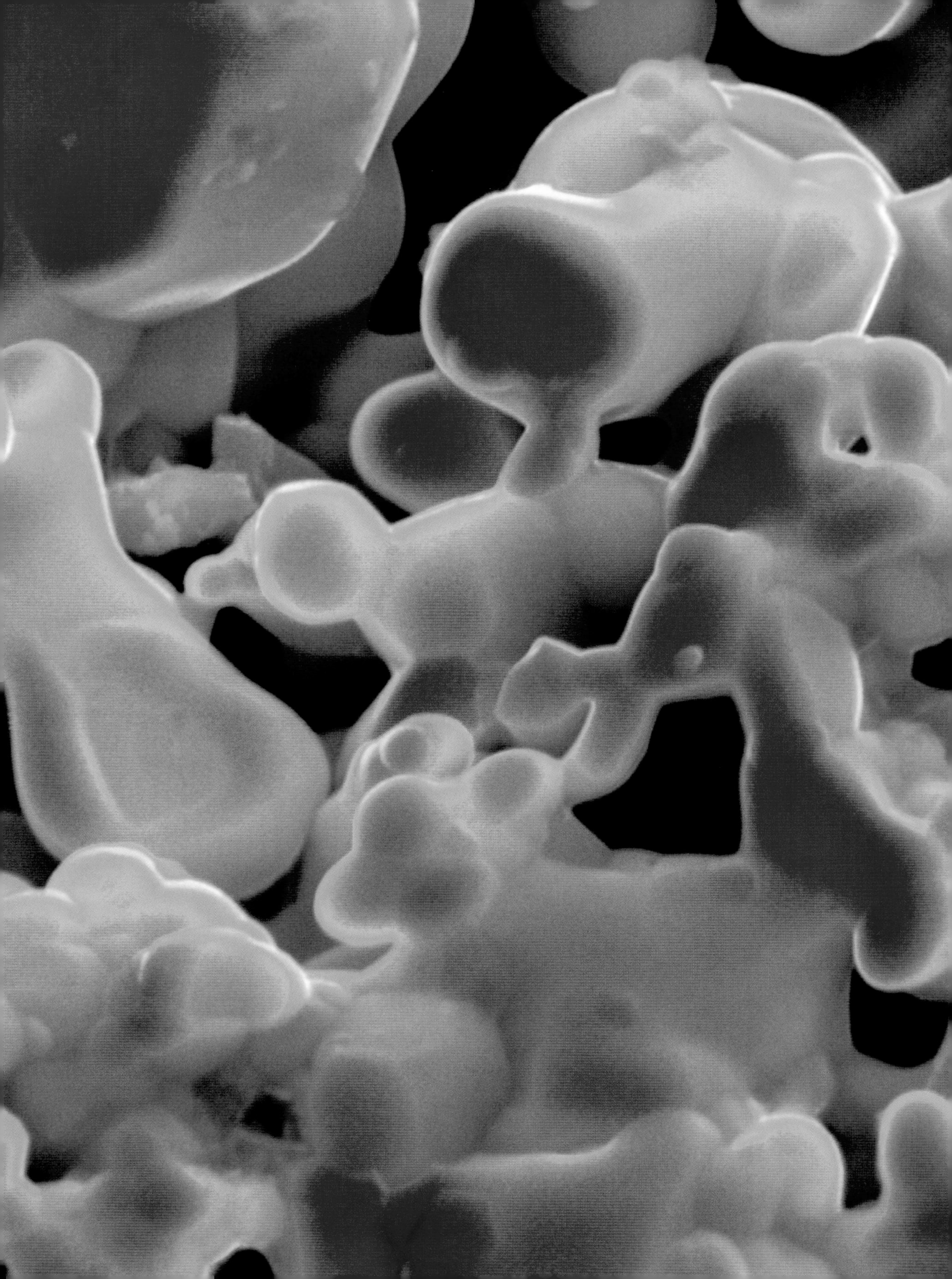

54 Superconductors

DEFINITION A SUPERCONDUCTOR IS A MATERIAL THAT HAS ZERO RESISTANCE TO ELECTRIC CURRENT

DISCOVERY SUPERCONDUCTIVITY WAS FIRST OBSERVED BY HEIKE KAMERLINGH ONNES, AT LEIDEN UNIVERSITY, IN 1911

KEY BREAKTHROUGH USE OF LIQUID HELIUM TO COOL MERCURY TO A TEMPERATURE AT WHICH ITS RESISTANCE VANISHED

IMPORTANCE A PLETHORA OF APPLICATIONS, INCLUDING POWER GRIDS, MEDICAL SCANNERS AND LEVITATING TRAINS

Drop the temperature of certain materials far enough and their electrical resistance – their ability to block electric current – vanishes. These materials are superconductors, through which electricity can flow unimpeded. They are used to make generators, imaging devices and particle accelerators – despite the fact that scientists still do not really understand how they work.

Between ten and 15 percent of the electricity generated by power stations is wasted: it is lost as heat from the copper cables that transport it. Superconductors eliminate this waste, carrying electric current over any distance with zero losses. Even the best ordinary conductors have some degree of electrical resistance, which puts the brakes on electric current, dissipating it as heat (this is the reason that all electrical appliances warm up when you leave them on). But, in 1911, the notion that electrical resistance is inevitable was dramatically overturned.

LEFT The fine structure of a superconducting material made from lanthanum, barium and copper-oxide. Superconductors conduct electric current with zero losses, meaning they can transmit electricity with 100 percent efficiency. However, at present, they only operate at extremely low temperatures. Research is under way to produce room-temperature superconductors.

Dutch physicist Heike Kamerlingh Onnes was studying the electrical properties of a wire made from mercury. To investigate the flow of electricity through conductors at ultra-low temperatures, he dropped the temperature of the mercury down towards absolute zero: –273.15°C. When he did this, Onnes noticed something rather unusual. He was about to make a discovery that would change the world.

Onnes noticed that the resistance of the wire was decreasing as the temperature fell. Some scientists had believed the opposite would happen, that current carriers in the metal would literally freeze solid, causing the resistance to rise. Onnes, on the other hand, had correctly predicted that the resistance would fall, but he had expected a gradual decline. So it came as a surprise when, at 4.2 degrees above absolute zero, the resistance of the mercury abruptly vanished. With nothing to block it, current now flowed through the wire freely – it had become a perfect conductor. Accordingly, Onnes referred to this new state of matter as 'superconductivity'. In 1913, he won the Nobel Prize in Physics for the discovery.

BSC theory

The hard part was explaining why. It took until 1957, when American physicists John Bardeen, Leon Cooper and Robert Schrieffer, at the University of Illinois, put forward the so-called BCS theory, after the first letters of their surnames. It won them the 1972 Nobel Prize in Physics.

Here is how the BCS theory works. In an ordinary conductor, negatively charged electrons escape their parent atoms to carry electric current through the material. Heat makes the lattice of atoms vibrate, and collide with the electrons, impeding the flow of current – this is electrical resistance. Cooling lowers the resistance by reducing the vibrations. But in a superconductor, resistance vanishes entirely. Here, electrons lock together into 'Cooper pairs' that slip freely through the lattice. Broadly speaking, each electron's negative charge attracts positive atoms, distorting the lattice to create a concentration of positive charge that pulls the next electron forward, and so keeps the current going.

'Giant superconducting loops are used to store electric current without loss for long periods – with no resistance to slow it down, the current keeps circling around forever.'

But even the best superconductors known at the time – using conventional elements and alloys – still demanded temperatures approaching –270°C, requiring costly ultra-low-temperature cooling fluids, such as liquid helium.

Getting warmer

In 1986, Johannes Georg Bednorz and Karl Alexander Müller, working at IBM in Switzerland, found a warmer alternative. They came up with a new superconducting material, a blend of ceramic and copper. Although still sub-zero, its 'critical temperature' – the upper temperature at which superconductivity begins – was now high enough that it could be reached using liquid nitrogen, a much cheaper coolant than helium.

The trouble was that they had no idea how these new 'high-temperature superconductors' actually worked – the materials were clearly doing something that was beyond the scope of the BCS theory. But that did not stop the discovery from bringing with it a raft of new applications.

Giant superconducting loops are used to store electric current without loss for long periods – with no resistance to slow it down, the current keeps circling around forever. Each of these super-batteries can hold around 20 megawatt-hours of electricity to top up power grids during times of peak demand. And in December 2003, Japan's MLX01 train achieved a blistering top speed of over 579km/h (360mph), due to powerful superconducting magnets that levitate the train clear of the ground – reducing frictional forces massively. This is possible because of a phenomenon called the Meissner effect, discovered in 1933 by German physicists Walther Meissner and Robert Ochsenfeld. They found that superconductors expel any magnetic fields within them. This means that a magnet placed above a superconductor will hover in midair as its field is pushed up and out of the material beneath.

Resistance is futile

Future applications include superconducting generators, dynamos that turn input force – like high-pressure steam from a nuclear reactor – into electricity with almost 100 percent efficiency. Physicists have calculated that if Europe alone incorporated these into its power stations, it would reduce annual carbon emissions by 53 million tonnes. Over the next decade, superconductivity looks set to become big business, with the worldwide superconductor market expected to be worth US$38 billion by 2020.

As of spring 2011, –138°C was still the highest critical temperature that had been achieved. The ultimate dream is to create superconductors that work at room temperature. But with no available theory to explain how higher-temperature superconductivity could actually work, no one knows whether this will ever be possible. Nevertheless, superconductivity remains a hotbed of scientific research – and one that could hold the solution to crucial world issues, such as climate change and energy production.

BELOW One property of superconductors is that they completely expel any magnetic fields passing through them – a phenomenon known as the Meissner effect. This means that an ordinary iron magnet placed above a superconductor will levitate, as shown here. The effect is exploited in Japan's MLX01 maglev train to increase speed and reduce journey times.

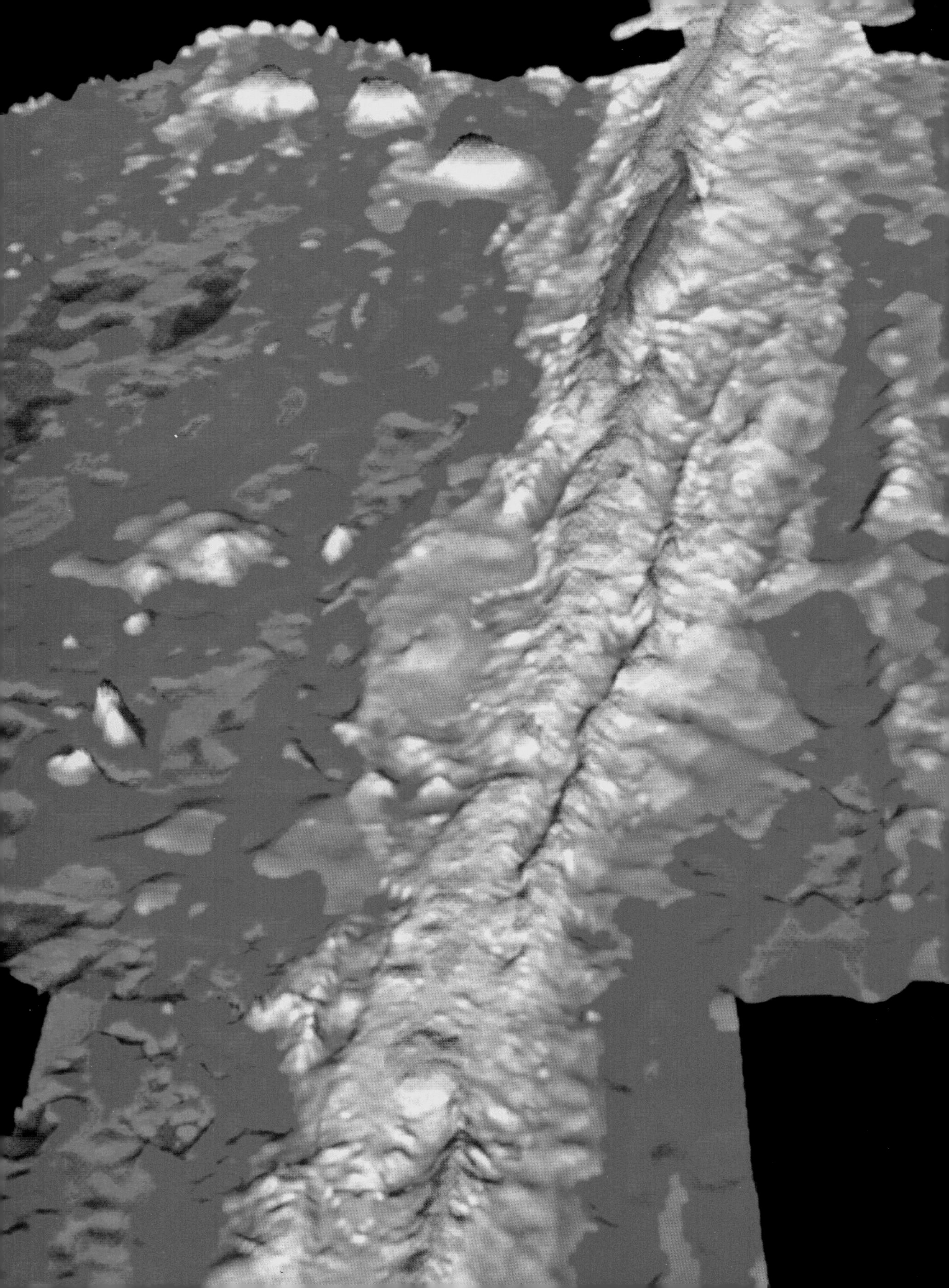

55 Continental drift

DEFINITION THE EARTH'S CRUST IS A NETWORK OF INTERLOCKING PLATES THAT JOSTLE TOGETHER AND CAUSE LANDMASSES TO MOVE

DISCOVERY THE THEORY WAS FIRST PUT FORWARD BY THE GERMAN GEOPHYSICIST ALFRED WEGENER IN 1912

KEY BREAKTHROUGH IDENTICAL FOSSILS DISCOVERED AROUND THE WORLD SUGGESTED THE LAND WAS ONCE A SINGLE CONTINENT

IMPORTANCE WEGENER'S THEORY FORMED THE BASIS FOR PLATE TECTONICS – THE DOMINANT FORCE IN SHAPING THE EARTH

Anyone who has looked at a map of the Earth can not have failed to notice that South America and Africa look as if they once fitted together. That is because they once did and moved apart in a phenomenon known as continental drift, first proposed in 1912 by Alfred Wegener. Today, continental drift is known to be a consequence of plate tectonics, which also underpins earthquakes and volcanic activity.

As recently as the start of the 20th century, geologists believed that the Earth's landmasses were fixed and immovable. They believe the continents had always existed in their present configuration – and always would do.

But with the discovery that the planet was a lot older than had previously been thought – due to the new field of radiometric dating (see page 212), developed in 1907 – some scientists began to wonder whether the form of our planet could really have been static for so long. One of them was a German geophysicist called Alfred Wegener. Born in Berlin in 1880, he was awarded a doctorate in astronomy by the University of Berlin in 1904. However, Wegener had always been more interested in the Earth sciences than space, and fascinated by issues such as weather and climate. And so after his PhD, he made a career change to become a geophysicist – securing a position at the University of Marburg.

It was here, one day during the autumn of 1911, that Wegener was browsing in the university library and came across a piece of research detailing instances of identical animal and plant fossils that had somehow been discovered on opposite sides of the Atlantic. Existing explanations put this down to long-

LEFT Contour map of a seafloor ridge in the Pacific Ocean. Blue represents the deepest regions and red the shallowest. Seafloor ridges are sites where the Earth is creating new crust from hot magma below. The seafloor spreads out from the ridge, and this movement drives continental drift. Hot spots may lead to underwater volcanoes.

sunken bridges linking these distant landmasses. But Wegener did not believe this. Noticing that the continents – especially South America and Africa – looked as if they might once have fitted together, he proposed that Earth's landmasses were once locked together as a single, giant 'supercontinent', which then subsequently drifted apart.

Continental drift

Wegener first presented his theory of 'continental drift' to the German Geological Society on 6 January 1912. But it did not receive a good response. The main problem was that he had no idea what physical mechanism might be causing the continents to drift. Wegener did not help his case by cobbling together a side theory involving centrifugal forces associated with the Earth's rotation – an idea that simple calculations could show was impossible.

'It was soon clear that the Earth's outer crust is made up of a network of plates: slabs of rock interlocking like pieces in a jigsaw puzzle. This view became known as the theory of plate tectonics.'

Wegener's theory of continental drift could have been consigned to the scientific scrap heap right there and then. But in the 1950s, new evidence emerged that brought about its resurrection. Different layers of rock in the Earth's surface represent different ages of planet – the deeper you go, the older the rocks you find. Many rocks preserve within them the direction of the Earth's magnetic field at the time they formed – a phenomenon called palaeomagnetism. But when geologists examined the direction of the field locked into rock beds they found that it changed depending upon the age of each rock layer – as if the entire rock bed was moving with respect to the Earth's magnetic poles.

The effect of paleomagnetism was soon backed up by undersea studies, which revealed oceanic floors to be continually on the move. At sites known as seafloor ridges, molten material in the planet's interior is welling up, forming new crust and pushing apart the existing seafloor. In other places, called subduction zones, the excess crust produced by this process is shuffling down beneath the planet's surface.

Fault lines

It was soon clear that the Earth's outer crust is made up of an intricate network of plates: slabs of rock interlocking like pieces in a jigsaw puzzle. This view became known as the theory of plate tectonics. The barriers where plates collide are called 'fault lines' and they can take a number of forms. Where plates are moving apart – as is the case at seafloor ridges – the faults are called 'divergent'. These faults are not just located at the bottom of the ocean – the Great Rift Valley is a spreading site on land, located in East Africa. Similarly, locations where plates are coming together are called 'convergent' faults. Subduction zones are an example, as are regions where convergent plates have caused the crust to ruck up, forming sharp mountain ranges – such as the Himalayas.

When the plates slide along past one another, the situation is called a 'transform' fault. Perhaps the most famous of these is the San Andreas Fault in California, which was responsible for the devastating San Francisco earthquake of 1906. Earthquakes are always likely to occur along fault lines, as the tectonic plates grate past one another. When earthquakes happen under the ocean the process can trigger huge tidal waves. The weakened crust along fault lines also makes them prime locations for the formation of volcanoes.

Restless planet

Planet Earth has seven major tectonic plates – African, Antarctic, Eurasian, Indo-Australian, North American, Pacific and South American – along with many smaller plates jostling in between. The plates are borne up on the semi-liquid mantle layer of the Earth's interior. The motion of the plates is now known to be driven by the churning motion of the underlying mantle as it transports heat upwards from the planet's interior. The viscous mantle material is like treacle – sticking to the overlying rock and pulling it along as it goes. Tectonic plates move at about the same rate fingernails grow, a few centimetres per year.

Wegener's notion of a supercontinent is now also well accepted. It has even been given a name – Pangea – and was thought to have existed around 200 million years ago. What is more, geologists believe that the continents may be converging again and heading towards forming another supercontinent in around 250 million years' time. This landmass has been named Pangea Ultima. Tragically, Wegener was not around to see his theory so gloriously reborn – having met his death on an ill-conceived expedition to Greenland during the winter of 1930. But his insights showed us what a dynamic and ageless world the Earth really is – continually resculpting its appearance as the aeons wheel by.

ABOVE This land rift in Iceland has been created as the tectonic plates that form the North American and Eurasian continents drift apart.

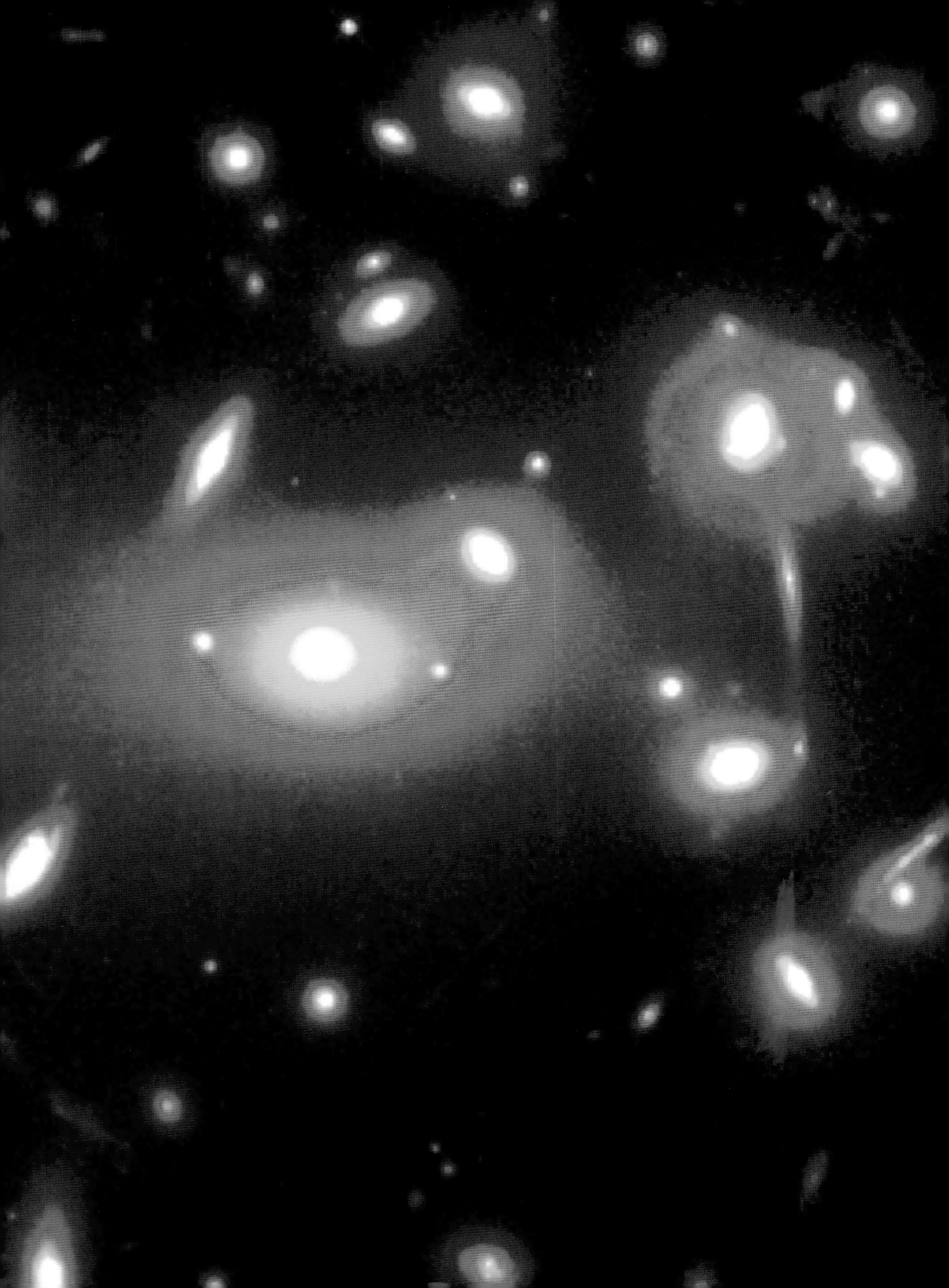

56 General relativity

DEFINITION GENERAL RELATIVITY IS A THEORY IN PHYSICS THAT EXPLAINS GRAVITY AS THE CURVATURE OF SPACE AND TIME

DISCOVERY FIRST PUBLISHED IN 1915 BY ALBERT EINSTEIN, THEN WORKING AT BERLIN'S KAISER WILHELM INSTITUTE FOR PHYSICS

KEY BREAKTHROUGH EINSTEIN FORMULATED AN EQUATION LINKING CURVATURE OF SPACETIME TO THE MASS AND ENERGY IT CONTAINS

IMPORTANCE GENERAL RELATIVITY SUPERSEDED NEWTON'S THEORY OF GRAVITATION THAT HAD HELD SWAY FOR OVER 200 YEARS

General relativity built upon the revolution that Einstein began with his idea of special relativity: it was a new theory equating the force of gravity with the geometry of space and time. It also provided the exact connection between geometry and the matter content of space, explained an anomaly in Mercury's orbit and predicted how light bends as it passes close to the Sun.

Despite the success of special relativity (see page 204), Einstein knew his theory was incomplete. The very reason it is called the 'special' theory is because it only applies to the special case of objects moving at constant speed. Einstein's next task was to generalize relativity to describe the motion of objects that are accelerating: the new theory was called 'general relativity'.

It was not just the theory of acceleration that Einstein planned to rewrite. He was also challenging Isaac Newton's law of universal gravitation, first put forward in 1687. According to this theory gravity travels between points in space instantaneously – in blatant contradiction of Einstein's special theory of relativity, which says that nothing can travel faster than the speed of light.

Early in the development of general relativity, Einstein realized that gravity and acceleration are equivalent. In what he described as 'the happiest thought of my life', it dawned on him that someone inside a sealed box with no windows who feels a force pushing them to the floor cannot tell whether that box is stationary in a gravitational field or accelerating upwards in space, away from gravity. Einstein called this the equivalence principle. But what form would the new theory take? A simple thought experiment in 1912 convinced Einstein that gravity manifests itself by curving the flat

LEFT Gravitational arcs in the galaxy cluster Abell 2218. The thin arc-like features are made by light from very distant sources, some five-to-ten times further away than Abell 2218. The cluster is so massive that light rays from background galaxies passing through it are deflected and magnified by the cluster's gravity.

spacetime of special relativity. He imagined a rotating disk. Rotation is caused by acceleration. Spin a weight on a string above your head and it is the tension in the string that accelerates the weight, stopping it flying off in a straight line and making it go round in a circle instead.

The outer edge of the rotating disk is moving at speed, and therefore according to special relativity its length should contract. But at the same time, the radius of the disk stays the same. The only way these two things can happen is if the disk is curved into a kind of dish shape. When Einstein saw this it was clear that curvature and acceleration go hand in hand – and, because of the equivalence principle, so do curvature and gravity.

'The two sides of Einstein's equation enabled physicists to determine how space and time are curved by mass and energy, and to compute the motion of objects across that curved stage.'

But how is this curvature determined by the mass and energy filling space? Answering that question pushed Einstein to the point of exhaustion. But in November 1915, he had the solution.

The field equation

The final equation linking matter and geometry is known as the 'field equation' of general relativity. On the left-hand side of the equation are mathematical quantities describing the curvature of space and time. And on the right are quantities describing its contents. Rather than just mass, which is the only source of gravity in Newton's theory, the right-hand side of Einstein's field equation factors in the energy present in space (for example, radiation) as well as the momentum and pressure. Clipped together, the two sides of the equation enabled physicists to determine how space and time are curved by mass and energy and to compute the motion of objects across that curved stage. For instance, the Sun can be imagined as curving space into a giant bowl shape, around the inside of which the planets circle like rolling marbles.

Mercury's riddle

Of course, the details are more complicated than that. One particular detail had been baffling astronomers for nearly 60 years. In 1859, French mathematician Urbain Le Verrier noticed that the elliptical orbit of Mercury seems to rotate around the Sun and traces out a rosette shape over time. At first astronomers blamed the gravitational pull of a new planet, orbiting closer to the Sun than Mercury. Searches for this new world, called Vulcan, drew a blank. Einstein recalculated Mercury's orbit using general relativity and predicted the effect that Le Verrier had seen. No new planet was required.

But the theory's big confirmation was yet to come. General relativity accommodates not just the deflection of massive objects by gravity, but light beams too. The degree of light bending around the Sun is tiny – about 0.0005 of a degree – so small that the beam would get lost in the Sun's glare. English astronomer Arthur Eddington got around this problem by observing the Sun during a solar eclipse – when its glare is blocked out by the Moon.

In 1919, Eddington travelled to the island of Principe, West Africa, where he took advantage of a solar eclipse to observe stars close to the Sun's position in the sky. Eddington's observations tallied with general relativity perfectly.

Einstein applied the theory of light bending in general relativity to the universe at large to predict the existence of 'gravitational lenses'. These form where the gravity of large galaxies and clusters magnifies the light from other celestial objects, revealing distant galaxies that might otherwise be invisible. The first real gravitational lens was discovered in 1979.

Ripples in time

Astronomers now seek evidence for the final untested prediction of general relativity: gravitational waves. These are ripples in space and time given out by moving sources of gravity, such as exploding stars. The waves are tiny – two points a metre apart see their separation distance fluctuate by just one billion-billionth of a centimetre when a gravitational wave passes by. If the ground-based detectors that hunt for these cosmic ripples succeed, they will give us a new window through which to observe the universe. Paradoxically, this could lead to general relativity's downfall. Explaining the Big Bang will ultimately require a theory of gravity that can accommodate the counter-intuitive laws of the tiny quantum world. General relativity refuses to do this, but probing the Big Bang through gravitational waves could show how the theory needs to be modified – pointing to the holy grail of physics, a theory of 'quantum gravity'. If physicists find that, it will be as much of a revelation to them as general relativity was to Einstein.

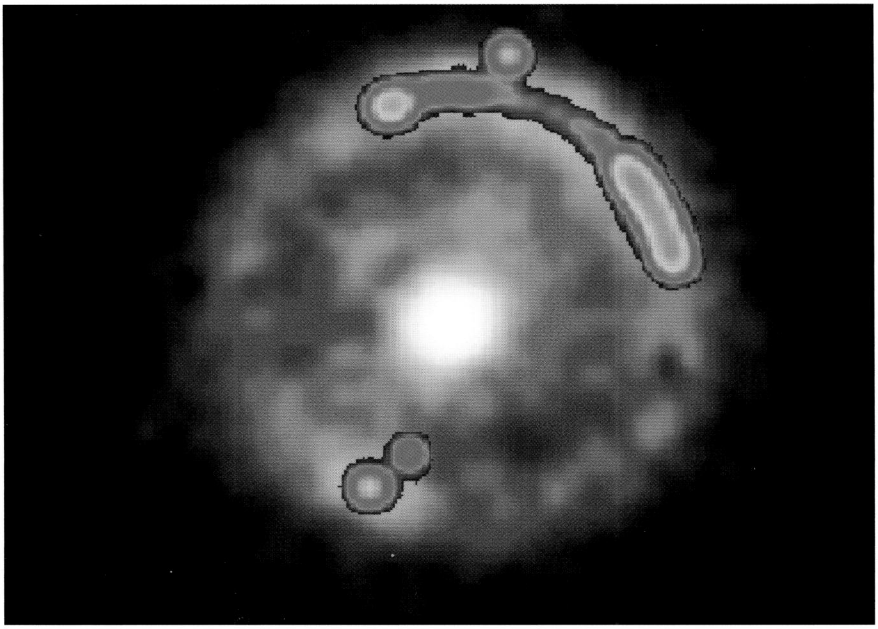

LEFT An 'Einstein ring' showing gravitational lensing. General relativity states that space is bent by the gravity of massive objects, and that this can curve light rays. In a gravitational lens, light from a distant galaxy is curved by the gravity of an intervening galaxy and magnified.

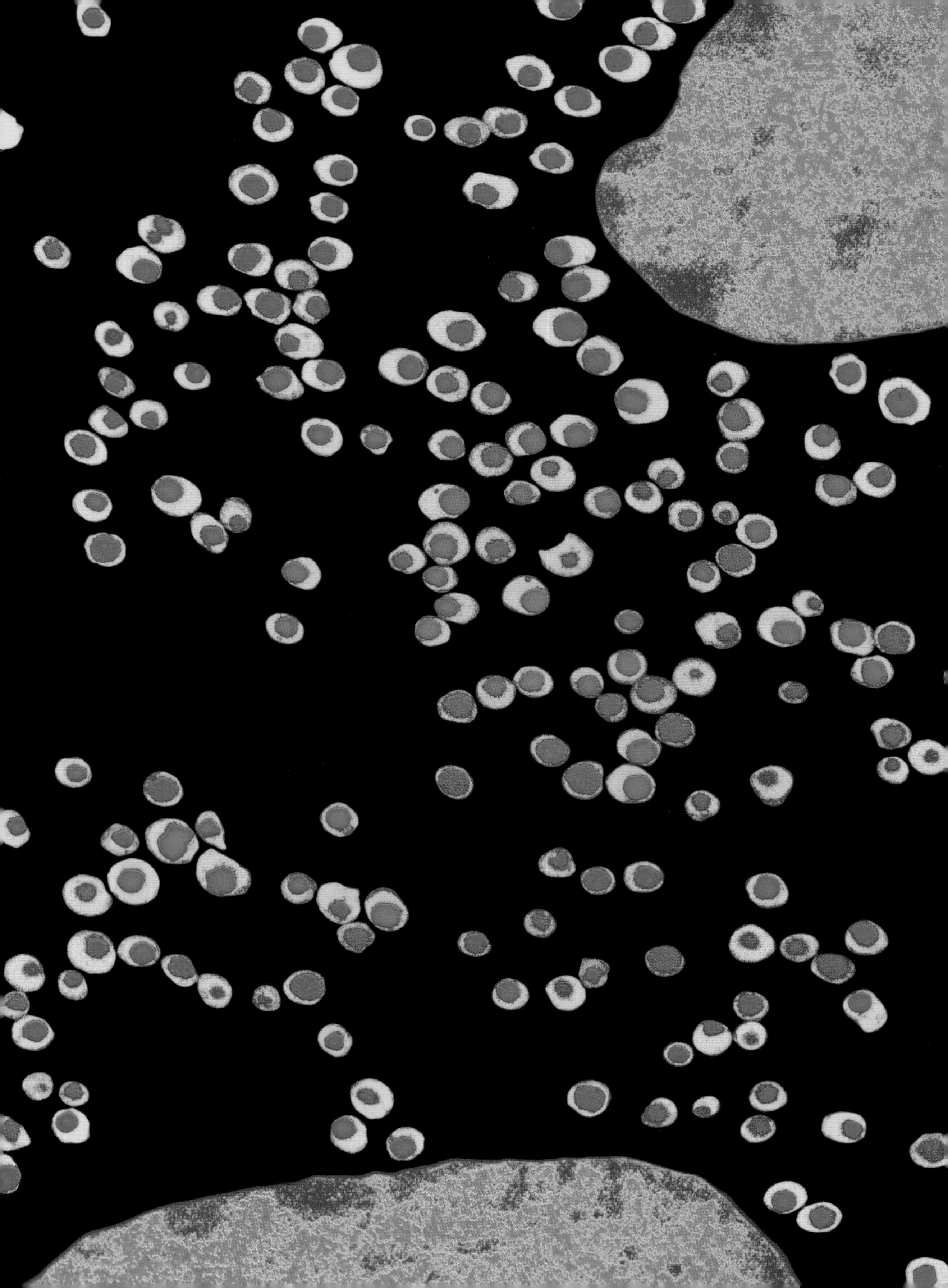

57 Insulin

DEFINITION A HORMONE PRODUCED BY THE PANCREAS TO REGULATE BLOOD SUGAR; ITS ABSENCE IS THE CAUSE OF TYPE 1 DIABETES

DISCOVERY FREDERICK BANTING AND CHARLES BEST WERE FIRST TO EXTRACT INSULIN AND TREAT DIABETES WITH IT, IN 1922

KEY BREAKTHROUGH BANTING AND BEST PREPARED AN EXTRACT FROM PANCREATIC CELLS AND USED IT TO TREAT DIABETIC DOGS

IMPORTANCE BEFORE THE DISCOVERY OF INSULIN, TYPE 1 DIABETES WAS AN INVARIABLY FATAL CONDITION

Diabetes is a condition in which the body is unable to produce enough of the hormone insulin to regulate its blood sugar. With no way to create reserves of energy, an early death used to be inevitable. But not anymore. In 1922, Frederick Banting and Charles Best used bovine insulin to treat diabetes sufferers, turning a terminal illness into a manageable condition.

Many people today suffer with diabetes yet are able to live more or less normal lives. But it has not always been this way. As recently as the early 20th century, diabetes was a killer illness. Some sufferers were able to extend their life through a strict diet that controlled the intake of sugars, but that would, at best, buy them a few extra years. The disease was even more painful to witness in younger diabetics since few would live as far as adulthood.

What is insulin?

Insulin is a hormone (a chemical messenger) that is normally produced in the pancreas of a healthy individual, and which helps to regulate blood glucose levels. We all need to eat in order to provide our bodies with energy – but since we cannot be eating continuously, our bodies store the energy we consume at meal times for use throughout the day. This is where insulin comes in. When we have had a meal and our blood sugar levels are high, insulin is steadily secreted from the pancreas, sending a chemical message to the body to convert some of the sugar in the blood into the energy-storage molecule glycogen, which is then packed away as a reserve supply in the liver and the muscles. When sugar levels later drop, insulin release is automatically halted and the glycogen is slowly converted back into energy. In sufferers of diabetes mellitus (to use its full name), this regulation process

LEFT Two hormone-producing cells, known as islets of Langerhans. The cell nuclei are at the top and bottom, although the dividing membrane between them is not visible. The small granules in between produce insulin (shown green).

goes wrong. The disease comes in one of two forms. In Type 1 diabetes, the pancreas is simply unable to produce enough insulin on its own. This is due to what is called an 'autoimmune' response, where the patient's immune system mistakenly identifies healthy pancreatic cells as foreign invaders and destroys them. The cause is not fully understood. Type 1 is the more serious of the two forms of diabetes and full insulin treatment is required to manage the condition.

Type 2 diabetes is less serious and often induced by an unhealthy lifestyle. Here, there is no reduction in insulin production – the body has just become resistant to the hormone, owing to factors such as obesity. It is usually treated with medication, plus a strict diabetic diet and regular exercise. Left untreated, a patient suffering from either type of diabetes will experience increased thirst and hunger followed by blurred vision, nausea, abdominal pain, coma and, ultimately, death.

Islets of Langerhans

The first step towards understanding this process was made in 1869 by German biologist Paul Langerhans. He was studying sections of pancreatic tissue under a microscope when he noticed small bunches of cells. It was suggested that these cell groups, which became known as 'islets of Langerhans', might secrete substances to aid digestion, since the pancreas was known to have ducts leading into the digestive tract.

'The first human test of insulin took place in January 1922 on a 14-year-old boy called Leonard Thompson, who was close to death as a result of his diabetes. He made an almost immediate recovery.'

In 1889, German physiologist Oskar Minkowski and German physician Joseph von Mering removed the pancreas from a dog, finding that the animal developed diabetes. However, when they simply tied off the ducts carrying pancreatic fluid into the intestine the animal only experienced some digestive trouble – diabetes did not result. From this experiment it was clear that the pancreas was the key to diabetes, but also that the disease was nothing to do with digestion – some other function of the pancreas was responsible. But what?

The answer to that was delivered by a surgeon working in Toronto, Canada, called Frederick Banting. He suspected that the islets of Langerhans were the vital part of the pancreas responsible for diabetes. In 1920, he had an idea to test the theory by carefully tying off the blood supply to disable all parts of the pancreas, except the islets. If no diabetes resulted, his theory would be proven correct.

But Banting needed financial resources in order to try out the idea. So he went cap-in-hand to physician Professor John Macleod, based at the University of Toronto. MacLeod was not convinced by the idea but, nevertheless, furnished Banting with a laboratory, a supply of dogs on which to experiment and an assistant – in the form of medical student Charles Best.

Extract of islet

When Banting and Best tied off the blood flow to the pancreas in one of the dogs, the organ automatically died back, as planned – leaving just the islets of Langerhans. And, just as Banting had suspected, the dog did not develop symptoms of diabetes.

ABOVE Charles Best played a key role in the discovery of the pancreatic hormone insulin, enabling the effective treatment of diabetes.

Next, they removed the isolated islets, ground them up and filtered them – to obtain a liquid extract. Then they injected this substance into the bloodstream of a dog in which they had induced diabetes by removing its pancreas. They found that regular injections of the extract kept the symptoms of the disease in check.

MacLeod was now listening and gave the pair more financial resources to continue their tests. He also suggested a name for their extract – 'insulin', after the Latin word *insula*, meaning 'island'. Banting and Best needed to upscale their insulin production and so began using pancreases taken from cattle. They also enlisted the help of a biochemist, James Collip, to help purify the resulting insulin – with a view to starting trials with human patients.

Clinical trials

The first human test of insulin took place in January 1922 on a 14-year-old boy called Leonard Thompson, who was close to death as a result of his diabetes. After receiving the insulin, he made an almost immediate recovery. A series of further successful clinical trials followed. One story tells of Banting, Best and Collip hurriedly administering insulin along a hospital ward of comatose diabetic children, who shortly after began to awake as if miraculously – to the astonishment of their families, who had been grieving just moments earlier.

Clinical insulin continued to be extracted from animals until the 1960s, when techniques were developed to synthesize it artificially. Today, most insulin is manufactured using bacteria that have been genetically modified to mimic the action of the islets of Langerhans. Banting and Macleod received the 1923 Nobel Prize in Physiology or Medicine for their work. They were angered that the Nobel committee had overlooked their colleagues and duly shared the prize money with Best and Collip. Their work has saved countless lives – and improved the quality of countless more.

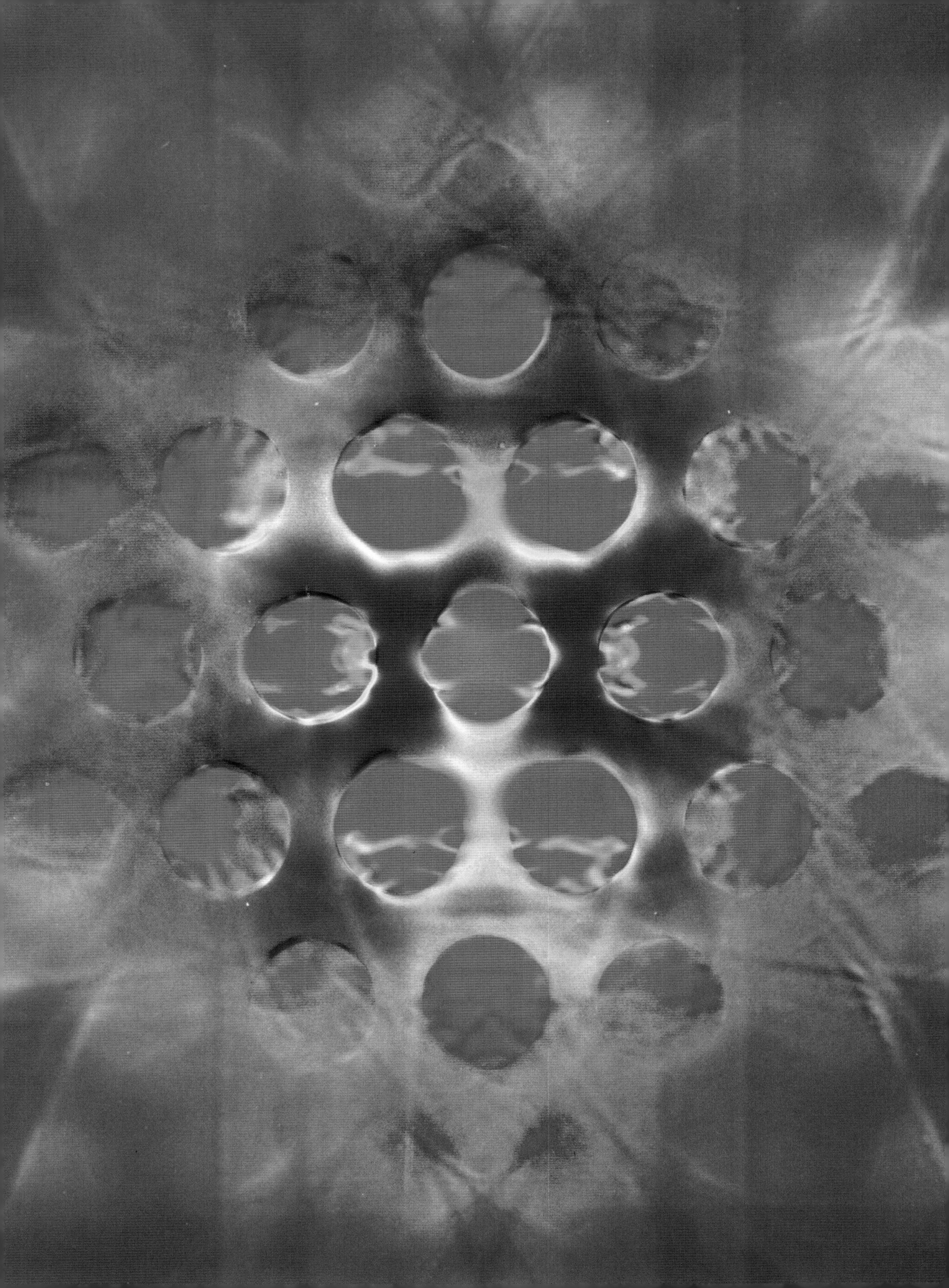

58 Wave-particle duality

DEFINITION SUBATOMIC PARTICLES OF MATTER BEHAVE LIKE WAVES, WHILE WAVES OF RADIATION BEHAVE LIKE SUBATOMIC PARTICLES

DISCOVERY THE THEORY WAS FORMULATED BY FRENCH PHYSICIST LOUIS DE BROGLIE IN 1924

KEY BREAKTHROUGH FINDING THAT ANY SOLID OBJECT IN MOTION HAS A CHARACTERISTIC WAVELENGTH GIVEN BY ITS MOMENTUM

IMPORTANCE IT LED TO THE MODERN INTERPRETATION OF QUANTUM THEORY, WITH PARTICLES DESCRIBED BY PROBABILITY WAVES

It was a centuries-old debate: is light made of particles or waves? Then, in 1924, Louis de Broglie came up with a simple yet seemingly impossible answer: it is both. Experiments soon proved his theory of 'wave-particle duality' to be correct. It prompted other physicists to develop a full wave theory describing the motion of subatomic particles – a theory that became known as quantum mechanics.

In the 17th century, the prevailing view was that light is composed of tiny particles, known as corpuscles. That was blown away at the start of the 19th century when the English physicist Thomas Young carried out a classic experiment proving that light is, in fact, a wave.

Or is it? Because then, in the early 20th century, Albert Einstein put forward his own explanation of the photoelectric effect – the way some metals give off electron particles when they are exposed to light. Einstein's explanation relied on light being made of particles, which could collide with the electrons to detach them from the metal. These particles of light – photons – were duly detected in 1923, by American physicist Arthur Compton.

So what on Earth was going on? Two experiments had presented two results that were seemingly in complete contradiction of one another – on the one hand light was behaving as a wave and on the other it was behaving like a stream of particles. The answer to this anomalous situation was to be provided by the French quantum physicist Louis de Broglie in his doctoral thesis, published in 1924 – the year after Compton's experiment. De Broglie was born in August 1892 in Dieppe. He had initially trained as a historian

LEFT Atomic structure of a titanium crystal created using the technique of electron diffraction. Electrons behave like waves as well as particles, which means that they are diffracted (spread out) as they pass through narrow openings – in this case, the gaps between adjacent titanium atoms. The form of the diffraction indicates the arrangement of the atoms in the crystal.

before turning his attention to science, obtaining his first degree in 1913. After serving as a radio operator during the First World War, he began a PhD at the University of Sorbonne, in Paris.

Photon momentum

In his thesis, de Broglie argued that, at the subatomic quantum level, particles can equally well be thought of as waves and as particles, and he backed his theory with a simple mathematical equation. It had previously been shown that the energy of a light photon is given by Planck's constant ($h = 6.62 \times 10^{-34}$) multiplied by the photon's frequency. Likewise, de Broglie was able to prove that a photon also has momentum – a quantity normally reserved for solid particles with mass – given by multiplying the particle's mass by its speed. For a photon, he showed this momentum is simply given by h divided by the photon's wavelength. Similarly, the equation could be turned on its head to show that a solid particle with momentum can also behave like a wave, and that this wave would have a characteristic wavelength given by Planck's constant divided by the momentum.

'A human being weighing 80kg (176lb) and walking at 1.4 metres (1.53 yards) per second has a de Broglie wavelength of 6×10^{-36} metres – a billion billion times smaller than the size of an electron.'

The equation meant that any moving objects – of any mass or size – behave as a wave to some degree. But the momentum of a macroscopic object is generally so large that its de Broglie wavelength is vanishingly small. For example, a human being weighing 80kg (176lb) and walking at 1.4 metres (1.53 yards) per second has a de Broglie wavelength of 6×10^{-36} metres – a billion billion times smaller than the size of an electron.

Electron diffraction

The dual nature of photons as both waves and particles had already been established. But in 1928, experiments were carried out to demonstrate that electrons – solid particles of matter that orbit in the outer regions of atoms – could also act like waves. Researcher Clinton Davisson and his doctoral student Lester Germer, working at Bell Labs in New Jersey, fired a stream of electrons at a piece of crystalline nickel.

The atoms in the nickel crystal were arranged in a well-ordered, rigid pattern. Davisson and Germer calculated that the de Broglie wavelength of the electrons should be similar to the spacing between the atoms in the crystal structure – meaning that the nickel should then act like a diffraction grating. These devices are normally used to split a light wave into its spectrum of component colours, and consist of a piece of glass scored with lines at a spacing comparable to the light's wavelength. The researchers realized that if the electrons were behaving like waves then they should also be diffracted by the atoms in the nickel, causing the narrow beam to fan out as it emerged on the other side. And this was exactly what the researchers saw. Electron particles were indeed behaving like waves – just as de Broglie had predicted.

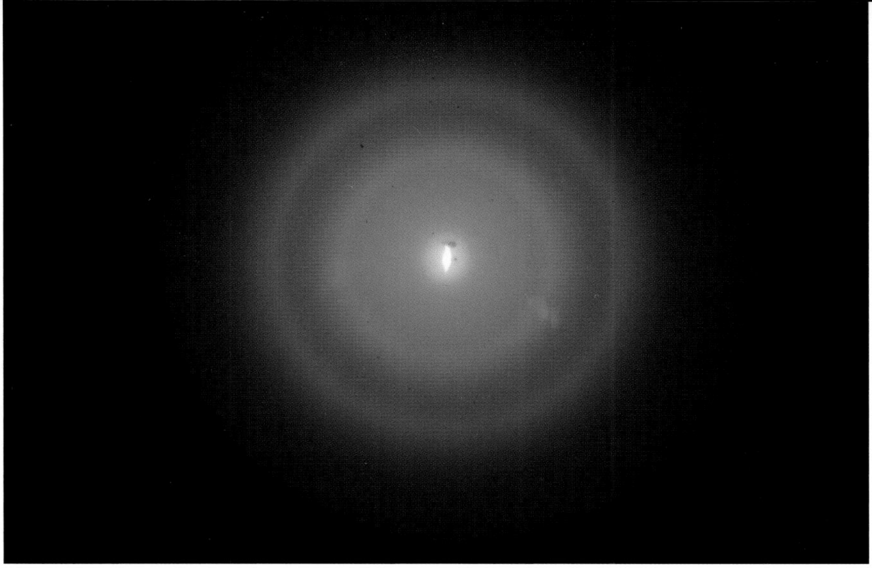

LEFT Diffraction rings produced by interacting electrons. Diffraction is a manifestly wave-like phenomenon, making this an irrefutable demonstration of wave-particle duality. Normally, diffraction happens when light waves pass through a grating with spacing comparable to the light's wavelength.

Electron diffraction is now used as a means to probe the structure of crystals – with the form of the diffraction pattern revealing the spacing between the atoms in the crystal.

Wave-particle duality, also known as the de Broglie hypothesis, had been verified. It earned de Broglie the 1929 Nobel Prize in Physics. But proving that a phenomenon exists is a different matter to actually explaining it. But the explanation finally came when Erwin Schrödinger built on de Broglie's work to describe the motion of solid particles in terms of waves of probability.

Young's experiment, revisited

The probability interpretation of quantum theory was illustrated beautifully in a revised version of Thomas Young's double-slit experiment, carried out by a French team in 1986. In Young's original 1801 experiment, light shone through two slits exhibited interference to produce a characteristic pattern of bright and dark bands on a screen. The new version used the same experimental set-up, with one cunning difference. Rather than shining a continuous light beam at the slits, photons were fired through the apparatus one at a time and the position of each on the screen was marked. Amazingly, the researchers found that as the pattern of dots on the screen accumulated, so the original interference pattern first seen by Young was recovered.

The only way this can happen is if each photon somehow passes through both slits to interfere with itself. And that is what was happening. A probability wave describing the photon's likely position passes through both slits. It interferes with itself to create a pattern of probability peaks on the screen giving the likelihood of where the photon will end up. When enough photons had passed through the apparatus, the pattern of dots mirrored this pattern of peaks, matching the positions of the bright bands that Young had seen 200 years earlier. Waves and particles were indeed the same thing.

59 Schrödinger's equation

DEFINITION AN EQUATION OF MOTION IN QUANTUM THEORY THAT TREATS SUBATOMIC PARTICLES AS WAVES

DISCOVERY FIRST FORMULATED BY AUSTRIAN PHYSICIST ERWIN SCHRÖDINGER IN 1926

KEY BREAKTHROUGH SCHRÖDINGER SAW THAT IF PARTICLES BEHAVE LIKE WAVES, THEN THEIR MOTION WILL OBEY A WAVE EQUATION

IMPORTANCE SCHRÖDINGER'S EQUATION IS TO QUANTUM THEORY WHAT NEWTON'S LAWS WERE TO CLASSICAL MECHANICS

Erwin Schrödinger set the jewel in the crown of quantum physics. Acting on the realization that particles can behave like waves, he came up with a wave equation to describe how subatomic particles moved. The equation explained many observed quantum phenomena and also ushered in the bizarre interpretation of quantum waves as waves of probability.

Schrödinger's principal contribution to science was the 'Schrödinger equation' – a wave equation governing the motion of solid particles of matter. It is one of the foundations of modern quantum mechanics, the theory that explains the physics of tiny subatomic particles.

Erwin Schrödinger was born in Vienna in 1887. He had broad interests in early life, but ultimately elected to study physics at the University of Vienna, graduating in 1910. Afterwards, he worked for the Austrian physicist Franz Exner before enlisting as an artillery officer during the First World War. After the war, Schrödinger had a number of university jobs across Europe, settling in 1921 at the University of Zurich where he remained for six years. It was here that Schrödinger carried out the work for which he is famous.

Schrödinger was led to this great achievement after hearing about the work of French physicist Louis de Broglie, who in 1924 had put forward his idea of wave-particle duality – that particles can behave like waves, and waves also behave like particles. If this was true, Schrödinger reasoned, then surely there must be a mathematical law of wave motion that describes how electrons and other subatomic particles move – similar, for example, to the equations that govern waves on a string or on the surface of the ocean.

LEFT Probability waves, also known as atomic orbitals, indicate where electrons orbiting in the outer region of an atom are most likely to be found. Schrödinger's equation is used to calculate the orbitals' form, determining the behaviour of quantum particles in terms of wave properties.

'When Schrödinger applied his new equation to the case of a hydrogen atom it was a resounding success, predicting exactly the quantized set of energy levels for the electron that had already been observed experimentally.'

To construct his equation, Schrödinger took elementary laws relating to the energy and momentum of bodies in ordinary mechanics, and into these substituted de Broglie's mathematical law linking the momentum of a particle to its wavelength (see page 236) and Planck's relationship between the energy of a photon and its frequency (see page 200). The result, which Schrödinger first wrote down in 1926, was indeed a wave equation.

When Schrödinger applied his new equation to the case of a hydrogen atom (consisting of a single electron moving in the electric field of a single positively charged proton particle) it was a resounding success, predicting exactly the quantized set of energy levels for the electron that had already been observed experimentally. Soon, Schrödinger was applying the equation to other problems in quantum theory – and again obtaining excellent agreement with his experiments.

But there was something missing. Even Schrödinger did not quite understand what exactly his equation was describing. It worked using a 'wavefunction' – a rather abstract quantity, but from which tangible physical parameters, such as energy, could be calculated. What did this wavefunction represent?

German physicist Max Born answered that question the very same year. He saw that the square of this weird wavefunction – which itself depends on the electron particle's position – gives the probability of finding the electron at that position. In this view, quantum theory is not a deterministic theory. It makes no hard and fast predictions about the behaviour of subatomic particles – it simply gives the probabilities of finding particles in one place or another when a measurement is made.

An analogy for this idea is a crime wave tearing through your neighbourhood – it does not mean you will definitely be the victim of a crime, but the likelihood increases. Similarly, the passage of a particle's probability wavefunction makes it more likely that you will find a particle if you go ahead and make a measurement.

Schrödinger's cat

The idea that particles act like probability waves until they are measured, at which point they become particles with a definite location, was called 'collapse of the wavefunction'. It was part of a view of quantum theory that became known as the Copenhagen interpretation, after a meeting between the key scientists responsible for it was held in Copenhagen in 1927, when the details of the theory were hammered out. The Copenhagen interpretation drew criticism from some physicists because it said that until a particle is measured, its state is spread across all possible states – meaning that it exists in all states at once.

Schrödinger himself disliked the Copenhagen interpretation, and was prompted to come up with his famous thought experiment to highlight its absurdity – Schrödinger's cat. He imagined a cat locked in a box with a vial of deadly poison linked to a radioactive source. If the source decays and emits a particle of radiation while the box is shut, then a hammer breaks the vial, releasing the poison and the cat dies. Otherwise, it lives. Because radioactive decay is a quantum process then, until the box is opened, the state of the source is governed by its probability wavefunction. According to the Copenhagen interpretation, the source is both decayed and not decayed at the same time. In which case, the cat is simultaneously alive and dead.

Einstein was another famous detractor of the Copenhagen interpretation, and is thought to have inspired Schrödinger's cat by writing to Schrödinger with an outline for a similar thought experiment in which a keg of gunpowder is made to exist in simultaneously exploded and unexploded states.

Quantum decoherence

Today, collapse of the wavefunction has been replaced in the philosophy of quantum physics with the phenomenon of 'decoherence'. Whereas collapse of the wavefunction seems to give special importance to the observer making a measurement, decoherence states that a system makes the transition from quantum (wavelike) behaviour to nonquantum (particle) behaviour as it interacts with its nonquantum surroundings. Quantum behaviour is like a house of cards – easily disrupted by the slightest perturbation, at which point it topples down and nonquantum behaviour takes over.

In the Schrödinger's cat experiment, the 'quantumness' governing decay (or not) of the radioactive source has decohered, and so become nonquantum, long before the observer opens up the box, meaning that the cat is always either alive or dead, but never both at the same time.

Schrödinger received the 1933 Nobel Prize in Physics for the equation. His prize was well deserved and the equation forms the absolute core of modern quantum mechanics. A few years later it was the starting point for 'quantum field theory' – an even more powerful discipline that uses quantum principles to try and understand the fundamental forces governing nature in terms of particles and their varying interactions.

BELOW The Schrödinger's cat thought experiment, in which quantum theory apparently permits an ill-fated feline to be both alive and dead at the same time. Erwin Schrödinger put forward this idea to illustrate why extreme care is needed in the interpretation of quantum theory's predictions.

60 Penicillin

DEFINITION THE FIRST INFECTION-FIGHTING ANTIBIOTIC DRUG, ISOLATED FROM A COMMON STRAIN OF MOULD

DISCOVERY THE DRUG POTENTIAL OF PENICILLIN WAS FIRST NOTED BY SCOTTISH BACTERIOLOGIST ALEXANDER FLEMING IN 1928

KEY BREAKTHROUGH FLEMING SAW THAT *PENICILLIUM* MOULD GROWING IN A DISH HAD KILLED ALL THE BACTERIA AROUND IT

IMPORTANCE THE DRUG WAS REFINED IN TIME FOR D-DAY, SAVING MANY THOUSANDS OF LIVES – AND MANY MORE SINCE

An old wives' tale tells of using mouldy bread as a treatment for infected wounds. Only, maybe it is not such a worthless tale after all. Alexander Fleming found that *Penicillium* mould – the sort you find thriving on an old loaf of bread – contains a substance that is toxic to bacteria. Fleming and others isolated, then mass-produced the substance, which he called penicillin. It has proved to be an indispensable treatment for infections, as well as other bacterial diseases – such as pneumonia and bronchitis.

The discovery of antiseptics in the late 19th century brought about a revolution in surgery, with a massive reduction in the number of deaths resulting from post-operative infections. But the problem of infected wounds was far from over. Injuries sustained in less-than-hygienic conditions – for example, on battlefields – soon turned septic, often with life-threatening consequences. And a number of diseases, such as tonsillitis, bronchitis and pneumonia, were also known to be caused by bacterial infections.

While traditional antiseptic chemicals were perfect for sterilizing surgical instruments and the exterior of open wound sites, they were generally too toxic to the human body to be ingested at the levels required to fight these major internal infections.

Stories had abounded in the folklore of many cultures of mouldy food being used to disinfect wounds. Joseph Lister, the father of antiseptics, had himself noticed that some moulds could kill bacteria – but discarded the effect as not powerful enough to be of any practical use. Lister died in 1912 – had he lived a while longer, he might have come to regret that conclusion.

LEFT A Petri dish culture of the fungus *Penicillium notatum.* This was the strain that Fleming found inhibited the growth of bacterial cells, and which led him to discover the world's first antibiotic drug.

Alexander Fleming was a Scottish bacteriologist at St Mary's Hospital in London. In 1928, he had been carrying out experiments on *Staphylococcus* bacteria, samples of which he had cultivated in Petri dishes. However, he failed to seal one of the dishes correctly and, as a result, it became contaminated with mould, which grew into large blotches on the surface of the bacteria colony.

Fleming was about to throw out the sample when he noticed something peculiar. Around the edges of the blue-green spots of mould on the plate were zones in which the bacteria appeared to have died. Fleming put the plate under his microscope, confirming that the bacteria near to the mould had indeed been killed. He surmised that the mould must be giving off some kind of substance that was toxic to the bacteria.

Fleming carried out further tests, breeding more of the mould and then filtering it to obtain a liquid extract. He then applied this extract to other bacteria samples, finding that they were also killed off by the substance. Further investigations identified the strain of the mould as *Penicillium*. He therefore named the substance he had extracted from it 'penicillin'.

Further tests confirmed penicillin's potency and that its toxic effects were minimal compared to antiseptics, leading Fleming to initially believe that the drug had great potential in the treatment of infection. However, his enthusiasm seems later to have waned after further experiments suggested that penicillin would not last long enough in the body for its bacteria-killing properties to take effect.

'Penicillin reduced the number of amputations required due to infection and saved many thousands of lives on the battlefield during the Second World War. And, of course, it has saved many millions more since.'

What Fleming may not have been aware of was that by this time the first clinical trial had already been carried out. In 1930, Cecil George Paine, a pathologist at the Royal Infirmary in Sheffield, successfully used a penicillin extract he had made to treat eye infections. He applied the penicillin treatment to a total of six patients – five of whom made recoveries as a result.

In 1939, a team at the University of Oxford in England – led by Australian Howard Florey and German-born Ernst Chain – began a programme of research to investigate the power of penicillin to treat infections in laboratory mice. They concluded that Fleming had been wrong – penicillin, it seemed, would remain active inside a creature's body for a sufficient time to fight infection. Human trials later confirmed this.

Mass production

The trouble was producing the drug in large enough quantities. In this, Florey and his team sought help from the USA. They were quickly referred to the Northern Regional Research Laboratory in Peoria, Illinois, where an advanced fermentation technique had been developed that could allow the mould to be grown in the interior of large, deep vats – rather than simply on the surface. The vats were filled with corn steep liquor through which air was continually pumped.

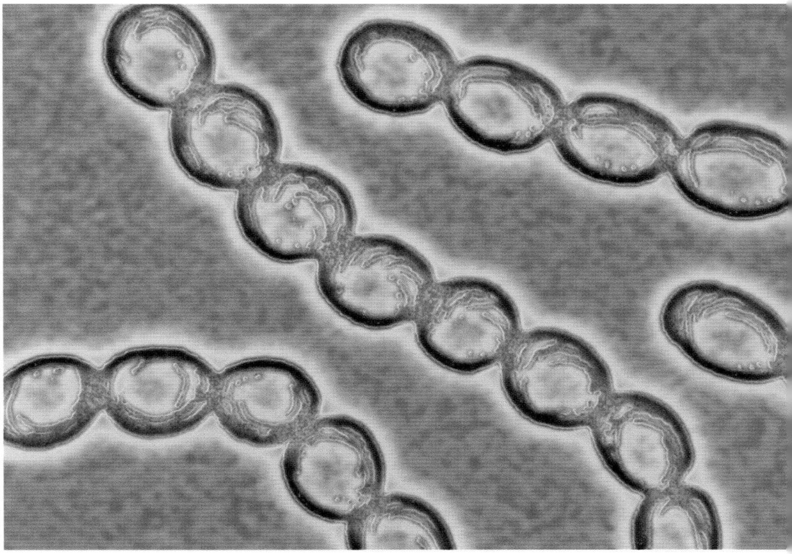

ABOVE Electron microscope image of *Streptococcus pyogenes* bacteria. 'Coccus' refers to the spherical cell shape, while 'strepto' indicates that the cells are joined in chains. This particular strain is responsible for throat, ear and skin infections, among other conditions, and is often treated with penicillin-like drugs.

Further clinical trials in humans followed. When these were also successful, the Peoria technique was put into mass production – to generate as many doses of penicillin as possible to treat injured Allied soldiers in the Second World War. By the time of the D-Day landings in 1944, approximately 2.3 million doses had been prepared. Penicillin greatly reduced the number of amputations required due to infection and saved many thousands of lives on the battlefield during this bloody conflict. And, of course, it has saved many millions more since.

American microbiologist Selman Waksman was responsible for naming these new medicines 'antibiotics'. The name was later used for a host of similar drugs, such as the modern forms of amoxicillin and methicillin. Fleming, Florey and Chain received the 1945 Nobel Prize in Physiology or Medicine for the discovery of penicillin, and for taking the steps needed to turn it into a usable medication.

Antibiotic resistance

Today, variations of penicillin are used, as most bacteria are immune to it in its original form. This phenomenon – called 'antibiotic resistance' – was first noticed as early as 1943. It is an example of evolution by natural selection, with bacteria continually evolving new defence mechanisms. Developing antibiotics is then rather like an arms race, with pharmacologists desperately trying to stay one step ahead of the ever-improving bacteria. Sometimes they succeed and sometimes they do not – which is when untreatable superbugs appear, such as MRSA (methicillin-resistant *Staphylococcus aureus*), which killed 17,000 people in the USA during 2005 alone.

The principal cause of antibiotic resistance is overuse of antibiotic drugs. Each time bacteria are exposed to antibiotics they gain more immunity to them. The only way to minimize this is to only prescribe the drugs when really necessary. To this end, doctors urge patients to only use antibiotics when the body's natural defences are unable to fight the infection. Otherwise, we risk letting the bugs gain the upper hand once more.

61 Game theory

DEFINITION A BRANCH OF APPLIED MATHEMATICS USED TO SELECT OPTIMAL STRATEGIES IN GAMES AND CONFRONTATIONS

DISCOVERY FIRST SCIENTIFIC PAPER ON GAME THEORY PUBLISHED BY HUNGARIAN MATHEMATICIAN JOHN VON NEUMANN IN 1928

KEY BREAKTHROUGH HE PROVED THE 'MINIMAX' THEOREM – PLAY FOR THE BIGGEST PAYOFF IN THE WORST-CASE SCENARIO

IMPORTANCE GAME THEORY HAS BEEN APPLIED TO ECONOMICS, SOCIAL PLANNING, MILITARY STRATEGY AND EVOLUTION THEORIES

Game theory is a branch of science which, as the name suggests, can help you select winning moves in a game of cards – or, equally, optimal ploys in a full-scale nuclear stand-off. It was first developed in the 1920s by mathematical genius John von Neumann and is now a sophisticated branch of applied mathematics that has delivered new insights into fields ranging from global economic theory to evolutionary biology.

The first person to experiment with game theory was the British diplomat and gambler James Waldegrave. In 1713, Waldegrave carried out a mathematical analysis of the card game *le Her*, to try to isolate winning strategies. However, the only record of his analysis was in an informal letter scribbled to a friend.

The first scientific study of gaming tactics was carried out by John von Neumann in 1928. He was born Neumann János Lajos at the end of 1903 in Budapest. He had a head for figures from an early age – reportedly memorizing telephone directories at the age of six. But it soon became clear that his abilities went way beyond rote learning – he had a natural aptitude for mathematics and science. In 1925, he obtained a PhD in mathematics from Péter Pázmány University (now Eötvös Loránd University).

After his PhD, Von Neumann taught at the University of Berlin. The rise of anti-Semitism, however, made life oppressive for Jews such as himself in continental Europe. Following the death of his father in 1930, his family emigrated to the USA, where János anglicized his name to John. His talent was soon spotted and John von Neumann was recruited as a professor at the Institute for Advanced Studies that had been set up in Princeton, New

LEFT In the 1920s, mathematicians began to analyze strategies in games such as poker and chess – founding the new field of science called 'game theory'.

RIGHT This phased-array radar at Otis Air Force Base on Cape Cod, Massachusetts, tracks intercontinental ballistic missiles aimed at the USA, as part of the country's defence against nuclear attack. One of the applications of game theory was analyzing cold war nuclear stand-offs.

Jersey, by educator Abraham Flexner and businessman Louis Bamberger. Indeed, he was one of the institute's first members – along with Albert Einstein and Kurt Gödel.

Parlour games

By the time he arrived in Princeton, Von Neumann's studies of game theory were already underway. In 1928, he published a paper in the German journal *Mathematische Annalen* entitled 'On the theory of parlour games'. In it, he looked at what are called zero-sum games – those in which one player's gains are equal to the other's losses.

A simple example of a zero-sum game might arise in cutting a cake. You and a friend have equally shared the cost of the cake. You get to make the cut that divides the cake into two pieces, and your friend gets first choice of which piece they have. You receive the other piece. What you lose, they gain – and vice versa – making a zero-sum game.

But what strategy maximizes the size of your piece? Von Neumann formulated a theorem called minimax for zero-sum games, which states that you should opt for the strategy that gives the biggest payoff in the worst-case scenario. Here, the worst-case scenario is that your opponent takes the biggest piece of the cake. Your optimum strategy is to minimize the size of the biggest piece – in other words, divide the cake exactly in half.

Von Neumann wrote up his findings in a book called *Theory of Games and Economic Behaviour*, co-authored with the economist Oskar Morgenstern, and published in 1944. They set out the fundamentals of game theory and applied them to economic theory. The book remains a definitive game theory text.

Nash equilibrium

In 1950, American mathematician John Nash extended Von Neumann's work to non-zero-sum games. He arrived at the concept of an equilibrium state – known as 'Nash equilibrium' – where all players in a game play the best strategy they can after taking into account the strategies of the other players. In a Nash equilibrium, no player can improve their lot by unilaterally changing their strategy. A classic example of this is the 'prisoner dilemma' game. Two men are arrested for a crime and are interrogated in separate cells. If both men refuse to cooperate with their interrogators they both get a minimal six-month sentence. If one man betrays his accomplice, he goes free while the accomplice gets ten years. Meanwhile, if both men

betray each other, then they each get five years. The best overall solution is to keep quiet. But the possibility of freedom means that both men can potentially improve their lot by cooperating with their interrogator. If they both do this then they each serve five years. And this is a Nash equilibrium – once the betrayal is done there is no way either man can improve his lot.

It is probably no coincidence that many results in game theory were derived during the cold war between the USA and Soviet Union, during which the world teetered on the brink of nuclear annihilation. In fact, mutually assured destruction – the strategy whereby a nuclear strike by either nation would provoke such a massive retaliation that both sides would be obliterated – is a game-theory Nash equilibrium. It is not ideal but, once in place, there is no unilateral strategy to improve the payoff for either side.

The cold war has been over for 20 years – highlighting the fact that Nash equilibria, and game theory's predictors as a whole, are not infallible. But countries do not always act out of pure self-interest. Sometimes they cooperate, working together to do what is best for the world (although getting nations to agree on curbing greenhouse gases seems to test this altruism to the limit).

Suicide bombers

The cold war underlines why it is crucial to understand the motives, and ascertain the rationality, of all players in a situation before game theory can be applied. For example, its predictions may hold good in negotiating with a hostage taker – who will want the best resolution of the situation for themselves, and almost certainly will not want to die. However, game theory is unlikely to furnish you with any winning strategies against the seemingly irrational resolve of a suicide bomber.

'Mutually assured destruction – the strategy whereby a nuclear strike by either nation would provoke such a massive retaliation that both sides would be obliterated – is a game-theory Nash equilibrium.'

A recent development is the application of quantum laws to game theory. This expands the discipline to games played via the exchange of quantum information (see page 323) – for example, when online poker or stock market trading is done on quantum computers connected up by quantum information channels. Quantum game theory could open up powerful new strategies for those gamblers and traders wily enough to exploit them.

62 Hubble's law

DEFINITION HUBBLE'S LAW IS A MATHEMATICAL EQUATION GIVING THE SPEED AT WHICH THE UNIVERSE IS EXPANDING

DISCOVERY AMERICAN ASTRONOMERS EDWIN HUBBLE AND MILTON HUMASON FIRST FORMULATED THE LAW IN 1929

KEY BREAKTHROUGH ASTRONOMICAL OBSERVATIONS REVEALED THAT THE RECESSION SPEED OF GALAXIES INCREASES WITH DISTANCE

IMPORTANCE THE DISCOVERY WAS ONE OF THE FIRST PIECES OF EVIDENCE FOR THE BIG BANG THEORY OF THE UNIVERSE

After years of observations of galaxies, Edwin Hubble and Milton Humason deduced that distant galaxies are moving away from us and that their speed increases the further away they get. This discovery became known as Hubble's law. Its key implication was that the universe is expanding – and this single breakthrough laid the foundations for modern cosmology.

When, in 1917, Albert Einstein applied his theory of general relativity to the universe at large, he was in for a big disappointment. For rather than the static, unchanging universe he was expecting, his equations predicted that galaxies should be rushing apart from one another – relativity was telling him that the universe is expanding.

This was way off the accepted view of the cosmos held by astronomers of the early 20th century, who still clung to the idea that the heavens on the largest scales are fixed and immutable. Einstein, uncharacteristically, bowed to popular opinion and took this to mean his theory must be wrong. So he introduced a correction term – called the 'cosmological constant' – that would alter the strength of gravity at long range and hold the universe still.

LEFT Images of Ultra Deep Field galaxies reveal the most distant objects ever seen. Each dot is a separate galaxy in the Fornax constellation, located about 13 billion lightyears away and 13 billion years old. Galaxies were first explained by Edwin Hubble, who stated that they are moving away from us, and that the further away they are, the faster they recede – a relationship called Hubble's law.

But in 1929, Einstein was in for an even bigger disappointment when an American astronomer called Edwin Hubble announced to the world that the universe is, in fact, expanding. Hubble, together with his assistant Milton Humason, had been studying galaxies with the 2.5-metre (2.7-yard) Hooker Telescope on Mount Wilson, California (at the time the biggest telescope in the world). In particular, they had been looking at the spectra of the light from the galaxies – obtained by splitting up the light into different frequencies and measuring the brightness at each frequency. Doing this reveals a pattern of peaks and dips caused by the presence of different chemical elements in the galaxies, which emit and absorb light at particular frequencies.

When Hubble and Humason did this for their distant galaxies, sure enough, they found the patterns of peaks and dips they were expecting – also known as 'lines' – corresponding to common chemical elements, such as iron, sodium and silicon. Only the peaks and dips were not where they should be. The pattern was there but all the lines were shifted to lower frequencies.

They realized that this meant the galaxies must be moving away. In 1842, Austrian physicist Christian Doppler had worked out that sound from a source that is moving away gets shifted to lower frequency. This 'Doppler effect' is why the pitch of an ambulance siren gets distorted as it moves past you. It was the same effect that was shifting the light from Hubble and Humason's galaxies to low frequency. Since low-frequency light is red in colour, the galaxies were said to be 'redshifted'. Each galaxy's redshift is directly linked to the speed at which it is moving away.

'Every point in space is moving away from every other point, as if space were the surface of a balloon that is being inflated.'

But was there a pattern to the redshifting? To answer that question, Hubble and Humason needed to measure the distances to the galaxies. They did this using a special kind of star called a Cepheid variable. These are stars that vary in brightness over a regular period of a few days to a few weeks, depending on their properties. In 1912, American astronomer Henrietta Leavitt had deduced that the period was determined by the star's average intrinsic brightness, or luminosity; this was called the period-luminosity relation.

The distance ladder

The period-luminosity relation furnished the astronomers with a way to gauge the distances to their galaxies. If they could find a Cepheid variable in each galaxy and measure the period of its variations, this would indicate the Cepheid's intrinsic brightness. If they also measured the apparent brightness of the star, this would tell them how much its light had been dimmed with distance and how far away the star and its host galaxy were placed.

When they applied the technique to galaxies in the sample, a mathematical relationship emerged – the redshift of each galaxy was proportional to its distance. Since the redshift is proportional to the speed at which each galaxy is moving this meant that the recession speed, v, is proportional to distance, d. Hubble wrote this as $v = Hd$. This is Hubble's law, where H is a constant of proportionality known as Hubble's constant. With v measured in km/s and d measured in megaparsecs (1 megaparsec or Mpc = 3.26 million lightyears), Hubble found H to be 500km/s/Mpc (311 miles/s/Mpc). Modern estimates made by probes sent into space give a more accurate value of 70km/s/Mpc (43 miles/s/Mpc). The upshot was that the universe is expanding after all.

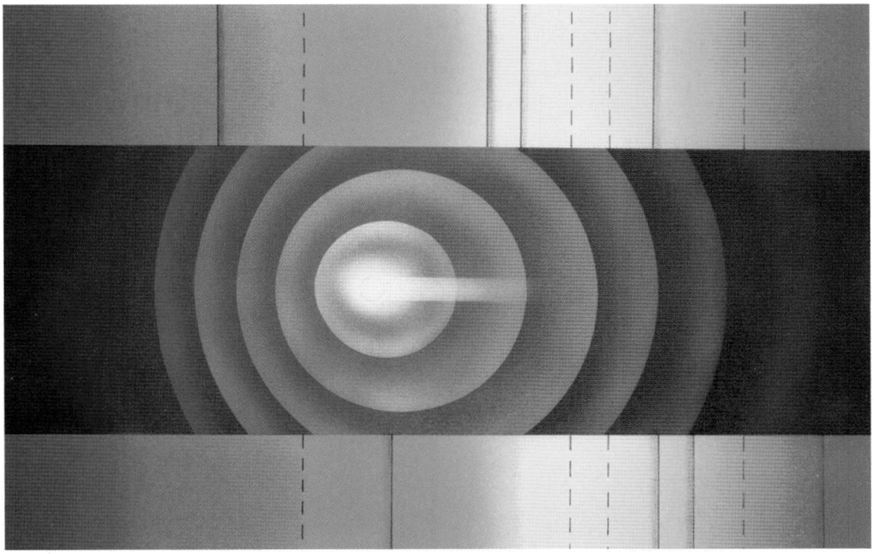

LEFT The redshift effect (right side of image), as studied by Hubble to deduce the recession speeds of galaxies. Redshifts arise where the wavelength of the light from a receding source becomes stretched out – literally shifted towards the red.

Every point in space is moving away from every other point, as if space were the surface of an inflating balloon. Galaxies, drawn as dots on the surface of the balloon, get further apart as the balloon expands – and the further apart any two galaxies are, the quicker they appear to recede from one another. When Albert Einstein learned of Hubble's discovery, he declared the cosmological constant to be the 'biggest blunder' of his professional career. However, in recent decades the idea has enjoyed something of a revival.

Cosmic inflation

Many cosmologists now believe the very early universe underwent a period of rapidly accelerating expansion – called 'inflation'. The idea neatly solves a number of problems with the standard Big Bang model. And it explains where the initial density irregularities came from around which the galaxies grew – essentially putting these down to microscopic quantum fluctuations, which were blasted up to astrophysical length scales by the rapid expansion. For inflation to happen the universe, shortly after the Big Bang, must have been filled with a kind of matter field mimicking the effect of the cosmological constant. But whereas Einstein's constant opposed the expanding motion of the universe, the field driving inflation served to accelerate it enormously.

Recent observations suggest that a similar kind of cosmological constant still pervades the universe today. In 1998, astronomers announced that the expansion of space on the very largest scales is accelerating – albeit not as rapidly as during inflation. The field of matter causing the acceleration is called 'dark energy' (see page 265). As of late 2010, cosmologists are still unsure what dark energy actually is. What is certain is that if Einstein were still alive today, he would be tearing his hair out.

63 Big Bang theory

DEFINITION THE BIG BANG THEORY SAYS OUR UNIVERSE EMERGED FROM A SUPERDENSE FIREBALL NEARLY 14 BILLION YEARS AGO

DISCOVERY THE THEORY WAS FIRST PUT FORWARD BY BELGIAN ASTRONOMER GEORGES LEMAÎTRE IN 1931

KEY BREAKTHROUGH LEMAÎTRE REALIZED THAT COSMIC EXPANSION IMPLIES THE UNIVERSE BEGAN AT A FINITE TIME IN THE PAST

IMPORTANCE THE BIG BANG THEORY IS THE FUNDAMENTAL MODEL ON WHICH ALL OUR UNDERSTANDING OF THE UNIVERSE IS BUILT

The universe grew from a hot, dense fireball of matter and radiation about 13.7 billion years ago. As the fireball expanded and cooled down, stars and galaxies condensed out of the primeval maelstrom. This is the Big Bang theory. It was first put forward by Georges Lemaître in 1931, in response to Edwin Hubble's discovery that space is expanding. The Big Bang theory remains today the best model for our universe's origin and evolution.

The Big Bang began in 1927. Not the actual bang, as such, but the theory stating such an event prompted the creation of our universe.

Belgian astronomer (and Catholic priest) Georges Lemaître came up with a range of solutions to Einstein's general theory of relativity (see page 229) that described the behaviour of various kinds of expanding universe. Lemaître realized that the key parameter was the average density of the universe. If it was equal to, or less than, a certain 'critical density' (around 20 grams per cubic centimetre) then the universe would continue to expand indefinitely – greater than this value and the expansion would one day halt and reverse.

Lemaître used his equations to derive Hubble's law (see page 253) for the universe's expansion – so that when Edwin Hubble announced the evidence for the law in 1929, Lemaître knew his theory was on sound footing.

LEFT An artist's impression of the Big Bang, from which our universe is thought to have begun 14 billion years ago. Only after the Big Bang, it is thought, did time and space – as well as matter and radiation – come into being.

Primeval atom

Lemaître's work so far was in itself an achievement. But he went further, using his new-found insight into the behaviour of the universe to make an astonishing inference. He understood that if galaxies are moving apart now, then if you run time backwards they get gradually closer. Keep winding back the clock and there comes a point where galaxies crash into each other and overlap. Go back further still and the mass in the universe eventually packs down into a hot, dense sphere – referred to as the 'primeval atom'. Lemaître supposed that the primeval atom was the seed from which our universe grew: an initial state from which matter, energy, space and time spontaneously exploded. Such a revolutionary theory for the origin of the cosmos might have gone unnoticed had it not been for the English astronomer Arthur Eddington. He had read Lemaître's work in a Belgian research journal, and arranged for it to appear in the widely read Monthly Notices of the Royal Astronomical Society – where it was finally published in 1931.

'During an interview for British radio in 1949, Hoyle famously declared that he refused to believe the universe began with what he mockingly referred to as a "Big Bang". The name stuck.'

Rival theory

Not everyone was convinced that the universe got going in the way Lemaître suggested. One opponent was the English physicist Fred Hoyle. He and colleagues Hermann Bondi and Thomas Gold put forward a rival 'steady state theory'. In it, they showed how it is possible for the universe to exist in a perpetual state of expansion without necessarily having a beginning. During an interview for British radio in 1949, Hoyle famously declared that he refused to believe the universe began with what he mockingly referred to as a 'Big Bang'. The name stuck.

But which theory was correct? One observational test was to look at the chemical composition of the universe. In 1947, Russian-born physicist George Gamow working with American Ralph Alpher deduced that if the universe had started out hot and dense – as the Big Bang theory required – then this would have left an imprint on the abundances of the elements hydrogen and helium. These are the lightest two elements. Hydrogen has a nucleus made of a single proton particle, whereas helium is made of two protons and two neutrons.

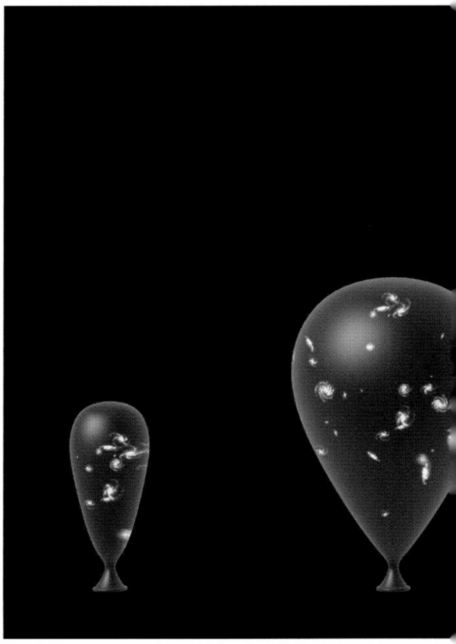

Gamow and Alpher analyzed the nuclear reactions that would have taken place as the early universe expanded and cooled. Their calculations showed that the universe created by a Big Bang should be roughly 75 percent hydrogen and 25 percent helium, with a smattering of heavier elements. Of course, nuclear reactions that have taken place since – in the cores of stars – will have generated more heavy elements. But when astronomers looked at the compositions of the oldest stars – they found exactly the 75:25 ratio of the two lightest elements that Gamow and Alpher had predicted.

Microwave echo

The observational prediction that convinced everyone beyond all doubt was made in 1948 – again by Gamow and Alpher, this time accompanied by physicist Robert Herman. They showed that if the universe started out as a hot fireball, which gradually expanded and cooled, then there should be a residual glow of that fireball still pervading space today. They worked out that the temperature of this residual radiation should be very small, –270°C, placing it in the microwave region of the electromagnetic spectrum. For this reason, it became known as the cosmic microwave background (CMB).

It was not until the 1960s that the CMB was detected – quite accidentally – at Bell Laboratories, New Jersey, by two radio astronomers called Arno Penzias and Robert Wilson. They were trying to eliminate unwanted microwave noise from their radio antenna. When theoretical cosmologist Robert Dicke, at Princeton University, learned of their plight he knew instantly what it was that they had found. Following their discovery, the Big Bang became the standard model upon which modern cosmology is based – and Penzias and Wilson subsequently received the 1978 Nobel Prize in Physics.

The Big Bang theory accurately explains the history of our universe – but what does it say about the future? As Lemaître had predicted in the 1920s, this hinges on the universe's average density. If this is less than the critical value, then space will continue to expand. Stars will burn out until there is nothing left to fill the ever-widening inky blackness. This scenario is known as the 'heat death'.

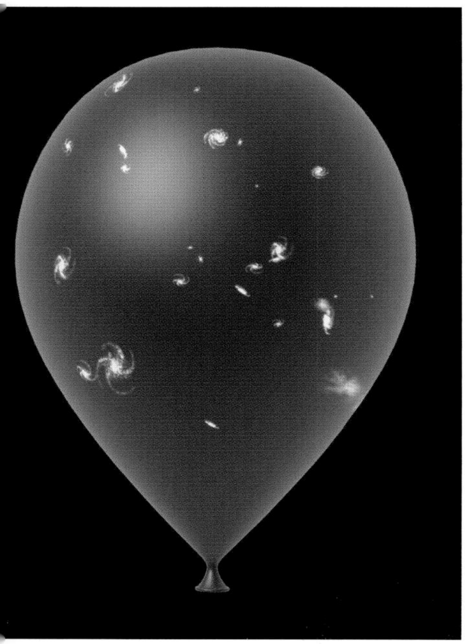

On the other hand, if the density is greater than critical then expansion will halt and turn around so that the universe collapses in on itself. Matter packs closer and closer together as space gets smaller and smaller until the universe returns to the superheated state in which it began and ultimately winks out of existence. This scenario is the 'big crunch'. Current observations favour the heat death theory that our universe will fade away rather than burn out. The good news is that it won't happen for a trillion years.

LEFT Computer artwork illustrating the spatial expansion of the universe. Blowing up a balloon is a good analogy for how space expanded after the Big Bang. Although we see space expanding in all directions we are not at the centre of the universe. If you draw dots on a balloon and then blow it up, each point recedes from every other point.

64 Antimatter

DEFINITION A PARTICLE OF ANTIMATTER HAS THE SAME MASS AS A PARTICLE OF MATTER BUT WITH THE OPPOSITE ELECTRIC CHARGE

DISCOVERY THE FIRST ANTIMATTER PARTICLES WERE DISCOVERED IN EXPERIMENTS BY CARL ANDERSON IN 1932

KEY BREAKTHROUGH USE OF A MAGNETIC FIELD TO SEPARATE PARTICLES FROM THEIR OPPOSITELY CHARGED ANTIPARTICLES

IMPORTANCE ANTIMATTER IS A POSSIBLE FUTURE SPACECRAFT FUEL, WITH 10,000 TIMES THE OOMPH OF NUCLEAR FISSION

Antimatter is another of those discoveries for which we must ultimately thank Albert Einstein. It was predicted in 1928 when physicists applied Einstein's theory of relativity to quantum mechanics – the emerging theory of the subatomic world. It was duly detected in experiments just a few years later, in 1932. Antimatter is now being investigated as a potential power source for future interstellar spacecraft.

A particle of antimatter has the same mass as a particle of ordinary matter but has other key properties – such as electric charge – reversed. For example, an electron has negative charge, but its antiparticle – the positron – has equal and opposite positive charge. Both particles have exactly the same mass.

When a particle of matter meets its antiparticle, all hell breaks loose as the two particles annihilate each other, converting their mass into energy. Indeed, matter–antimatter annihilation is the most efficient way to convert mass into energy – with the amount of energy liberated per kilogram of fuel being several orders of magnitude higher than that produced in nuclear reactions.

LEFT The tracks of electrons and their antimatter counterpart, positrons, in a cloud chamber detector. The air in the chamber is saturated with vapour, which passing particles cause to condense, creating a visible trail. A magnet applied to the chamber makes the paths of positrons and electrons – with opposing electrical charges – curve in opposite directions.

Antimatter was first predicted by the English theoretical physicist Paul Dirac. In 1928, Dirac was trying to develop a version of Schrödinger's equation that was consistent with Einstein's special theory of relativity. Schrödinger's equation is a wave equation in quantum theory that describes the motion of solid particles, such as electrons. Schrödinger formulated the equation using relationships linking the energy and momentum of bodies taken from Newton's theory of motion. Einstein, however, had shown that Newton's theory breaks down as an object approaches lightspeed. And so Schrödinger's equation only described slow-moving electrons. Dirac realized that he could extend the equation to fast-moving particles by reformulating it using the energy-momentum relations that Einstein had derived in his special theory. But when he did this he found something unusual. As well as a solution of the equation describing the electron, a second solution appeared describing the motion of a particle with a positive electric charge. Dirac initially wondered whether the solution corresponded to the proton – the positively charged particle in the nuclei of atoms. But the proton was known to be much heavier than the electron, while his solution was implying that the new particle should have the same mass.

Dirac's equation had actually predicted the existence of the positron. It was detected experimentally in 1932 by the American physicist Carl Anderson. He had been investigating cosmic rays – high-energy particles from space that regularly crash into the planet. The cosmic-ray particles themselves never make it through the atmosphere, but the fragments from their collisions with atoms and molecules in the air filter down via a cascade of secondary collisions that can be picked up by particle detectors on the ground.

Anderson was using a detector called a cloud chamber – a sealed glass tank filled with alcohol vapour. As particles pass through the chamber they cause alcohol droplets to condense to form observable trails in the tank. Anderson had also placed a magnetic field across his cloud chamber. Electrically charged particles in a magnetic field move in a circle – either clockwise or anticlockwise, depending on whether their charge is positive or negative. The radius of the circle is fixed by the particle's mass.

'When a particle of matter meets its antiparticle, all hell breaks loose as the two particles annihilate each other, converting their entire mass into energy.'

In this way, Anderson detected plenty of electron particles in his chamber. But occasionally he saw a circular particle track with the radius corresponding to the mass of the electron, but curving in the opposite direction. He had found Dirac's positron – a discovery that would earn him the 1936 Nobel Prize in Physics.

Antiatoms

Since Anderson's discovery, physicists using particle accelerators – machines with powerful magnets that accelerate subatomic particles to near lightspeed and then collide them – have produced other antiparticles, including antiprotons and antiquarks (quarks are the particles from which protons are made). In 1995, a CERN laboratory team made the first antiatom when they combined a positron and an antiproton to make antihydrogen.

Could antimatter ever be used as a source of energy? After all, if matter–antimatter annihilation kicks out 10,000 times the power of nuclear fission reactions then surely it could solve all of our energy woes? The trouble is that whereas the fuel for nuclear fission can be mined out of the ground, antimatter

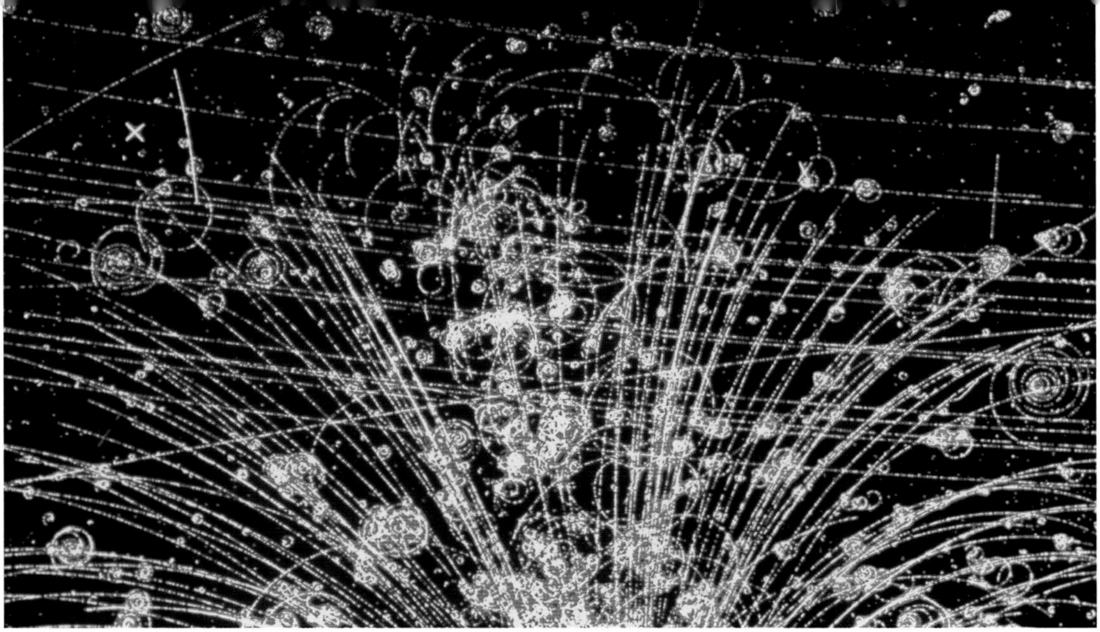

does not occur naturally on Earth. Instead, it has to be manufactured inside particle accelerators. And the process is inefficient – requiring about ten billion times as much energy to create as is released when the antimatter is annihilated. Even making one gram of the stuff would take several hundred million years and about $60 trillion – though in the future this could change.

ABOVE Shrapnel of subatomic particles generated by the impact of a high-energy cosmic ray from space. The cosmic ray's energy has been converted into new matter and antimatter particles. The two types of particle are separated using a magnetic field.

Small amounts of antimatter exist naturally in space. If James Bickford, of the Draper Laboratory, in Massachusetts, has got his sums right then around four tonnes of the stuff should waft into our solar system from interstellar space every year. Calculations suggest that this material should get snagged by the magnetic field of the giant planet Jupiter – in much the same way that the magnetic field of our own planet accumulates charged particles from the Sun, to form the so-called Van Allen radiation belts.

Superfuel

A spacecraft equipped with magnetic storage tanks, which could hold the antimatter while preventing collisions with any matter, could conceivably retrieve this superfuel and return it to Earth for us to use.

Some space scientists have even suggested that antimatter could be the ultimate power source for interstellar spacecraft. The perennial problem in rocketry is the colossal mass of fuel required in order to accelerate, which also has to be accelerated, requiring more fuel – and so on. But antimatter offers a fuel source with an appealing power-to-weight ratio. Rocket scientists have worked out that an antimatter engine would be 2000 times more efficient than a space shuttle's main engine. Perhaps Jupiter will be the location of humanity's first interstellar spaceport – where spacecraft from Earth stop to fuel up before departing for the stars.

65 Dark matter

DEFINITION DARK MATTER IS HIDDEN MASS THAT, TOGETHER WITH DARK ENERGY, MAKES UP 95 PERCENT OF OUR UNIVERSE

DISCOVERY FIRST OBSERVATIONAL EVIDENCE NOTED IN 1934 BY FRITZ ZWICKY AT THE CALIFORNIA INSTITUTE OF TECHNOLOGY

KEY BREAKTHROUGH SEEING GALAXIES IN THE COMA CLUSTER MOVE TOO FAST FOR THE CLUSTER'S GRAVITY TO HOLD IT TOGETHER

IMPORTANCE DARK MATTER IS ESSENTIAL FOR UNDERSTANDING THE EVOLUTION AND ULTIMATE FATE OF THE UNIVERSE

We cannot see it, but its existence is indisputable. Invisible material makes up 95 percent of our cosmos. We know it is there because of its gravity: distant stars and galaxies move as if in the grip of a much stronger gravitational field than can be accounted for by the stars and galaxies usually seen by astronomers. Dark matter (and its cousin dark energy) is an energy field that will determine our universe's ultimate fate.

The idea that there might be more to the universe than meets the eye was first put forward in the 1930s by Swiss astronomer Fritz Zwicky. While researching at the California Institute of Technology, Zwicky studied the Coma Cluster – a bundle of over 1000 galaxies, 320 lightyears away in the constellation of Coma Berenices. Zwicky was puzzled: he had measured how fast the galaxies were moving, and the fact that they were not flying off into space showed that there must be enough gravity to hold them together. But when Zwicky totalled up the gravitational pull of the cluster by estimating the mass of the galaxies and applying Newton's law of gravitation, he was surprised. It seemed that the galaxies were moving hundreds of times too fast. The only way to explain the figures was if the cluster contained hundreds of times more mass than could be seen. Zwicky called this mass 'dark matter'.

No one believed him. In the 1970s, American astronomer Vera Rubin was studying the 'rotation curves' of spiral galaxies – how the speed at which a galaxy turns varies with the distance from its centre. Close to the centre, her observations agreed with the theoretical predictions. But in the outer regions of a galaxy – the halo – the theory and observation went their separate ways. The theory predicted that in the halo the rotational speed should gradually

LEFT Artistic interpretation of dark matter – invisible material that accounts for almost all matter in the universe. Computer simulations reveal that dark matter forms a three-dimensional, weblike structure that threads through space – and around which visible matter, such as bright galaxies, gathers under the force of gravity.

decrease, dropping to zero at large distances. But Rubin found the rotation speed to be roughly constant, as if the whole halo was turning as a rigid mass. The only way this could happen was if the halo contained hidden dark matter. Rubin's observations suggested that the parts of spiral galaxies that we can see make up as little as five percent of their total mass.

More evidence for dark matter emerged through studies of 'gravitational lensing'. This happens when the light from a distant galaxy gets bent by the gravity of a galaxy cluster on the line of sight from Earth, distorting and magnifying the image of that distant galaxy. Astronomers apply mathematical techniques to the image to infer the mass of the lensing cluster, providing results that agree with Zwicky's measurements of the Coma Cluster and indicate the clusters are 100 times more massive than observations suggest.

BELOW A cluster of galaxies, known as a gravitational lens. The gravity of this cluster focuses the light from objects behind it. The behaviour of a gravitational lens allows astronomers to calculate the mass of the lensing cluster.

Vital ingredient

Dark matter is now a key ingredient in theories of how galaxies and clusters first formed – without its extra gravity to start the process, it is unlikely these structures would have appeared. And we would not be here.

It soon became clear that dark matter is just the tip of the cosmic iceberg. In the late 1980s, Saul Perlmutter and colleagues at Lawrence Berkeley National Laboratory in California, began an 11-year study of distant supernovae – explosions marking the deaths of stars. The aim was to understand the large-scale behaviour of the universe. In the 1920s, Edwin Hubble had discovered that the universe is expanding. Perlmutter's team hoped their research would tell them how fast and, crucially, the rate at which the expansion is being slowed by gravity.

Antigravity

To the team's bewilderment, when they collated their data they found cosmic expansion is not slowing down at all – it is speeding up. The universe's gravity was doing the opposite to what they expected. If their mathematics was to be believed, this repulsive 'antigravity' must be generated by a sea of exotic energy filling the whole of space. And this 'dark energy', as it became known, weighed more than everything else in the universe put together – including dark matter. On 8 January 1998, Perlmutter's team presented their findings to the world at a meeting of the American Astronomical Society. The discovery was voted 'Breakthrough of the Year' by prestigious journal *Science*. It is now regarded as one of the most significant scientific advances of the 20th century.

Final confirmation – as if it were needed – came in the first decade of the 21st century with the launch of dedicated space missions to accurately measure the parameters of the universe from the cosmic microwave background, the microwave echo of the Big Bang. These spacecraft confirmed the existence of dark matter and dark energy, showing that our universe is composed of around 23 percent dark matter, 72 percent dark energy and just five percent atoms (the 'ordinary matter' making up you and me, and all the stars, planets and galaxies).

But what is this dark matter and dark energy made of? The embarrassing truth is that no one really knows. The consensus view is that dark matter is probably exotic subatomic particles – of the sort predicted by extensions to the standard model of particle physics, such as supersymmetry (see page 333). But this is yet to be confirmed.

Dark energy, on the other hand, is thought to be energy locked away in the structure of empty space. Generally, we imagine the energy of empty space to be zero, but there is no reason why this should be so – and indeed studies of particle processes in the first moments after the Big Bang suggest that the 'vacuum energy', as it is called, could be extremely large.

'It seemed the galaxies were moving hundreds of times too fast. The only way to square the figures was if the cluster contained hundreds of times more mass than could actually be seen. Zwicky called this mass "dark matter".'

End of days

Dark energy will decide the universe's ultimate fate. Depending on its characteristics, space could expand forever or recollapse back on itself in a so-called 'big crunch'. In a third, more dramatic scenario cosmologist Robert Caldwell suggested in 2003 that if dark energy takes a particularly extreme form, called 'phantom energy', it could cause cosmic expansion to accelerate so fast that all the matter in the universe – from galaxies to atoms – would be torn to bits. He called this event the 'big rip'. Even if the theory is correct, however, this would not happen for tens of billions of years.

Cracking the enigma of dark energy is one of the most pressing priorities for 21st-century science. In 2020, NASA launches the Wide Field Infrared Survey Telescope (WFIRST) – a space probe that, amongst other things, will investigate the detailed physical properties of dark energy. Researchers hope it will finally shed some light on this dark corner of space – and tell us once and for all what the bulk of our universe is actually made from.

66 Ecology

DEFINITION THE STUDY OF THE INTERACTION BETWEEN ORGANISMS AND THEIR NATURAL ENVIRONMENT

DISCOVERY ECOLOGY AS A SCIENTIFIC DISCIPLINE WAS FOUNDED BY ENGLISH BOTANIST ARTHUR TANSLEY IN 1935

KEY BREAKTHROUGH TANSLEY PUT FORWARD THE CONCEPT OF AN INTERACTING NETWORK OF ORGANISMS, TERRAIN AND CLIMATE

IMPORTANCE REGULATING THE IMPACT OF HUMAN ACTIVITIES ON THE EARTH'S COMBINED ECOSYSTEM IS NOW AN URGENT PRIORITY

Overfishing, deforestation, pollution – it seems we have little regard for our environment. Throughout his life, Arthur Tansley popularized the view of the natural world as a self-sustaining network, or 'ecosystem'. His work established ecology as a science and brought the realization that our ecosystem is a finely tuned mechanism that should be handled with care.

Our understanding of the natural world has emerged gradually over the course of scientific history. To begin with, we considered nature on a piecemeal basis. In the 18th century, Carl Linnaeus classified organisms according to his taxonomic scheme (see page 61), treating each species as an isolated individual unit. But then, in the 19th century, Charles Darwin's theory of evolution by natural selection made everyone sit up and see the natural world in a wholly different light.

Up until this time, species of plants and animals were regarded as God-given – no one really understood where they came from or what made them the way they were. Darwin's theory changed all of that, arguing that life forms are shaped by the environment in which they live. Natural selection causes organisms to adapt to their surroundings, to assume the form that maximizes their chances of survival. Darwin's suggestion meant that nature was no longer all about individual species – it was about interactions that take place between species and their environment.

LEFT Deforestation in the Amazon. The dark green areas indicate intact forest, while the light green and brown regions have been cleared for raising livestock. Ecology brings scientists the means to quantify the damage that deforestation causes.

In 1866, German naturalist Ernst Haeckel came up with a name for this new view: ecology. Danish botanist Eugen Warming built on this reasoning further, in a way extending Darwin's purely biological theory of natural selection to embrace abiological entities as well, such as the landscape and climate. Other researchers at around this time were realizing the importance of food chains – how, for example, plants feed off the nutrients in soil, grazing animals then eat the plants and predators ultimately eat the grazing animals. In fact, rather than a chain, the process is more of a cycle with waste products – urine, faeces and the bodies of dead organisms – being returned to the ground in the form of nutrients so that the process can begin again.

A picture was gradually emerging of life on Earth as a kind of self-organizing hierarchy. But the real work in developing this idea, and popularizing it with scientists and the general public, was done by Arthur George Tansley.

Field trips and academia

Tansley was born in London in 1871. He was interested in botany as a child and this became his principal subject when he went away to university – reading botany with zoology at Trinity College, Cambridge. He graduated with first-class honours and shortly after became an assistant professor at University College London. Early in his career, he was able to make a number of foreign field trips – visiting Ceylon (modern-day Sri Lanka), Malay (now Malaysia) and Egypt to study the indigenous plant life.

'Tansley coined the term "ecosystem" as it is presently understood, to mean a self-supporting unit in the natural world comprising biological life forms, landscape and climate all working together.'

Tansley had been influenced by the work of Eugen Warming and other early ecologists, including Andreas Schimper and Francis Oliver. And this, combined with his field studies, strengthened the picture in his mind of the Earth's ecology as a delicately balanced system that we should treat with care and respect. Tansley recognized that while it can take an ecology many years to form, the removal of just one species from the food chain instantly disrupts the hierarchy of natural cycles – bringing the whole ecology crashing down almost overnight.

Tansley propounded this view wherever he went – and through his various duties as lecturer in botany at Cambridge University (from 1907), founder and editor of the journal *New Phytologist*, as first President of the British Ecological Society and as editor of the *Journal of Ecology*.

Earth's ecosystem

Tansley put forward his views and theories formally in a number of scientific papers published over the course of his career. One of the most important of these appeared in the journal *Ecology* in 1935. In this paper, Tansley coined the term 'ecosystem' as it is presently understood, to mean a self-supporting unit in the natural world comprising biological life forms, landscape and climate all working together. He would later also adopt the term 'ecotope' to describe different ecological classifications of landscape and terrain. Tansley understood how small ecosystems, for example, the cycles of life in a pond, all mesh together to form progressively larger systems, culminating with the biggest ecosystem of all – our planet's biosphere. It was a view that would turn out to be extremely prescient as the decades rolled by, and concern gradually mounted over the impact upon our global ecosystem of pollution, habitat destruction and the over-exploitation of the planet's natural resources.

Tansley was, almost from the start, the single dominant figure in ecological research in Britain – and, by the end of his career, had also left his stamp on the world stage, transforming the way we think about organisms and their interconnectivity with each other and the rest of the planet. He was knighted for his services by King George VI in 1950.

The Gaia hypothesis

In the 1960s, Tansley's theories were extended in a controversial direction by the English environmental scientist James Lovelock. Lovelock proposed a view of the Earth known as the Gaia hypothesis, whereby the planet's ecosystem functions like a single giant organism in its own right. In Lovelock's view, the Earth is able to regulate its vital parameters, such as atmospheric composition and ocean salinity, in order to sustain life on its surface.

The mechanism responsible for this is called homeostasis, named after the process of the same name that enables organisms such as you and me to automatically regulate our temperature (by perspiring) or our blood sugar levels (via insulin release). Although some environmentalists consider Lovelock's arguments compelling, his views are far from universally accepted by mainstream science.

BELOW In this map of Earth's biosphere, dark green represents the greatest density of plant matter, while yellow marks the least dense regions. In the oceans, darker pinks mark the greatest concentrations of plant life, with lighter shades delineating the less dense areas.

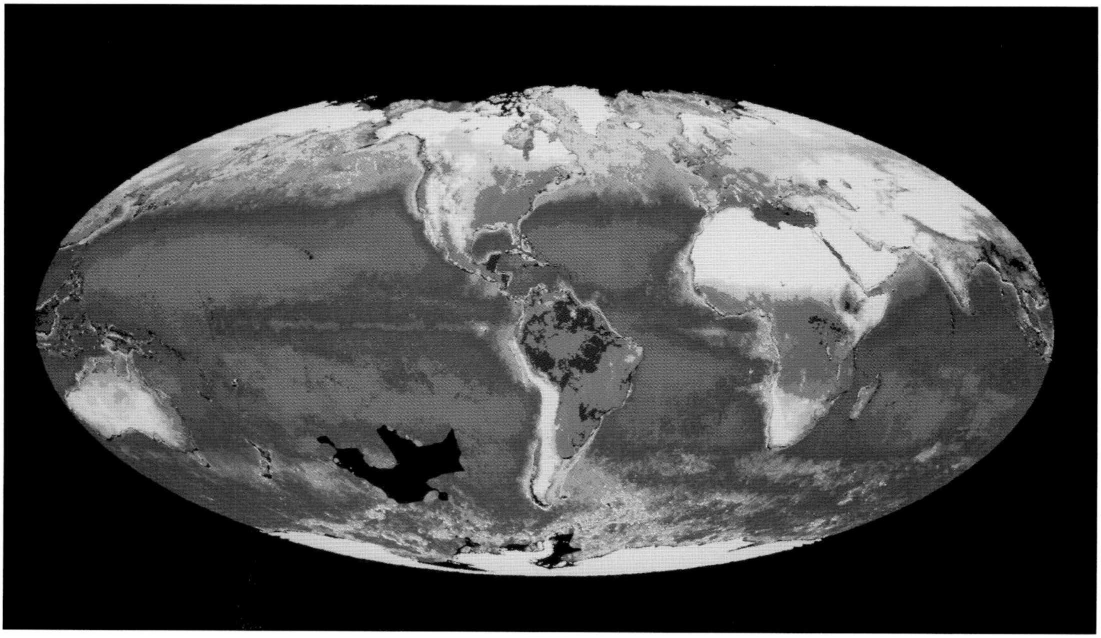

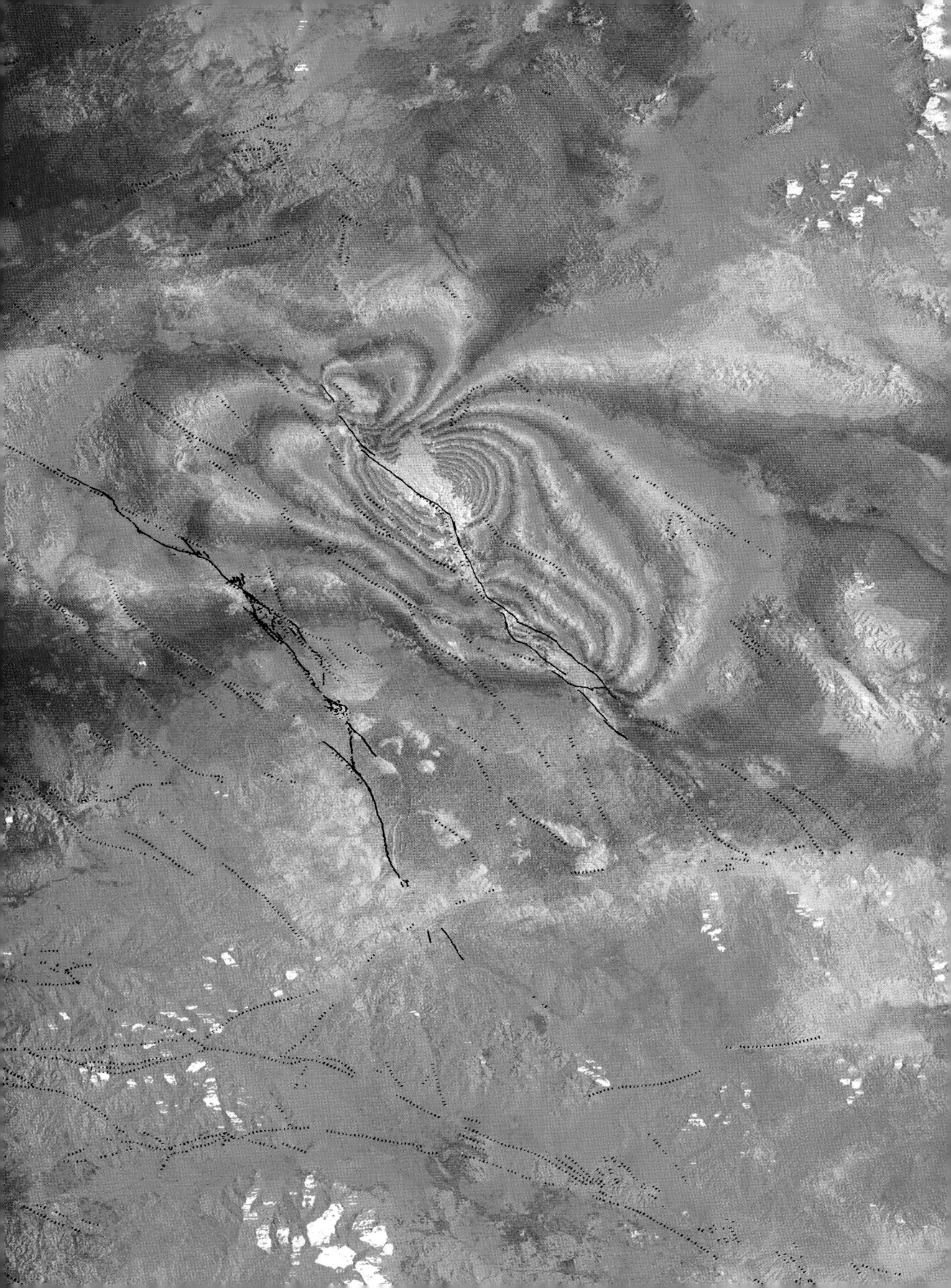

67 Radar

DEFINITION A WAY TO DETECT DISTANT AIRCRAFT BY SENDING OUT A BEAM OF RADIO WAVES AND LISTENING FOR AN ECHO

DISCOVERY FIRST PUT INTO PRACTICE AS A MILITARY EARLY-WARNING SYSTEM BY ROBERT WATSON-WATT IN 1935

KEY BREAKTHROUGH WHILE DEVELOPING NEW WEAPONRY, WATSON-WATT SAW THAT RADIO BEAMS COULD BE USED FOR DETECTION

IMPORTANCE RADAR IS NOW ESSENTIAL FOR AIR-TRAFFIC CONTROL AND HELPED THE ALLIES WIN THE SECOND WORLD WAR

Life was pretty dangerous for early aviators – unaware of anything that they could not see from the cockpit. Then in the 1930s, a Scottish engineer called Robert Watson-Watt discovered radar: a way of using radio signals to detect distant aircraft. As a warning system, it enabled the Royal Air Force (RAF) to repel Nazi bombing raids and is now an essential link in air-traffic control. It has even contributed to the development of convenience foods.

During the First World War, London and other British cities had been bombed by German Zeppelin airships. In the 1930s, with war clouds gathering over Europe once more, concern was mounting amongst military leaders over whether the country could defend itself against new attacks made by faster-moving, high-altitude planes. These aircraft would be hard to hit with conventional anti-aircraft guns, and would fly so high that attempting to intercept them following a visual sighting would take too long.

While the British Air Ministry was pondering this problem, rumours began to reach them that the Germans had developed an energy beam weapon, nicknamed the 'death ray'. The Air Ministry was simultaneously concerned that this fearsome-sounding device could be used against Britain, but also intrigued that if such a thing was possible it might be an effective weapon against the Luftwaffe's heavy bombers.

In 1935, the Air Ministry tasked electrical engineer Robert Watson-Watt with finding out if he could develop a British equivalent of the mysterious death ray. Watson-Watt, himself a descendent of steam-engine inventor James Watt, was born in 1892 in Brechin, Scotland. He graduated in 1912 from the

LEFT Satellite radar image of a Californian earthquake. The colours correspond to contours, each representing a ground displacement of 10cm (3.9in). At the quake's epicentre, the contours are closer together, indicating more movement. Radar can calculate distance by bouncing radio waves off objects, and measuring how long it takes them to return.

University of Dundee with a degree in engineering. During the First World War, he worked for the British Meteorological Office where he helped to develop a technique for detecting thunderstorms, using a radio receiver to pick up the distant electromagnetic bursts givens off by lightning strikes. His device, which worked using directional antennas, proved extremely useful in alerting pilots to storm hazards.

After the war, Watt remained as a government-employed radio researcher – ultimately becoming Superintendent of the Radio Department of the National Physical Laboratory in Teddington, England. And it was here that he was working when he received the request from the Air Ministry to investigate the death-ray rumours. A quick calculation soon showed these to be false – with the stupendous power requirements of such a device rendering it impractical. However, the calculations of Watson-Watt and his colleagues showed that it might well be possible to use an electromagnetic ray to detect enemy aircraft – potentially giving the RAF enough advance warning of an air raid to scramble fighter interceptors.

It works in the same way to shouting at a mountain and waiting for the echo to come back. If you have a stopwatch and you know the speed of sound in air, then by timing how long the echo takes to reach you, you can calculate how far away the mountain is. Similarly, Watson-Watt envisaged sending out beams of radio waves in the direction from which bombers were expected to approach, then looking out for the electromagnetic echo produced when the waves were reflected by incoming aircraft – using radio-receiver equipment.

'Watson-Watt and his colleagues showed that it might well be possible to use an electromagnetic ray to detect enemy aircraft – potentially giving the RAF enough advance warning of an air raid to scramble fighter interceptors.'

The Air Ministry was interested and asked for an immediate demonstration. Watson-Watt needed to hack something together quickly so, rather than build his own transmitter, he used the existing radio signal from the BBC's shortwave radio station based at Daventry. He set up two receiving antennas about 10km (6.2 miles) away from the broadcasting mast, and successfully used these to pick up reflected signals from a test aircraft that the ministry had arranged to fly around the site.

Homeland defence

Watson-Watt was soon working full-time on the development of the fledgling technology. In 1937, the first three experimental radio towers were built in a network called Chain Home that would ultimately provide early warning protection to the south and east coasts of the British Isles. By the time war broke out in 1939, there were 19 Chain Home stations in place, and this number had risen to 50 by the time the war was over in 1945. The technology was soon shared with the Allied powers. In 1940, scientists in the US Navy coined an acronym for it – radar – short for 'radio detection and ranging'. Radar played a big part in helping the vastly outnumbered RAF win the

Battle of Britain – enabling them to ambush attacking formations of Nazi German planes. This was helped in no small part by Watson-Watt's subsequent development of small radar sets for aircraft, which enabled fighter pilots to home-in on enemy aircraft, even at night.

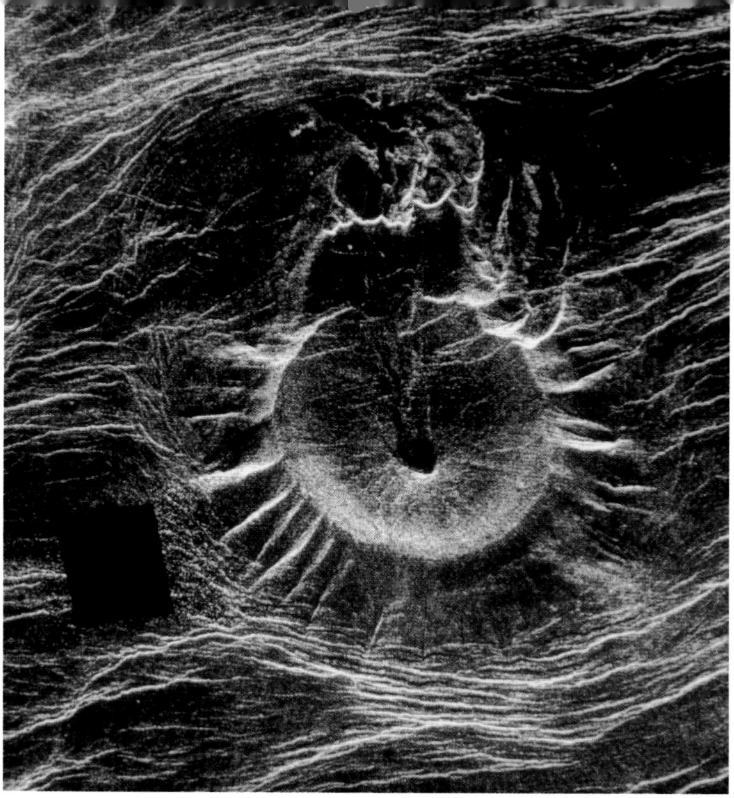

ABOVE Volcano on the planet Venus, imaged using radar apparatus on the NASA Magellan space probe. Radar is one of the few imaging techniques that can penetrate the dense, gloopy atmosphere that smothers Venus.

Radar operators were able to track signals, enabling both speed and direction to be calculated accurately, thereby allowing fighter planes to be vectored in on their targets. Also, combining signals from multiple receivers allowed the size of the radar contact to be estimated. After the war, radar technology was employed in the civilian sector, where it soon became essential in both aviation and shipping, as well as the scientific fields of meteorology and astronomy.

From astronomy to microwave ovens

Some of the most stunning images of the planet Venus – a world whose surface is hidden from view by a thick, murky atmosphere – were returned by the Magellan space probe in the 1990s. Magellan beamed radar signals through the murk and down to the planet's surface as it orbited. Timing the return of these signals enabled an accurate topographic map of the surface to be compiled. A similar relief map of the Earth was constructed using data from a dedicated space shuttle mission in 2000.

Radar research has even given us one of our most popular modern convenience gadgets. In 1945, Percy Spencer, an electronics engineer working for Raytheon, in Waltham, Massachusetts, was working on a military radar set when he noticed that a chocolate bar in his pocket had melted. When he turned the radar on other foods he found that they were cooked by the beam, too. The specific radar set he was using gave off microwaves – electromagnetic radiation with a wavelength of around a thousandth of a millimetre. He soon developed his discovery, adding a metal box to trap and concentrate the microwave radar beam within a confined space. Spencer had invented the microwave oven.

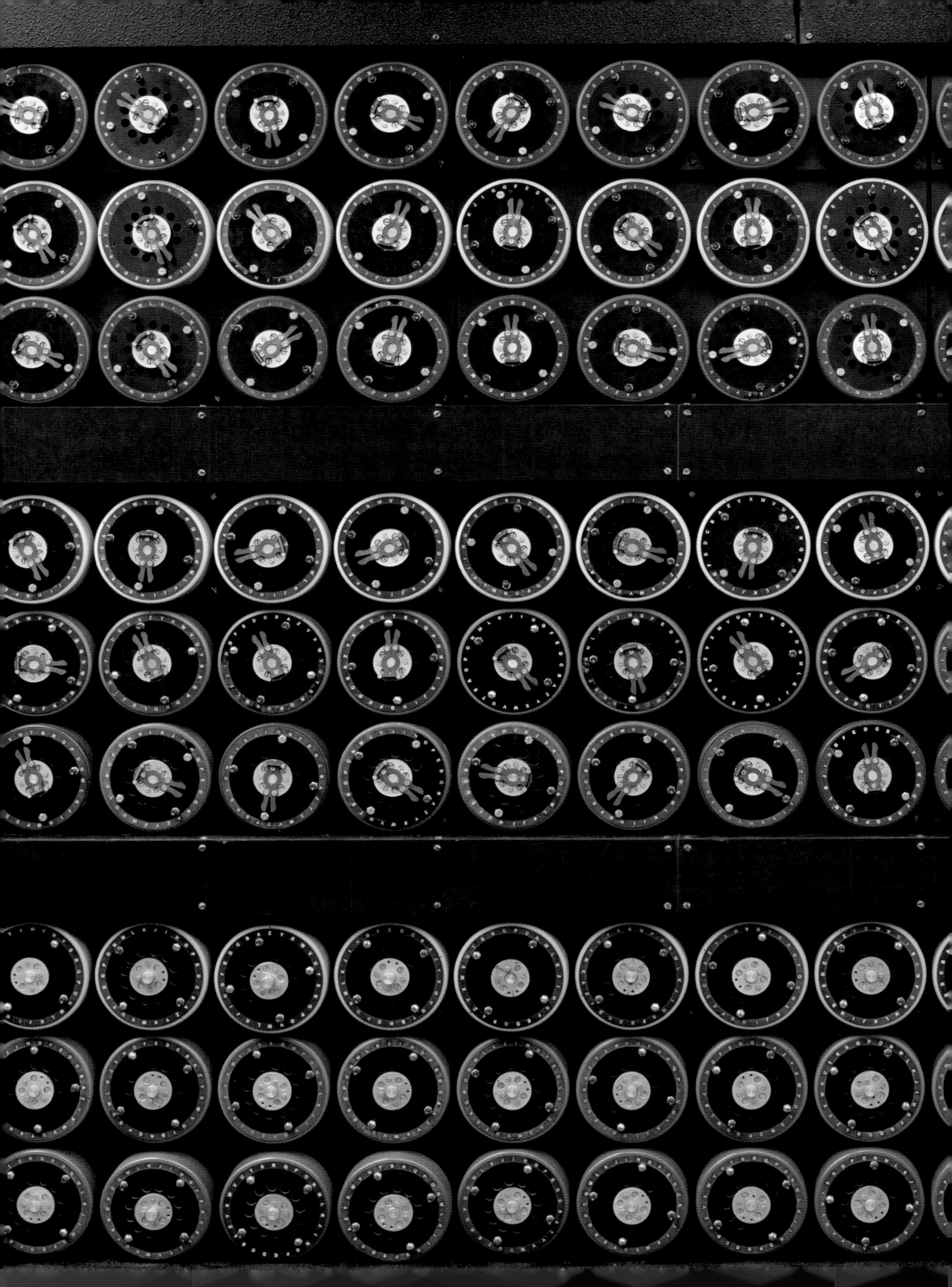

68 Computers

DEFINITION MACHINES THAT PERFORM COMPLEX CALCULATIONS BY FOLLOWING AN INSTRUCTION SET, OR 'ALGORITHM'

DISCOVERY THE THEORY OF THE PROGRAMMABLE COMPUTER WAS DEVELOPED BY ENGLISH MATHEMATICIAN ALAN TURING IN 1936

KEY BREAKTHROUGH REALIZATION THAT A COMPUTER'S PROGRAM COULD BE FED INTO IT JUST LIKE THE INPUT DATA IT PROCESSES

IMPORTANCE TURING'S WORK UNDERPINS EVERY MODERN COMPUTER IN EXISTENCE TODAY

Most people have a computer in the home; many have one in their pocket. Anyone who has ever used a computer in any form can thank Alan Turing. In the 1930s, he came up with the theoretical framework underpinning modern digital computers. During the Second World War, he applied his theory to decipher the German Enigma code.

The computer regularly tops polls for the greatest invention of all time. And with good reason. Computers were initially the preserve of universities and research institutions, where their ability to perform tortuously difficult mathematical calculations brought revolutions to all fields of science.

An astute observer might well have foreseen these developments. What was less obvious, however, was the way in which computers would pervade the lives of ordinary people. The home computer arose as a phenomenon in the 1970s and 1980s, as computer power was harnessed for communication, education and entertainment, making established tools, such as the typewriter, redundant. Today, with the internet, personal computers are the basis for social and economic networking, for watching movies and listening to music, and for shopping. The smallest computers fit in your pocket and they pack more processing power than the early mainframes of the 1950s that used to fill entire rooms. So how did the computing revolution begin?

LEFT Turing's Bombe computer was used during the Second World War to decrypt enemy messages. Each drum corresponds to a rotor setting on the German Enigma machine. By mimicking the action of several Enigma machines linked together, the Bombe removed inconsistent code settings.

Primitive computers, machines capable of performing calculations to order, have been built throughout history. The Antikythera mechanism is a complex arrangement of cogs and gear wheels that was recovered from a shipwreck off the Greek island of Antikythera, and has been dated to 150–100 BC. It is

believed to have been a device for computing the positions of celestial objects in astronomy.

Computers do not have to be electronic devices – anything that is able to process input information according to logical rules to give an output is a type of computer. Consider two cogs meshing together – one with 20 teeth, one with ten teeth. Turn the little cog twice and the big cog goes round once. In general, turn the little cog x times and the big cog turns $x/2$ times – it is a computer for dividing the number of input turns by two.

ABOVE A room full of computer processors from the TERA-10 supercomputer – the largest of its kind in Europe. They process results from the Laser Megajoule – an experimental nuclear fusion device based in France. The TERA-10 offers 50 teraflops (50,000 billion operations per second) of computing power.

In 1837, English mathematician Charles Babbage built an elaborate mechanical computer, known as the analytical engine. The machine could take numbers fed in via punched cards – stiff paper with a pattern of holes cut in it – upon which it could then perform arithmetic operations.

Up until the 1930s, the development of computers had been a purely practical pursuit. But now this changed with the birth of computer science. At the cutting edge of this new discipline was a young English mathematician called Alan Turing. He studied at King's College, Cambridge, and so impressed his tutors that he was awarded a fellowship by the college to stay on and carry out further research.

It was here that he came up with the idea for what has become known as the 'Turing machine'. This was a hypothetical device that can carry out any calculation for which there is a well-defined algorithm – that is, a set of instructions by which the calculation can be performed. Turing imagined data being fed on a ticker tape into the machine, which would then apply its algorithm accordingly. Output would be directed to another ticker tape. For example, the algorithm might contain instructions like 'if input number equals 1, then move tape two steps left and change machine to state 2' and 'if machine is in state 2, then write input number to output tape'.

Turing (and, independently, the American mathematician Alonzo Church) showed in 1936 that any computation for which there is a well-defined algorithm can be computed on a Turing machine. Turing went further still to argue that the instructions upon which a Turing machine operates can themselves be encoded as a stream of input data and fed into a grander device, called a Universal Turing Machine, that is able to simulate the action of any other Turing machine. In other words, a Universal Turing Machine can be 'programmed' to mimic any other Turing Machine and thus perform any computation for which there is an algorithm. What Turing had conceived was the blueprint for the modern programmable computer.

Enigma busters

During the Second World War, Turing got to put his theoretical ideas into practice when he worked at the British code-breaking centre at Bletchley Park, deciphering transmissions encrypted in the Germans' fiendish Enigma code. Turing designed a computer called the 'Bombe', an electromechanical device to generate thousands of possible encryption keys and then automatically perform a check to reject those keys that were logically inconsistent. This reduced the Enigma's billions of possible keys to a few thousand, which could then be checked by hand – still a formidable job, but not impossible.

Turing later oversaw the construction of an improved code-breaking machine – Colossus. This was the world's first fully electronic digital computer, and was operational at Bletchley Park in time for the D-Day landings in 1944. The work of the British code-breakers, empowered by Alan Turing's computer technology, shaved an estimated two years off the duration of the war and saved hundreds of thousands of lives. Having seen his visions of the 1930s implemented using the relatively new technology of electronics, Turing was inspired by what computers might ultimately be capable of. After the Second World War, working at the National Physical Laboratory in Teddington, he set about building the first electronic Universal Turing Machine – a fully programmable digital computer. The result was the Automatic Computing Engine (ACE), which ran successfully for the first time in 1950.

> **'The work of the British code-breakers, empowered by computer technology, shaved an estimated two years off the duration of the war.'**

In 1952, Alan Turing – who was homosexual – was arrested and subjected to a government programme of chemical castration. (At this time, it was against British law to be a practising homosexual.) His public and private life were destroyed and he committed suicide two years later.

Legacy of innovation

In the final years of his life, Turing took the first steps in the field of artificial intelligence – the construction of computers that are able to think – by devising the 'Turing test' for assessing a computer's intelligence. He is also said to have written the first algorithm for a chess-playing program although, at the time, there was no computer powerful enough to run it.

In 2009, British Prime Minister Gordon Brown issued a public apology for the government's treatment of Turing. In his intellectual endeavours, as well as his private life, Alan Turing was a man living ahead of his time.

69 Splitting the atom

DEFINITION DIVIDING THE NUCLEUS OF AN ATOM INTO TWO PIECES AND RELEASING ENERGY; ALSO KNOWN AS NUCLEAR FISSION

DISCOVERY FIRST CONFIRMED BY OTTO HAHN, LISE MEITNER AND FRITZ STRASSMANN AT THE KAISER WILHELM INSTITUTE

KEY BREAKTHROUGH BOMBARDING URANIUM NUCLEI WITH NEUTRONS AND CREATING NUCLEI OF THE LIGHTER ELEMENT BARIUM

IMPORTANCE NUCLEAR FISSION IS THE PRINCIPLE UPON WHICH ALL MODERN NUCLEAR POWER PLANTS ARE BASED

Atoms were thought to be the fundamental components of matter. But in the early decades of the 20th century it became clear that their dense nuclei are made of even smaller particles, and can thus be split apart. The process is called nuclear fission and was discovered by a team of German physicists in 1938. Fission gives off colossal amounts of energy – of which nuclear electricity and atomic bombs are both applications.

In the first decades of the 20th century, the physics of the very small underwent something of a revolution. Physicists knew that matter on the tiniest scales was made from building blocks called atoms. They believed atoms were like billiard balls – smooth, featureless and, above all, indivisible.

By the 1930s, physicists had a better idea what an atom looks like up-close – and it is nothing like a ball. The emerging field of quantum theory – the new radical physics of the particle world – had told them that an atom is made of a cloud of negatively charged electron particles orbiting around a central positively charged 'nucleus'.

The nucleus had been discovered in 1909 by a team led by physicist Ernest Rutherford, working at the University of Manchester in Britain. They fired 'alpha particles' – tiny positively charged chunks of matter – at a sheet of gold foil, finding that most of the particles passed straight through, but very occasionally one would bounce back in the direction it had come. In 1911, Rutherford used the results to infer that the positive charge of an atom is concentrated into a small volume at its centre – and this is the nucleus. It is tiny – tens of thousands of times smaller than the atom in which it resides.

LEFT Mushroom cloud from an atom bomb test in the Nevada desert, 1953. The yield of this explosion was equivalent to about 61 kilotonnes of TNT, nearly five times the force of Hiroshima. These early atom bombs were powered by nuclear fission – splitting in half the nuclei of heavy atoms, such as plutonium, to liberate energy.

What lies within?

But what is the nucleus made of? It was Rutherford again who provided the answer ten years later in 1919. This time, he fired alpha particles into a cloud of nitrogen gas. He found that occasionally a particle would be absorbed by a nitrogen nucleus – and when this happened, the nitrogen spat out a nucleus of the lighter chemical element hydrogen. Rutherford concluded (correctly, again) that the hydrogen nucleus must be a fundamental component of all atomic nuclei – he called it the proton. Each proton has positive electrical charge, equal but opposite to the charge of the electron.

Protons, it also turned out, determine the particular chemical element in the periodic table (see page 169) to which an atomic nucleus belongs. One proton makes a hydrogen nucleus, two makes a helium nucleus, three is lithium and so on. The number of protons in a nucleus is often referred to as its 'atomic number', denoted by the label Z.

Let's split

The discovery of protons threw up a new problem. If the protons were positively charged, and bunched together as tightly as Rutherford's gold-foil experiment suggested, then what stops the nucleus from flying apart due to electrical repulsion? The answer came in 1932, when English physicist James Chadwick discovered a second component of the nucleus. Called the neutron, it has no electric charge and forms a kind of 'packing' between the protons in the nucleus, lessening the electrical forces between them.

'The neutrons had actually cracked open the uranium nuclei, splitting them in half. But that was not all. When the mass of the barium and krypton nuclei was added up, it came to less than the initial mass of uranium. The mass difference had been converted into a massive quantity of energy.'

But neutrons could have a disruptive influence, too. In 1938, German physicists Otto Hahn, Lise Meitner and Hahn's assistant Fritz Strassmann began experiments to fire neutrons at the heavy atomic nucleus of uranium – an element with atomic number 92. They were trying to make chemical elements heavier than uranium. Their idea was that every so often a uranium nucleus would capture a neutron, which would then convert, via a process called beta decay, into a proton, thus raising the atomic number of the nucleus.

But instead they created small amounts of the lighter elements barium (atomic number 56) and krypton (atomic number 36). The neutrons had actually cracked open the uranium nuclei, splitting them in half. But that was not all. When the mass of the barium and krypton nuclei was added up, it came to less than the initial mass of uranium. The mass difference had been converted into a massive quantity of energy. Albert Einstein's famous formula $E = mc^2$ says that mass (m) can be converted into energy (E) by multiplying by the speed of light (c) squared. And that is exactly what had happened to the mass lost in splitting the uranium nucleus. Hahn, Meitner and Strassmann had discovered nuclear fission.

ABOVE Sellafield nuclear reprocessing plant, Cumbria, England. This facility serves all British nuclear power stations by processing spent nuclear reactor fuel into weapons-grade plutonium and preparing radioactive waste for disposal.

But energy is not all that is given off in the process. Each time a uranium nucleus splits in half, it emits more neutrons. And each of these neutrons can then go on to split another uranium nucleus, repeating the process and setting up a self-perpetuating 'nuclear chain reaction'.

One thing leads to another

Hungarian physicist Leo Szilard had realized the possibility of such chain reactions in 1933, having read H. G. Wells's 1914 novel *The World Set Free*. In it, Wells imagines that scientists discover a way to speed up the rate of natural radioactivity – the way some chemical elements decay, giving off particles and energy over time. The scientists in the novel then use it to build atomic bombs. Szilard saw that Wells's vision could be made reality if each radioactive decay was able to trigger the decay of another nearby nucleus. But at the time he could see no way to make the idea work.

With the discovery of Hahn's group, Szilard suddenly saw the mechanism he had been looking for. With war brewing in Europe, he approached Albert Einstein. The two of them famously wrote to President Roosevelt, warning of the danger that Germany could use nuclear chain reactions to build a new, terrible weapon. The letter led directly to the Manhattan Project and the development of the world's first atomic bombs by America, two of which were used against Japan in 1945. The bombs operated on exactly the principle that Hahn, Meitner and Strassmann had discovered.

But there were peaceful applications, too. Nuclear fission underpins all modern nuclear power stations. The world's first commercial nuclear power plant – Calder Hall, in England – opened in 1956. And today, even some prominent environmentalists see nuclear power as the most efficient and environmentally friendly way to generate electricity. Hahn received the 1944 Nobel Prize in Chemistry for the discovery of nuclear fission. Many believe Meitner should have shared the prize with him.

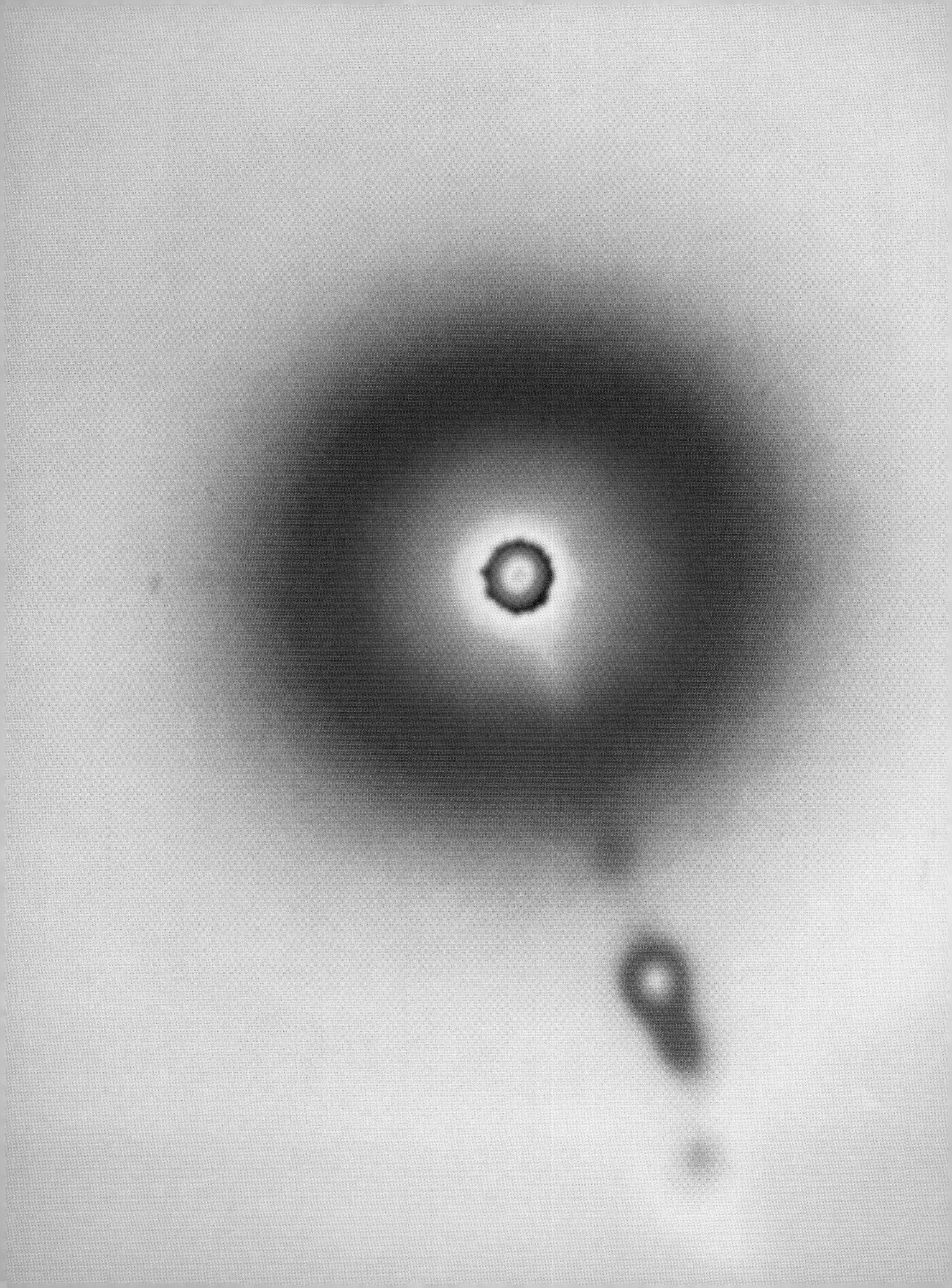

70 Black holes

DEFINITION AN OBJECT WHOSE GRAVITY IS SO POWERFUL THAT NOT EVEN LIGHT CAN GET AWAY FROM IT

DISCOVERY PREDICTED TO EXIST IN NATURE BY AMERICAN PHYSICISTS ROBERT OPPENHEIMER AND GEORGE VOLKOFF IN 1939

KEY BREAKTHROUGH SHOWING THAT THERE IS NO NATURAL FORCE THAT CAN STOP THE HEAVIEST STARS COLLAPSING TO A POINT

IMPORTANCE BLACK HOLES ARE A TESTING GROUND FOR STUDYING GRAVITY UNDER THE MOST EXTREME CONDITIONS

Black holes are places in space that are so small and dense that not even rays of light can escape their immense gravitational pull. They are believed to form when a massive star runs out of fuel and collapses in on itself at the end of its natural life. And even these fierce gravitational powerhouses pale next to the monstrous black holes thought to reside inside most galaxies.

Robert Oppenheimer is best known as the father of the atomic bomb – he led the Manhattan Project to develop America's nuclear capability. But six years before the devastation of Nagasaki and Hiroshima, he made a startling contribution to astrophysics. Oppenheimer and his colleague George Volkoff demonstrated how black holes – regarded as nothing more than a theoretical curiosity – could form when stars reach the end of their lives.

Many people associate black holes with Einstein's theory of general relativity, but in fact they were first put forward by English scientist John Michell in 1784 – using only Isaac Newton's law of universal gravitation (see page 57).

LEFT The elliptical galaxy M87 in the Virgo galaxy cluster. This galaxy has a supermassive black hole at its core, weighing an estimated seven billion times the mass of the Sun. Extending from the galaxy is a colossal jet of material, 5000 lightyears long, which is thought to have been ejected by the black hole.

Michell considered a feature of Newton's theory called 'escape velocity'. Throw a ball up in the air and it falls back to the ground. Throw it a little harder and it travels a little higher before falling back. But throw it hard enough and, in theory, it will be able to escape the pull of the Earth's gravity. The minimum speed needed to do this is called the Earth's escape velocity – it is just over 11km per second (6.8 miles per second). Other planets have their own escape velocity that increases with the planet's mass and decreases with its size. The escape velocity from the surface of Mars is just under half that of Earth; the escape velocity from Jupiter is over five times that of

Earth. Michell wondered whether objects might exist with an escape velocity greater than the speed of light – so that not even light can get away from them. He prophetically named these objects 'dark stars'.

After Einstein's formulation of general relativity, German physicist Karl Schwarzschild derived a solution of its equations for the gravitational field around a planet or star. His findings echoed the work of Michell, showing that when the size of a star of mass M was less than $2GM/c^2$ (where G is Newton's gravitational constant and c is the speed of light), nothing could escape from the resulting object's gravity. This outer surface, with a radius given by $r = 2GM/c^2$, from which not even light can escape, was named the 'event horizon' by American physicist David Finkelstein in 1958. Anything unfortunate enough to fall over the event horizon of a black hole can never escape and is doomed to be crushed out of existence at a point of infinite density at its centre – known as a 'singularity'.

But few physicists thought that such objects existed. The sceptics included Albert Einstein, who believed that the rotation possessed by all celestial objects stabilized them against gravitational collapse: a rotating object rotates faster as it shrinks (like an ice-skater pulling her arms in), and an outward 'centrifugal' force always accompanies the rotation (the same effect that makes your clothes stick to the inside of the washing machine as it spins).

Heart of darkness

Ordinarily, stars are supported against gravity by thermal pressure. Nuclear reactions in the star's core generate heat. This makes the atoms and molecules inside bounce around more, creating pressure – and this is what holds the star up. But when the star reaches the end of its life and runs out of nuclear fuel in its core, it cools, the internal pressure drops and it collapses under gravity.

'Anything unfortunate enough to fall over the event horizon of a black hole is doomed to be crushed out of existence at a point of infinite density at its centre – known as a "singularity"'.

Rotation alone, it turned out, is not sufficient to halt the collapse. But quantum theory – the physics of atoms and molecules – was providing forces that could. In the 1930s, Indian physicist Subrahmanyan Chandrasekhar showed how electrons squeezed together exert a pressure on one another through a law called the exclusion principle. The principle says that quantum particles do not like to share quantum states and resist attempts to squash two or more particles into the same state. Stars supported by this so-called 'electron degeneracy pressure' are called white dwarfs. They are extremely dense, packing the mass of the Sun into a sphere the size of Earth. Chandrasekhar's calculations suggested that the maximum mass of a white dwarf is about 1.4 times the mass of the Sun, beyond which it collapses.

A white dwarf exceeding its Chandrasekhar limit would collapse down to become an even denser object, a neutron star, cramming down a solar mass

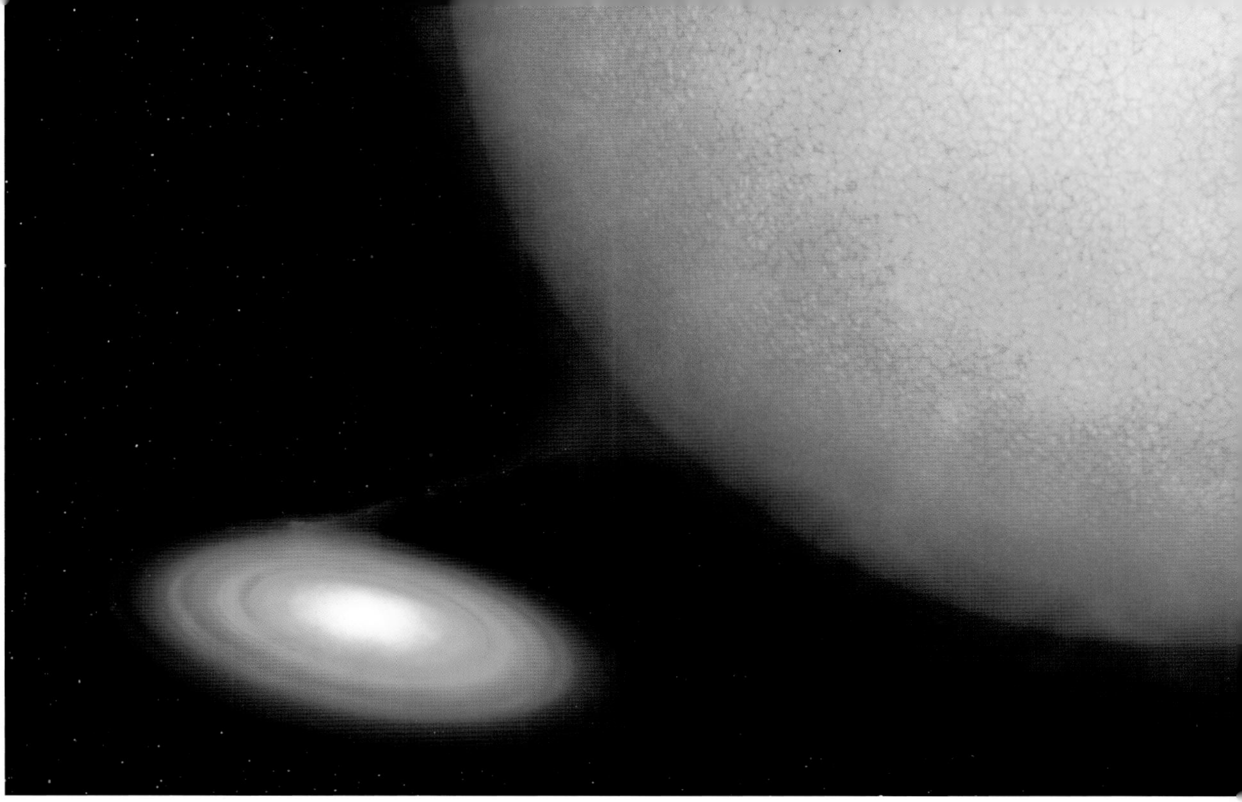

to just 20km (12.4 miles) across. And this is where Oppenheimer and Volkoff took up the story. They calculated how heavy a neutron star can be and still hold itself up against gravity. Their equations give the maximum mass of a star to be about three times the mass of the Sun. There is no known force that can prevent a neutron star heavier than that from collapsing to a point. The discovery of neutron stars in the 1960s led to more interest in the work of Oppenheimer and Volkoff. In 1967, the American physicist John Wheeler referred to these objects as 'black holes', because they neither reflect nor give off any light. The term quickly entered physics parlance.

ABOVE Artwork of Cygnus X-1, the first identified black hole. Here, the black hole is seen orbiting around a supergiant star. The hole pulls material away from the star, which then encircles it.

Night vision

Black holes attract matter, which circles around their equator, rubbing against other accreted matter and heating up so much that it gives off X-rays. And this is how the first black hole was found, in 1964. It is called Cygnus X-1 – astronomical shorthand for the brightest X-ray source in the constellation Cygnus – and it is part of a binary system, where two stars orbit one another. Analyzing Cygnus X-1's motion enabled astronomers to deduce its weight: 8.7 times the mass of the Sun, way above Oppenheimer's limit.

Today, numerous stellar-mass black holes are known, as well as monstrous 'supermassive black holes' that lurk at the centre of most galaxies, including our own Milky Way. The heaviest is in the galaxy M87 and weighs in at a colossal 6.2 billion times the mass of the Sun.

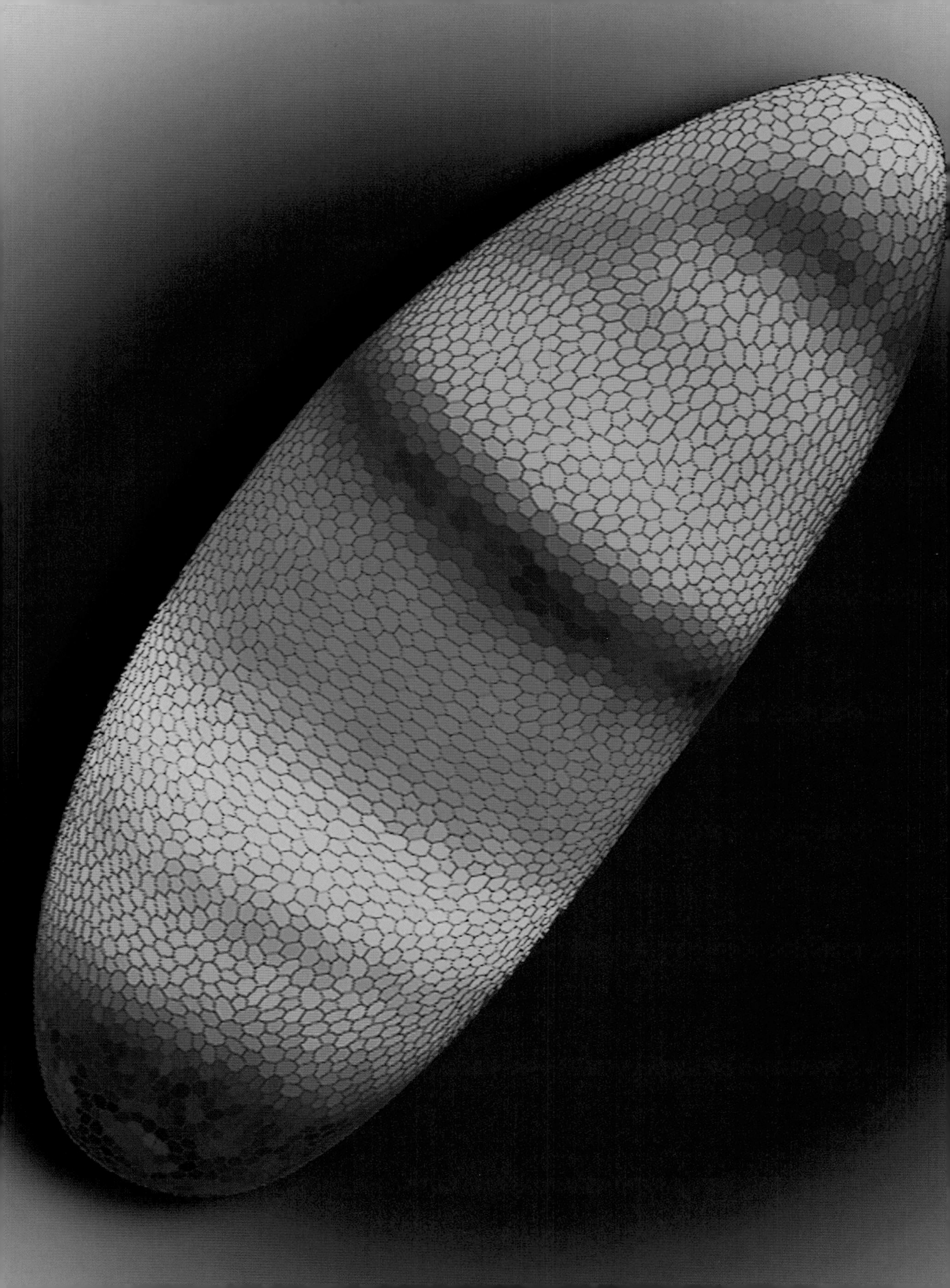

71 DNA carries genes

DEFINITION DNA IS A MOLECULE FOUND INSIDE CELLS; GENES ARE BIOLOGICAL DATA ENCODING THE TRAITS THAT WE INHERIT

DISCOVERY FIRST EXPERIMENTAL EVIDENCE PROVIDED BY OSWALD AVERY, COLIN MACLEOD AND MACLYN MCCARTY IN 1944

KEY BREAKTHROUGH REMOVING THE DNA FROM BACTERIA SHOWED THAT THE BACTERIA WERE THEN UNABLE TO PASS ON THEIR GENES

IMPORTANCE ISOLATING THE CHEMICAL CARRIER OF GENES PAVED THE WAY FOR GENETIC ENGINEERING AND GENE THERAPY

We pass on our characteristics to our children via chunks of biological information known as genes. This much was known in the late 19th century. But it took another 80 years to figure out how nature achieves this feat – when ingenious experiments revealed that genes are encoded into a chemical molecule called DNA.

'She's quiet – just like her dad.' Parents pass on their traits to their children, and the resemblance is often uncanny. It is not just physical characteristics, such as eye colour – but intelligence, temperament and all aspects of a person's being are handed down through the generations. Humans, and in fact all living organisms – from minke whales to rhododendrons – do this with the help of biological envoys called 'genes'.

The term was coined in 1909 by Danish botanist Wilhelm Johannsen – after the Greek word *genos*, meaning 'birth'. Johannsen had rediscovered the work of Gregor Mendel who, in 1866, had put forward the idea that traits are passed between generations in discrete units (see page 161).

LEFT Map of fruit-fly embryo showing how genes are 'expressed'. In all life forms, each cell contains copies of every gene. But in any part of the body only a selection of genes are expressed – or switched on – and this is what differentiates between tissue types. Gene expression is controlled by regulator genes, in this case, Knirps (green), Kruppel (blue) and Giant (red).

Your genes make up a library that holds all the facts and figures needed to build exact copies of every single cell in your body. Most genes work by encoding the information needed to make particular proteins. These are large, complex molecules that maintain the structure and healthy function of living cells. Every cell stores a copy of your entire gene library from which it can pick and choose in order to make the proteins needed for its particular cell type. Many genes are encoded on long, spiral-shaped molecules found in the centre of cells. The molecules are commonly known as DNA, short for 'deoxyribonucleic acid'. DNA is built from more basic chemical units known as nucleotides. And the precise sequence of the nucleotides is what stores genetic data, rather like the binary 1s and 0s of computer code.

Mendel's laws, published in 1866, set out the basic rules of genetic inheritance. However, they were not adopted by the wider scientific community until the early 20th century. DNA, on the other hand, was first discovered inside the nuclei of biological cells back in 1868, by Swiss biologist Friedrich Miescher. But it took until 1944 for the direct connection with genes to be established: this landmark breakthrough was achieved jointly by American scientists Oswald Avery, Colin MacLeod and Maclyn McCarty, based at the Rockefeller Institute for Medical Research, New York City.

Mice and men

The first step was a landmark experiment carried out in 1928 by the English microbiologist Frederick Griffith. Griffith had studied two different kinds of pneumonia bacteria – one fatal and one non-fatal. First of all, Griffith injected laboratory mice with the two different strains of the bacteria. As expected, the mice injected with fatal pneumonia died, while those injected with the non-fatal variety lived. Next, Griffith killed off a sample of the fatal pneumonia bacteria by exposing it to extreme heat. Mice injected with this altered sample also survived. But then he injected another group of mice with both the dead, fatal bacteria and the live, but non-fatal variety. And this was where the surprises began – because these mice also died. The virulence of the fatal (but dead) pneumonia had somehow been transmitted to the live, but supposedly fatal, strain of pneumonia. But how?

Avery and his team decided to find out the answer to Griffith's question by repeating the experiment – this time using chemical techniques to selectively destroy different parts of the fatal bacterial cells in each individual trial. Some biologists had suspected that cell proteins were the messengers that carry genes from one organism to another. So the team first applied an enzyme that destroyed all the protein in the cells. If the protein really was responsible for the transmission, then destroying it should prevent the safe bacteria from becoming fatal. When they carried out the experiment, however, the mice still died. It was only when the team applied a DNA-digesting enzyme to the fatal bacteria that the test mice survived. Avery realized immediately what this meant: DNA was the molecule that was passing the fatal traits between the bacteria.

'It was only when the team applied a DNA-digesting enzyme to the fatal bacteria that the test mice survived. Avery realized immediately what this meant: DNA was the molecule that was passing the fatal traits between the bacteria.'

Viral validation

If any doubt remained, it was blown away in 1952 in a follow-up experiment by the American biologists Alfred Hershey and Martha Chase. They looked at phage viruses – disease-causing agents that infect bacteria cells in order to

reproduce. During the reproductive process, a virus must pass on its genes. But does it do this via proteins – or DNA, as Avery's team had found with pneumonia bacteria?

To find out, the researchers labelled the DNA in the viruses with a radioactive form of phosphorous – a chemical element that is found in DNA but not in protein. The team were then able to show that the radioactivity was passed on to the cells. Next, they inserted radioactive silicon into the viral proteins. Conversely, silicon is found in protein but not in DNA, and in this case no radioactivity was passed from the viruses to the bacteria cells. The evidence was conclusive: genes are carried by DNA, just as Avery's team had found. Avery was awarded the Copley Medal of London's Royal Society in 1945 for the discovery though, inexplicably, he was passed over for a Nobel Prize.

ABOVE Different dog breeds – from left to right, Beagle, Wheaten Terrier, Rhodesian Ridgeback, Red Siberian Husky, Irish Setter, Golden Retriever, Boxer and Sheltie. Different breeds are created by selective breeding: the genetic blueprint of each breed is encoded in its DNA, ensuring that its traits are passed to subsequent generations.

Instructions included

Since the initial breakthrough, scientists have discovered that there is even more to genes than just DNA. Living cells come equipped with an extra genetic code that tells them how to interpret the information on the DNA – a sort of cellular instruction manual. Every single cell contains a copy of your entire DNA. But it is the instruction manual that pulls out the particular genes needed to build each different cell type – such as skin, bone, muscle or internal organs. This selection process is called 'gene expression'.

It seems possible that some of the rules in this instruction manual can change over the course of an organism's life due to factors such as nutrition and stress. In other words, life experiences can cause physiological changes in a parent that may be passed to offspring. Called 'epigenetics', this theory is a new spin on an old idea put forward by the 19th-century French scientist Jean-Baptiste Lamarck. Evidence for epigenetics is mounting. In 2006, an international team of researchers conducted a study revealing that lifestyle factors – such as poor nutrition and smoking – in fathers impacts upon the health of sons and grandsons many years down the line.

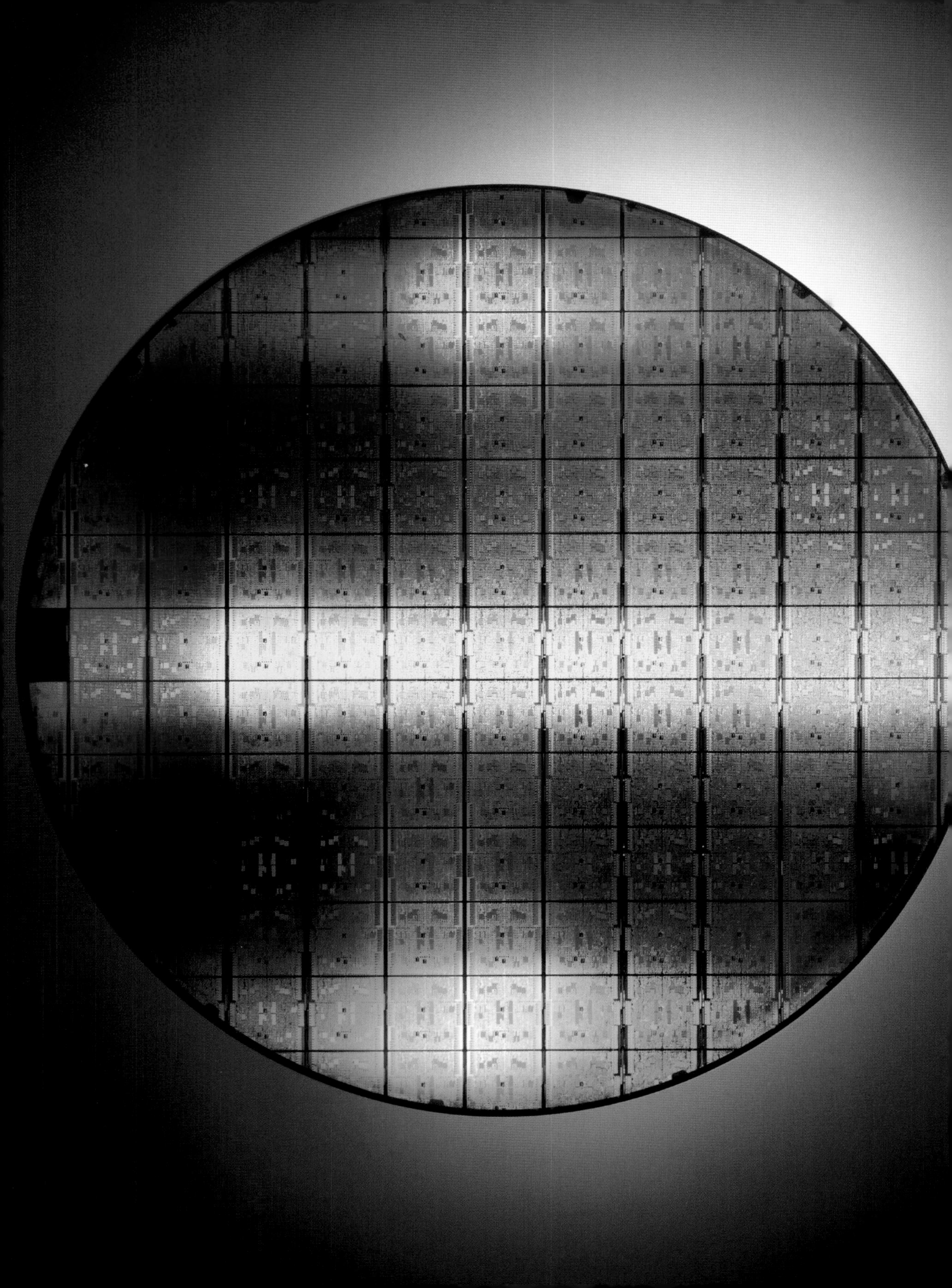

72 Semiconductor electronics

DEFINITION DEVICES FOR REGULATING ELECTRIC CURRENT, MADE FROM PARTIALLY CONDUCTING MATERIALS

DISCOVERY THE FIRST TRANSISTOR WAS BUILT BY WILLIAM SHOCKLEY, WALTER BRATTAIN AND JOHN BARDEEN IN 1947

KEY BREAKTHROUGH CONTROLLING CURRENT THROUGH A SANDWICH OF DIFFERENT SEMICONDUCTORS BY APPLYING A VOLTAGE TO IT

IMPORTANCE PROMPTED THE MINIATURIZATION OF ELECTRONICS THAT GAVE US THE MICROCHIP

It is arguably the greatest invention of the 20th century. When the first practical semiconductor device – the transistor – was built at the Bell research laboratory of the American Telephone & Telegraph Company in 1947, it kick-started the whole field of micro-electronics.

It was the discovery he had been striving towards for two years, but William Shockley was furious. In 1945, Shockley began developing a new electronic amplifier using semiconductors – materials that do not quite conduct electricity well enough to be called conductors, and yet do not insulate enough to be called insulators.

Brilliant yet aloof, Shockley chose to work from home, leaving his colleagues Walter Brattain and John Bardeen to get on with their research unsupervised at the Bell Labs in New Jersey. Shockley was not expecting them to press ahead and make the discovery without him, but that is what they did.

In November 1947, a chance observation by experimentalist Brattain gave theoretical physicist Bardeen a critical new understanding of how electricity behaves on the surface of a semiconductor. Brattain had patched together a crude device from plastic, gold foil and semiconducting germanium, then tested it. The amplifier worked, controlling a large electric current from a tiny input voltage. It was the first 'transistor': from 'trans' (meaning 'to change') and 'resistor' (a device to regulate current). Bell Labs announced the discovery in June 1948, but its importance was not realized until 1951, when Shockley improved the design. Shockley, Bardeen and Brattain jointly received the 1956 Nobel Prize in Physics.

LEFT Semiconducting silicon wafer used to make microchips. Tiny electrical components are imprinted on the wafer by depositing insulating layers; then selected areas of silicon are etched away to leave precise conducting pathways. The wafer is then divided up, ready for insertion into microchips.

N-type, P-type

Semiconductors work by the movement of both negatively charged electrons – which are what carry electric current through ordinary conductors – and also positively charged 'holes' in the material that the movement of an electron leaves behind. Negatively charged semiconductors, with an excess of electrons, are known as N-type, while those that are positively charged, owing to an excess of holes, are called P-type.

These two types of semiconductors are made by 'doping' a sample of a pure semiconductor – that is, by adding impurities to it. Making an N-type semiconductor simply means adding an impurity with an excess of electrons – such as arsenic. On the other hand, a semiconductor can be turned P-type by simply adding a doping agent that is deficient in electrons – gallium is a common example.

The existence of semiconductor materials – the most common being the metallic elements silicon and germanium – had long been known. But their applications in electronics were not immediately obvious.

The very first semiconductor device was the diode – an electronic component through which current only flows in one direction. The first semiconductor diode was built at the end of the 19th century by German physicist Karl Ferdinand Braun. But this 'cat's whisker', as it was known (because it used a whisker-like wire in contact with a semiconducting crystal), was unreliable.

Bardeen's team were onto something much bigger. They had realized that multiple slices of N- and P-type materials can be brought together to form a semiconducting junction, the overall conductivity of which can be controlled by varying an external voltage applied to the junction. And this was the train of thought that led to the transistor.

So how did it work?

Transistors are essentially a sandwich made from alternating layers of N-type and P-type material. In an NPN transistor, increasing the voltage of the middle (P-type) layer increases the number of electrons there. This makes the middle layer better at conducting electrons between the sandwich's two outer layers – and so the current through the transistor increases. In a PNP transistor, on the other hand, the process is similar but the current is regulated not by electrons, but by the number of positive holes in the middle layer.

> 'Fixed groups of semiconducting transistor switches made logic gates – electronic circuits that process information. And logic gates were the building blocks of something amazing: digital computers.'

At their most basic level, transistors are electronic amplifiers. They allow a minuscule voltage to control the flow of a very much larger current – for example, magnifying a feeble radio transmission so it can be heard over a loudspeaker. That in itself was nothing new. Amplifiers had been around since 1907, when electrical engineers had built primitive circuits based on vacuum tubes, or 'valves'. The tubes used an arrangement of metal terminals inside a sealed glass bulb to regulate the flow of current. But these were fragile, expensive, unreliable – and very bulky. The Bell Labs transistor, however, measured just 1.3cm (0.5in) across, and would soon trump valves on all their other shortcomings, too.

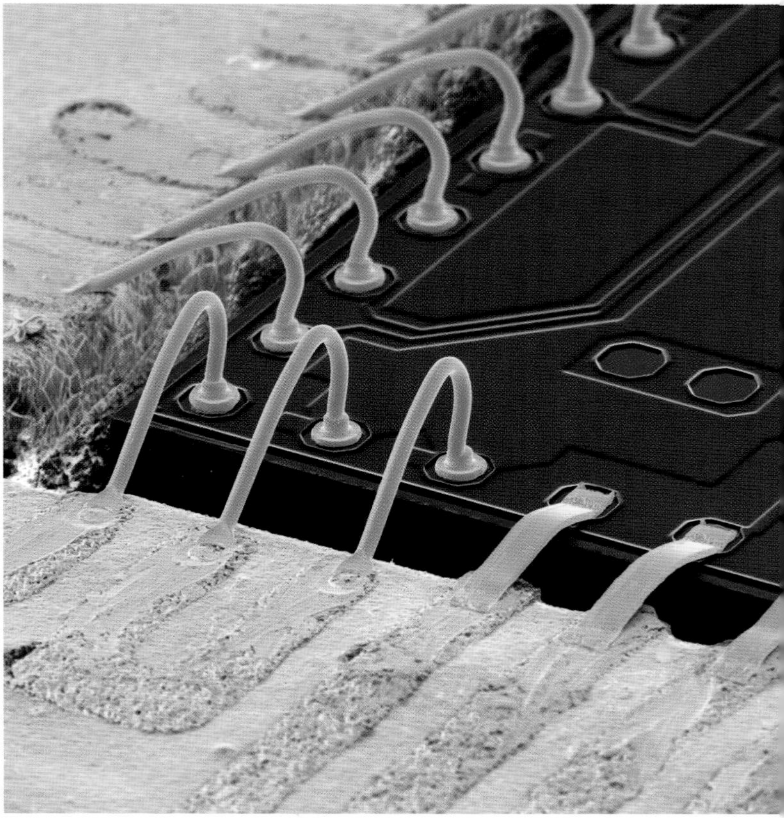

ABOVE Close-up of a microchip. Clearly visible are wires (often made from gold) linking the silicon wafer in the centre to its external casing. These lead to the pins that connect the chip to other components in the circuit board.

The first commercial transistor went on sale in 1949. And hot on its heels came a truly iconic invention: the transistor radio. It was released in 1954, priced $49.95 (about $400 today). The radio did not catch on immediately – that had to wait until the 1960s, when manufacturing and retail prices fell dramatically and it became massively popular.

But the big application was still to come. The transistor was not just an amplifier; it was also a switch, able to flick quickly and reliably between 'on' and 'off' settings in response to an input voltage. Vacuum tubes could also do this to a limited degree, but transistors were much more efficient. Fixed groups of semiconducting transistor switches could make logic gates – electronic circuits that process information. And logic gates were the building blocks of something amazing: digital computers.

Less is more

The first transistor computer was built at the University of Manchester, England, in 1953, and more followed during the 1950s and 1960s. A key advantage was their small size, being more compact than their vacuum-tube cousins. Computers got smaller still with the invention, in 1958, of the semiconductor microchip, also known as the silicon chip. This could fit many transistors onto a single tiny wafer of semiconducting silicon, and so began the miniaturization of electronics.

Today, a semiconductor microchip can hold billions of transistors, each so small it could fit on the surface of a blood cell – and so cheap that $1 buys you 50 million of them. And that's why transistors, and other semiconductor components, are now in every electronic device from phones to space rockets.

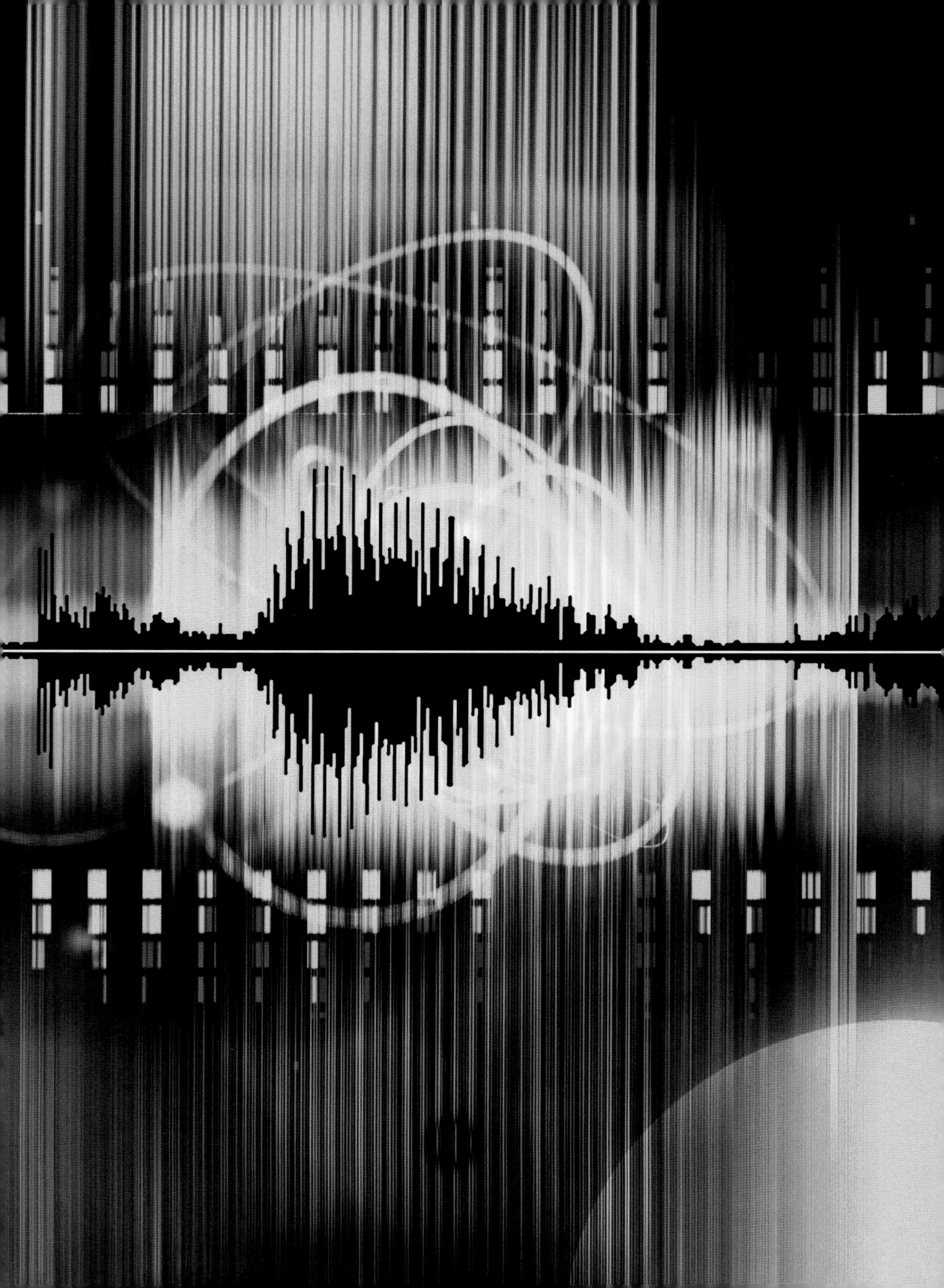

73 Information theory

DEFINITION A MATHEMATICAL THEORY GOVERNING THE TRANSMISSION OF INFORMATION IN THE PRESENCE OF NOISE

DISCOVERY PUT FORWARD IN 1948 BY CLAUDE SHANNON

KEY BREAKTHROUGH SHANNON REMOVED REDUNDANT DATA AND ADDED KNOWN DATA TO GAUGE THE EFFECTS OF NOISE

IMPORTANCE USED IN ALL COMMUNICATION DEVICES AND COULD HOLD CLUES AS TO THE NATURE OF BLACK HOLES

Information might seem like a very simple concept – maybe one that is too simple to merit its own theory. But in 1948, Claude Shannon published a startling piece of research setting out how information could be quantified mathematically – and how that formalism could maximize the efficiency of information transmission and reduce the effects of noise.

The foundations of information theory were laid down by American electronics engineer and mathematician Claude Shannon. Shannon was born in 1916, in Michigan. He was a bright child, showing talent at scientific subjects from an early age and spending his spare time on technical projects, such as radio-controlled models and building a telegraph system between his house and a friend's. He entered the University of Michigan in 1932, taking two degrees – mathematics and electrical engineering – before moving to the Massachusetts Institute of Technology to begin work on his PhD.

Right from the start, Shannon was interested in the communication of information. One of his first projects was the theory of Boolean logic – processing information using logic gates (such as AND and OR) – and how it could be used to simplify the operation of switches in telephone exchanges. Then he turned his solution around to show that the operation of relay switches could be used as a calculating device to solve problems in Boolean logic. His work formed the basis for the construction of computers.

During the Second World War, Shannon worked on cryptography (the science of encrypting military communications) and fire-control systems (primitive computers used for aiming artillery guns). He co-authored one

LEFT Computer depiction of patterns present in a human voice. Like all forms of communication, a voice signal is a stream of data that can be analyzed electronically – for example, as a form of identification. Voice data is inherently imprecise – it can be muffled or distorted – but this can be compensated for using error-correction techniques from information theory.

scientific paper that was prescient of his future work – examining the operation of fire-control systems when there was noise in the communication channel between the computer and the gun. In this case, he asked, how could the operator be sure that the gun would receive the intended signal sent by the computer?

Binary digits

Shannon continued this line of enquiry after the war. He was fascinated by the idea of how information could be communicated efficiently – and in the presence of noise. This research culminated, in 1948, with Shannon publishing a seminal research paper entitled 'A Mathematical Theory of Communication' in the Bell System Technical Journal.

In the paper, Shannon identified the 'bit' as the fundamental unit of information. Bit is short for binary digit – a simple 1 or 0, which could correspond to the yes/no answer to a question, a true/false input to a logic problem or the on/off state of a switch inside a computer. Just as the ordinary decimal digits that we use can combine to make bigger numbers, so bits can also combine to make longer binary numbers called bytes. For example, a 2-bit byte can encode all the numbers between 0 and 3 – 0 (00), 1 (01), 2 (10) and 3 (11). Similarly, the numbers from 0 to 7 can be represented with 3 bits, while 0–127 takes 7 bits. In general, *n* bits can encode numbers from 0 to (2^n – 1).

'Next time you create a ZIP file to minimize the size of the attachments you send with an email, you can thank Claude Shannon – the compression relies on exactly his principle of information entropy.'

Shannon came up with two ways to improve the efficiency by which information could be transmitted as a stream of bits. The first was called 'source coding'. This involved removing as much redundant information as possible from a signal before it was sent. For example, imagine transmitting the results of tossing a coin 100 times. The result of each coin toss can be represented by a bit – let's say 0 = tails and 1 = heads. For an evenly weighted coin, the results could be transmitted as a stream of 100 bits. Shannon then asked how many bits would be required if the coin was biased, finding that it was much less. To see why, consider a very biased coin that only comes up heads one time in a hundred. On average, in 100 flips there will be one head in a long sequence of tails. Specifying its position will require a number between 1 and 100, which takes 7 bits (6 bits only gives 0–63, which is not enough). Therefore, the information content of the signal is reduced from 100 bits to just 7 bits in the case of the biased coin.

Information entropy

Shannon came up with a formula for the minimum amount of information needed to transmit a message. He called this quantity 'information entropy', after noticing that the formula was similar to the formula for calculating entropy in thermodynamics (see page 138).

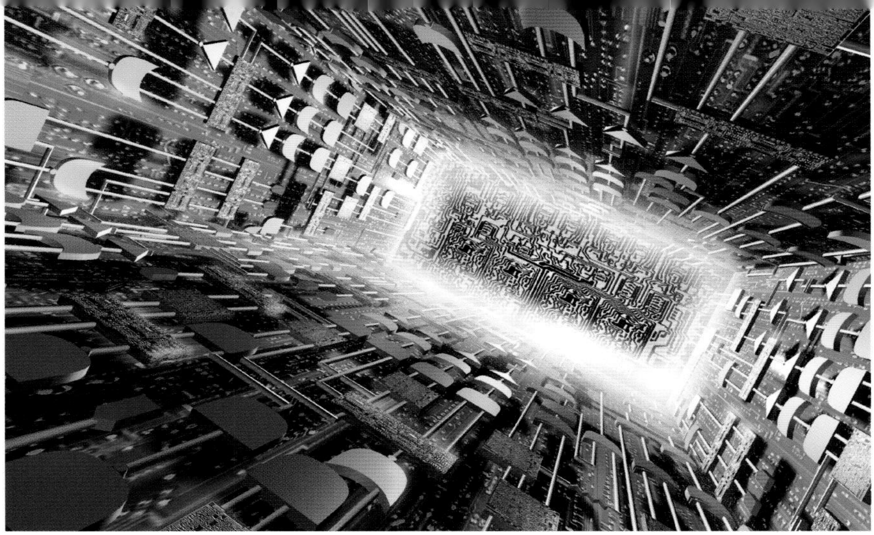

LEFT Computer artwork of electrical components joined together to form an electronic circuit. The coloured symbols represent logic gates that process signals. These gates deliver particular outputs based on certain inputs. Claude Shannon, the father of information theory, used logic gates to design telephone exchanges.

Next time you create a ZIP file to minimize the size of the attachments you send with an email, you can thank Claude Shannon – the compression relies on exactly his principle of information entropy. Creating a ZIP file is a 'lossless' form of source coding, because what you get out when you decompress the file is exactly what went in. Nothing is lost. Other kinds of compression are 'lossy' – they actually throw away information that is deemed not to be important. Audio MP3s are an example of lossy compression, and work by removing audio frequencies that are beyond the range of human hearing. An MP3 is typically about one-eleventh the size of the same song at CD quality.

Channel coding

Shannon also tried to develop ways of sending data over noisy communication channels – and this was his theory of 'channel coding'. Whereas source coding involves removing as much redundancy as possible, channel coding reintroduced a small degree of redundancy to gauge the amount of noise. For example, by splicing a known sequence of bits in amongst the signal, it is possible for the receiver to tell how that known sequence has been scrambled by the noise, and to then correct the actual message accordingly.

This is the basis of modern error correction codes. They enable you to have a conversation over a badly distorted mobile-phone connection, to play a scratched CD or to scan the barcode on a crumpled packet of food at the supermarket checkout.

Shannon's work has proved crucial to the development of modern communications and computer technology. Today, information theory is essential in all electronic communication and is also making its presence felt in other areas of science. Researchers suggest that it plays a role in the way genes organize inside biological cells and may even establish a link between the physics of black holes and the large-scale structure of our universe.

74 Quantum electrodynamics

DEFINITION A QUANTUM THEORY OF THE ELECTROMAGNETIC FIELD AND ITS INTERACTION WITH ELECTRICALLY CHARGED PARTICLES

DISCOVERY FORMULATED IN A WORKABLE FORM BY SIN-ITIRO TOMONAGA, JULIAN SCHWINGER AND RICHARD FEYNMAN IN 1948

KEY BREAKTHROUGH FIXED VARIOUS PROBLEMS THAT HAD BEEN PLAGUING THE THEORY IN ITS ORIGINAL FORM

IMPORTANCE MOST ACCURATE THEORY IN PHYSICS AND THE BLUEPRINT FOR OTHER 'QUANTUM FIELD THEORIES'

Quantum electrodynamics is the quantum theory of the electromagnetic field. An early version of the theory was developed by Paul Dirac and colleagues in the 1920s and 1930s, but it turned out to have fatal flaws. In 1948, physicists Sin-Itiro Tomonaga, Julian Schwinger and Richard Feynman reformulated the theory, removing the flaws and creating what many regard as the greatest theory ever devised in physics.

In 1928, English physicist Paul Dirac applied Einstein's theory of special relativity to Schrödinger's quantum-wave equation for particles to yield a mathematical expression describing the motion of electrons travelling at close to the speed of light. That was, itself, a momentous achievement.

But Dirac's new theory was to bring some unexpected discoveries. It accounted naturally for the phenomenon of quantum spin – which describes the rotational properties of quantum particles. And it predicted the existence of antimatter, particles with opposite fundamental properties such as electric charge. The first antimatter particle – the positron, or anti-electron – was discovered just a few years later.

Dirac and other giants of quantum theory, such as Wolfgang Pauli and Werner Heisenberg, developed the equation further. They showed that Dirac's theory not only described fast-moving electrons but also encapsulated an accurate quantum description of light. Light, of course, was just an electromagnetic wave. And so in general terms, the Dirac equation was offering a quantum mechanical description of the electromagnetic field and its interaction with charged particles such as the electron.

LEFT Artwork showing high-energy subatomic particles smashing into one another. Quantum electrodynamics (QED) was one of the first theories for predicting what happens during such collisions. QED models the interaction between electrically charged particles and electromagnetic waves – for example, when particles of matter and antimatter annihilate, leaving only radiation.

Force carriers

It was already known that electromagnetic waves could also be thought of as particles – called photons. According to the Dirac equation, these particles were also the quanta of the electromagnetic field, used to represent the tiniest increments of field energy. When two electrically charged particles interacted with one another, the motion of each is influenced by the electric field generated by the other. And, according to the Dirac equation, they are doing this by the mutual exchange of photons. Photons came to be thought of as the 'force carriers' of the electromagnetic field.

Dirac and his colleagues modelled all this using mathematical entities called creation and annihilation operators – also governed by the Dirac equation – to describe the appearance and disappearance of photons throughout space, which in turn accounted for the undulations of the electromagnetic field. They had constructed a quantum theory of electromagnetism and electric charge, which became known as quantum electrodynamics.

Long-hand solutions

It soon became clear, however, that all was not well with the theory. Because it was so complicated, solving the equations mathematically was not always possible. Instead, physicists would perform a 'perturbative expansion' – writing the solution as a long series of mathematical terms. The idea was that the dominant contribution should come from the first few terms in the series – the later terms, being negligible in size, could be discarded to a good approximation.

'This revised quantum electrodynamics is widely regarded by physicists now as the most accurate physical theory ever constructed, with experimental tests verifying its predictions to an astonishing 29 decimal places.'

At least, that was the idea. In quantum electrodynamics as it stood, the later terms in the perturbative expansion not only got bigger but actually became infinite in some cases. And there was worse news to come. Improved experimental technology was beginning to show up cases where the predictions of the theory blatantly disagreed with observations. For example, fine-scale microwave observations were revealing a small shift in the energy levels of the hydrogen atom (now called the 'Lamb shift') that quantum electrodynamics was unable to account for.

A resolution to these anomalies came in 1948, with work by Sin-Itiro Tomonaga (Tokyo University of Education), Julian Schwinger (Harvard University) and Richard Feynman (Cornell University), all working independently. Tomonaga and Schwinger had an idea to remove the 'divergences' – the way perturbative expansions gave infinite answers. Called renormalization, it was a process for dumping infinite terms in the expansion. Some regarded it as a fudge: Feynman himself called it 'hocus pocus', but the plain fact was that it worked. More recently, physicists have warmed to renormalization – finding a way to interpret it physically as a kind of variation with particle energy of the parameters governing the theory.

Feynman's main contribution to the story was his 'path integral' formulation. This is a way to calculate the probability of, say, an electron travelling from A to B in the presence of an electromagnetic field by adding up the probability of it getting there by all possible paths. It was a complex procedure, though Feynman, who was known for cracking intractable problems by thinking in picture form, offered an ingenious tool – Feynman diagrams. These were a way of showing graphically all the possible paths that a particle could take. Each diagram has a unique mathematical representation, so the calculation involved simply noting down the relevant diagrams and adding up the relevant mathematical terms.

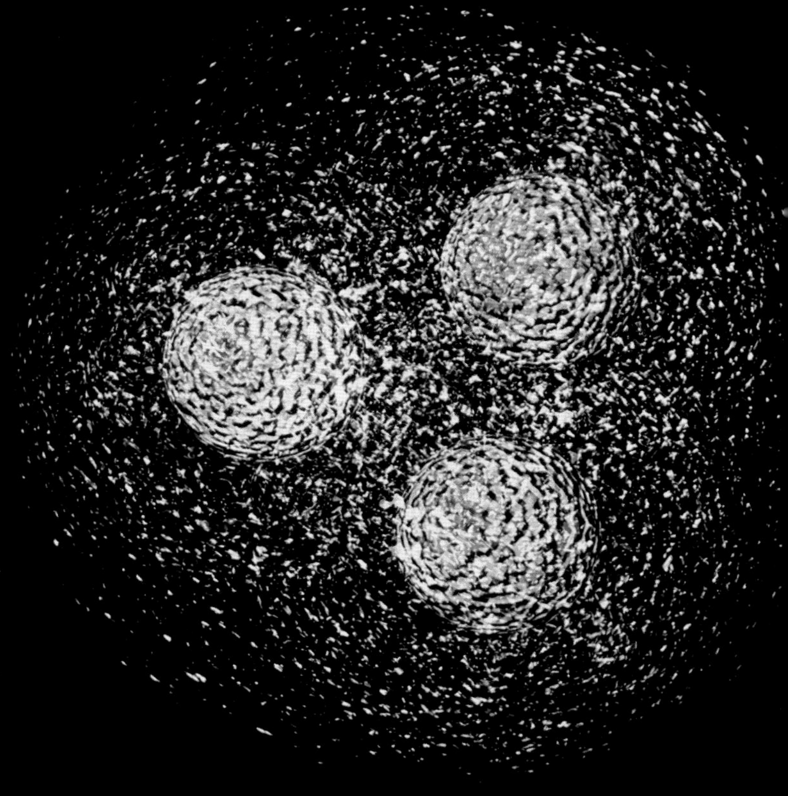

ABOVE Internal structure of a proton particle, made up of three smaller quarks: two 'up' quarks (blue-green) and a 'down' quark (red). The behaviour of quarks is governed by a branch of quantum theory called quantum chromodynamics.

The resulting reboot of quantum electrodynamics was a resounding success. The theory resolved the issue of divergences and corrected inaccuracies such as the Lamb shift. Indeed, this revised quantum electrodynamics is widely regarded as the most accurate physical theory ever constructed, with experimental tests verifying its predictions to an astonishing 29 decimal places. Tomonaga, Schwinger and Feynman received the 1965 Nobel Prize in Physics for their work.

Quark theory

The formulation set the standard for further theories, in particular quantum chromodynamics – a quantum theory of the 'strong force' that presides inside nuclei. It works by splitting the components of the nucleus (protons and neutrons) into smaller particles known as quarks, which possess a quantity known as 'colour' – analogous to electric charge in electrodynamics (nothing to do with colour in its everyday sense). Colour charges generate fields of particles called gluons, which are the force carriers of the theory – like photons in quantum electrodynamics. Quarks, and the other predictions of quantum chromodynamics, have since been verified in particle accelerators.

In the 1970s, quantum electrodynamics was unified with the weak force, which operates alongside the strong interaction inside atomic nuclei, to form the electroweak theory. This theory's predictions have also been verified (with the exception of Higgs boson, see page 333) and together, electroweak and quantum chromodynamics form the standard model of particle physics.

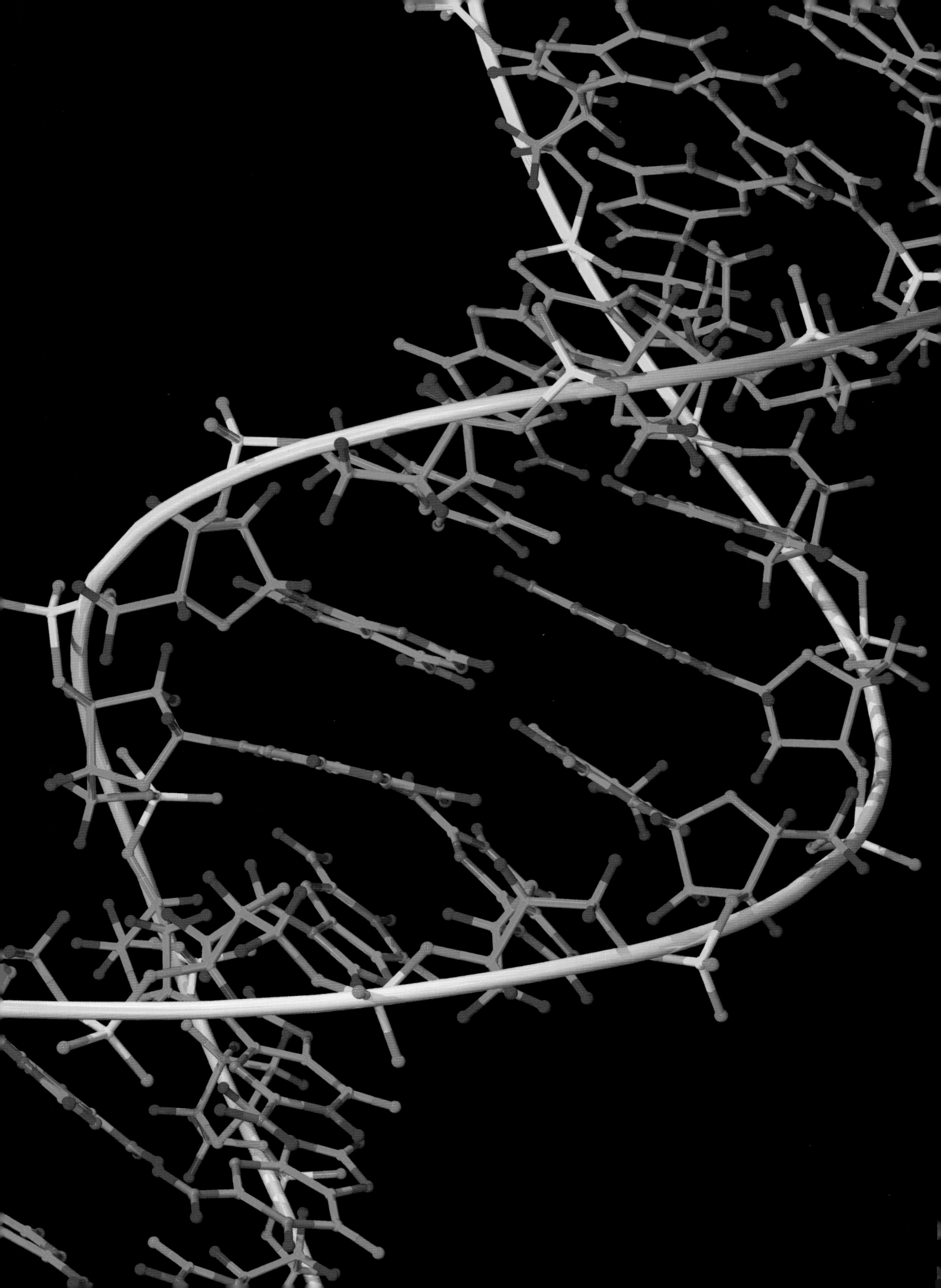

75 The double helix

DEFINITION THE FUNDAMENTAL MOLECULAR STRUCTURE OF DNA CONSISTS OF TWO ENTWINED STRANDS

DISCOVERY DEDUCED BY JAMES WATSON AND FRANCIS CRICK, AT THE UNIVERSITY OF CAMBRIDGE, IN 1953

KEY BREAKTHROUGH X-RAY IMAGES OF DNA MOLECULES HINTED AT A HELICAL STRUCTURE; WATSON AND CRICK FILLED IN THE DETAILS

IMPORTANCE UNDERSTANDING THE STRUCTURE OF DNA, AND HOW IT WORKS, LAID THE FOUNDATIONS FOR MOLECULAR BIOLOGY

Nine years after Oswald Avery's discovery that DNA is the carrier of genes, James Watson and Francis Crick took the revolution to the next level. They unpicked the structure of DNA and, ultimately, found the mechanism by which it turns genes – units of biological data that encode the traits and characteristics of life forms – into living, breathing organisms.

It is often said that a picture is worth a thousand words. One day in 1953, during a visit to King's College London, American molecular biologist James Watson would have agreed with this statement. He had just been shown an X-ray image of a DNA molecule, taken by biological physicist Rosalind Franklin. As Watson stated in his 1968 book *The Double Helix*, 'The instant I saw the picture, my mouth fell open and my pulse began to race.'

Watson and his English colleague Francis Crick, working at the University of Cambridge's Cavendish Laboratory, had been trying to deduce the structure of DNA (deoxyribonucleic acid). This followed ground-breaking work by the American scientist Oswald Avery and his colleagues who, in 1944, had proved that DNA provides the route by which genetic information is transmitted (see page 289). The quest was now on to find out what DNA looked like – and how it really worked.

LEFT Double helix structure of DNA, consisting of two long strands of sugar phosphates (pale blue). Stretched between these are pairs of nucleotide bases consisting of the molecules guanine, cytosine, thymine and adenine – the order of which encodes the body's genetic data.

List of ingredients

DNA had been already discovered inside the nuclei of biological cells in 1868. Further research had revealed that it is made up of four base chemical components – thymine, guanine, adenine and cytosine, usually labelled by their first initials, T, G, A and C. These chemicals, known as nucleotides, are

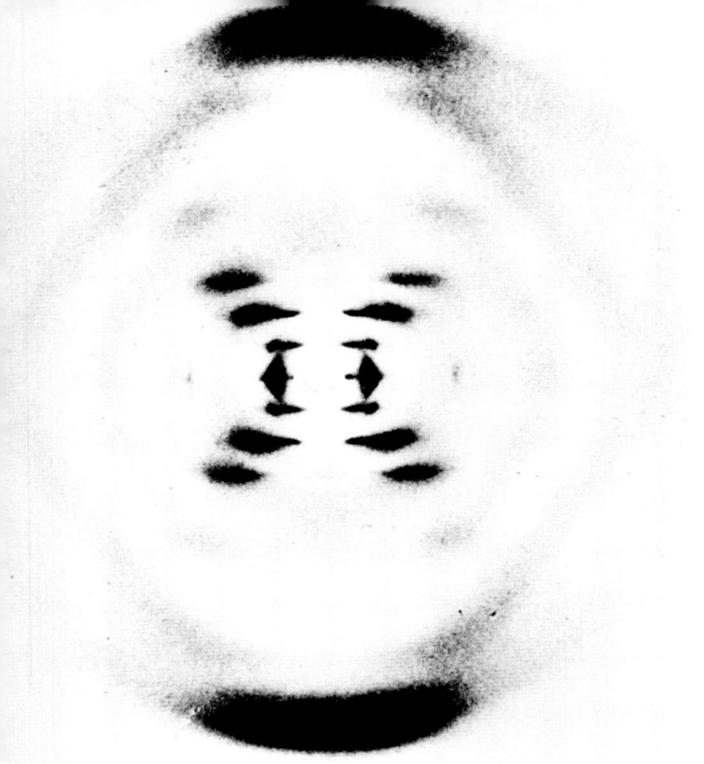

ABOVE X-ray diffraction image taken by Rosalind Franklin. This was one of the biggest clues in unpicking the structure of DNA. As a side-on view, it shows the two outer phosphate strands of the double helix crossing over.

made up from oxygen, hydrogen, nitrogen and carbon, bonded together in different combinations.

One of the first clues about DNA's physical structure came from the biochemist Erwin Chargaff, based at Columbia University, New York. He carried out chemical tests revealing that within the DNA of any organism the quantity of A and T were about equal, as were the quantities C and G. For example, A and T each make up about 30 percent of human DNA, while C and G each make up 20 percent. It was as if the A and T bases, and the C and G bases, were paired up with one another. This became known as Chagraff's rule.

But how did this pairing go to make up the overall shape of the DNA molecule? That is where Franklin's X-ray image came in. She had been making crystals from DNA and shining X-rays through them to probe the crystal structure. Just as ordinary light shone through a narrow grating spreads out, or 'diffracts', to produce a characteristic pattern, so X-rays, with their much higher frequency, are diffracted by the much narrower grating formed by the regularly spaced molecules inside a crystal. And the patterns produced by this X-ray diffraction can be reverse-engineered to produce images of the molecules themselves.

Photo 51

That is exactly what Franklin had done for DNA. And her X-ray diffraction picture, called 'Photo 51', showed the molecule forming an X-shape. There was already speculation that DNA would have a helical, 'corkscrew' structure – the American chemist Linus Pauling had shown that this was the case with some proteins. Photo 51 showed that in DNA the structure is a double helix with two long molecular strands entwined around one another. What the photograph clearly showed was a side-on view of the two strands, making an X-shape where they appeared to cross over.

In addition, Photo 51 showed the bases paired-up across the two helices, linking them together rather like rungs in a long, twisted ladder. The pairings obeyed Chagraff's rule. So, for example, opposite a T base on one helix would be an A base on the other. Similarly, a C base on one helix corresponded to a G base on the other. As a direct result of Photo 51, Watson and Crick were able to build the first accurate model of DNA's molecular structure. The base pairs formed the rungs linking the two helices, while the helices themselves were made of alternating molecules of phosphate and deoxyribose sugar. Franklin's X-ray picture revealed that the radius of the DNA helix was about 1 Angstrom unit, 1 ten-millionth of a millimetre. Watson and Crick's model measured almost 2 metres (2.2 yards) high.

Their findings were published in the journal *Nature* on 25 April 1953. In recognition of the discovery, Watson, Crick and Maurice Wilkins (who collaborated with Rosalind Franklin on her X-ray diffraction work) received the 1962 Nobel Prize for Medicine or Physiology. Franklin had died from ovarian cancer in 1958 – and Nobel prizes are not awarded posthumously.

DNA's code

Crick went on to further develop the model of DNA and its function. In 1958, he set out how DNA encodes genetic information in the sequence of bases along its structure. Each digit in the genetic code of an organism is written as a sequence of three successive bases, called a codon. So a codon of TAG on one helical strand – corresponding to ATC on the other, because of Chagraff's rule – represents a different genetic 'number' than, say, CGA (or GCT on the other strand).

Crick deduced how DNA is able to replicate itself, copying the genetic sequence and passing it on to new cells. It does this by unzipping the double helix, so that new bases can then attach themselves to the two single strands (again, in accordance with Chagraff's rule) to create two new double helices.

'Watson and Crick were able to build the first accurate model of DNA's molecular structure. The base pairs formed the rungs linking the two helices, while the helices themselves were made of alternating molecules of phosphate and deoxyribose sugar.'

Protein provider

Crick also worked out how DNA makes proteins, the main components of cells. It does this using a related molecule called RNA (ribonucleic acid). Also made of long chains of nucleotides, RNA is able to take an imprint of the DNA in a cell's nucleus in a process called 'transcription'. It then ferries this copy, rather like a fax, out of the nucleus and into protein-making factories called ribosomes in the outer portion of the cell. Here, the process of 'translation' takes place, in which the RNA code is read off and its sequence used to bolt together amino acids – the building blocks of proteins.

The discovery of the double helix structure of DNA, and Crick's follow-up work on its function ushered in a new field of science: molecular biology. This is the study of how the appearance, physiology and even behaviour of organisms are determined by the structure of their DNA. Molecular biology is arguably the most important area of biological science. It has led to gene therapies that promise treatments for incurable illnesses, including HIV. Meanwhile, genetic engineering has yielded crops that are both pest-resistant and more nutritious. And in 2010, researchers even created a new form of life – by literally engineering a custom-designed DNA molecule from scratch.

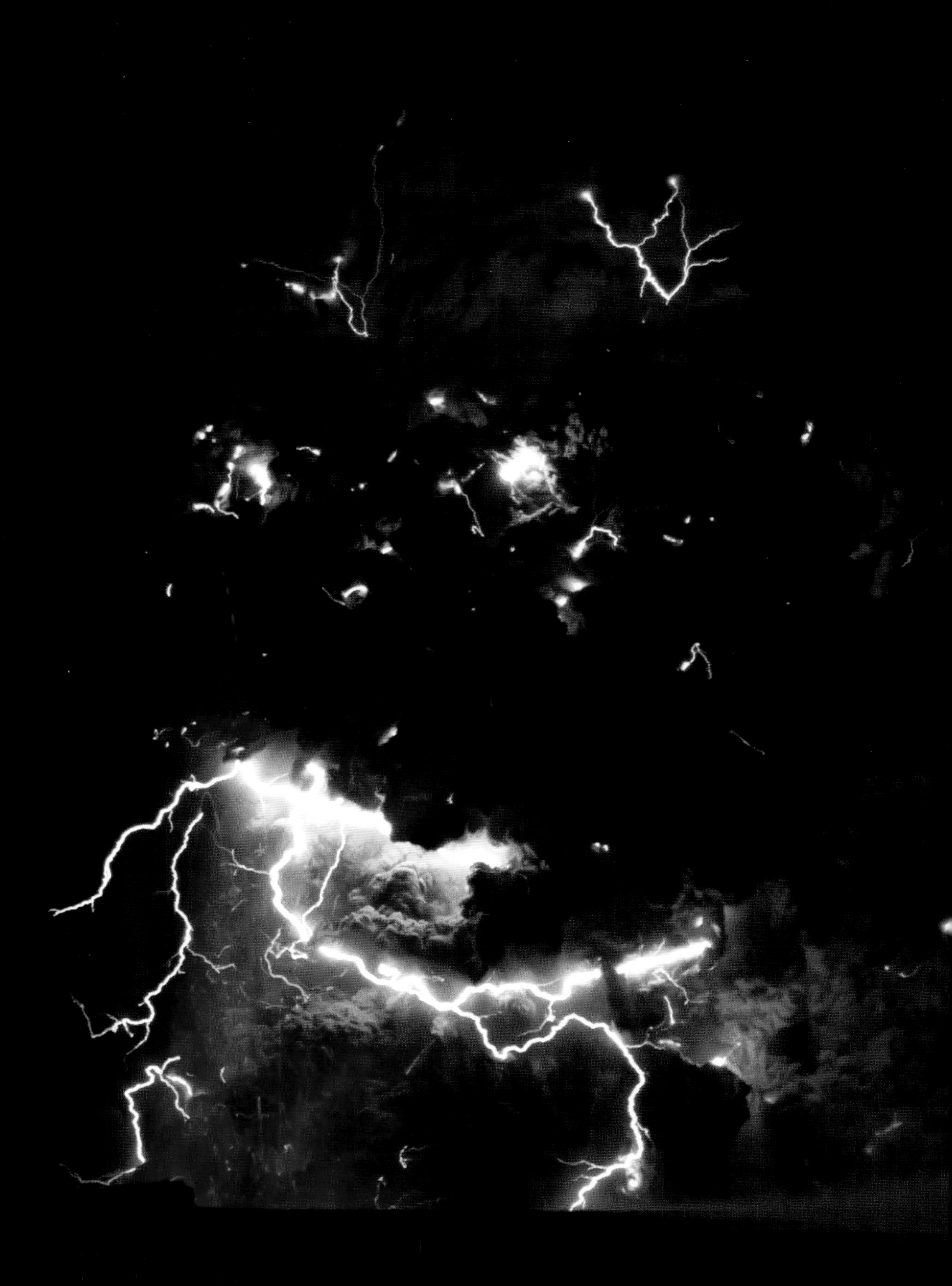

76 Miller-Urey experiment

DEFINITION AN EXPERIMENT TO DEMONSTRATE HOW ORGANIC COMPOUNDS FORMED ON THE YOUNG EARTH

DISCOVERY EXPERIMENT WAS FIRST CARRIED OUT IN 1953 AT THE UNIVERSITY OF CHICAGO BY STANLEY MILLER AND HAROLD UREY

KEY BREAKTHROUGH AFTER ONE WEEK, THE EXPERIMENT HAD CREATED AMINO ACIDS – LIFE'S BUILDING BLOCKS

IMPORTANCE EXTENDED DARWIN'S THEORY OF EVOLUTION, SHEDDING LIGHT ON THE ORIGIN OF LIFE ON EARTH

About 3.5 billion years ago, something special happened on planet Earth – the first life appeared. No one knows for sure how it happened, but in 1953 two chemists came up with a tantalizing clue. They zapped chemicals thought to have existed in the atmosphere of the early Earth with electricity to simulate the action of lightning bolts. The result was amino acids – the chemicals from which all life is constructed.

By the mid-20th century, thanks to new discoveries in genetics, Charles Darwin's theory of the evolution of life by natural selection was well established. It was accepted that new species were created as old ones adapted to new environments. Take a colony of tree-climbing monkeys and introduce them to an open savannah and, over the course of many generations, they will lose their tree-climbing traits and acquire new ones that optimize them to their new surroundings.

In this way, the evolutionary tree of life could be recreated, tracing each species back to the species that went before it, and then to the species before that and so on, until . . . until what? There inevitably comes a point at which there is no species that went before – just raw chemicals which, for one reason or another, combined to create the first-ever life forms on planet Earth. But how?

LEFT Volcanic lightning on the Eyjafjallajökull volcano in Iceland, 2010. The lightning occurs when electric charge builds up on rising clouds of ash. If results from the Miller-Urey experiment are correct, this process could have prompted the generation of life on Earth.

In 1922, the Russian biochemist Alekandr Oparin put forward a theory for the origin of life. He believed that life was a fairly unremarkable biochemical process. New astronomical observations had revealed the presence of methane and other gases in the atmospheres of the giant planets, such as Jupiter, and this led Oparin to believe that these gases would have been common on the early Earth. He theorized that in the harsh conditions presented by the young planet's volcanism and electrical storms, these elements could have been readily cooked up into the precursor chemicals for life.

Amazingly, it was many years before anyone thought to test Oparin's ideas. But in 1953, American chemists Harold Urey and Stanley Miller at the University of Chicago conducted a landmark experiment that supported Oparin's theory as to how life on Earth first got going.

Recipe for life

Miller and Urey mixed up a combination of water, methane, ammonia and hydrogen. Following Oparin's original brief, they deliberately did not include oxygen, which, judging from the atmospheres of the giant planets, he believed would have been absent. Next, the researchers heated their mixture to produce a vapour, that was representative of the atmospheric conditions on the early Earth.

The gases rose up from the vessel in which they were heated and passed along a glass tube into another vessel – a spark chamber. Here, high-voltage electric current was used to create arcs between two electrodes, simulating the effects of electrical storms in the atmosphere of our embryonic world. Finally, the vapour leaving the spark chamber was fed through a condenser, turning it back into liquid, which was then piped back into the heating vessel so that the process could be cycled over and over.

'After running their apparatus for one week, Miller and Urey found the liquid was full of organic compounds, including amino acids which are essential in creating proteins, a major component of all life.'

What they found was incredible. After running their apparatus for just one week, Miller and Urey found the liquid was now full of complex organic compounds, including amino acids which are essential in creating proteins, a major component of all life on Earth. Although life itself had not been created in the experiment, the implication was that if these chemicals had been brewed up on the early Earth and allowed to slosh about in its oceans, then they would have combined in the right way in order for primitive life to have emerged.

DNA bases

The work served as an inspiration to other researchers. In 1961, Spanish biochemist Joan Oró repeated the experiment but added hydrogen cyanide to the list of ingredients – a chemical that has been detected in interstellar clouds, and thus could well have been present on the early Earth. Oró found that including hydrogen cyanide in the cocktail of chemicals led to the production of the molecule adenine. This is a component of both DNA and RNA – essential in storing, accessing and copying life's genetic code – and is used to build the molecule adenosine triphosphate (ATP), which is essential in transferring energy between cells.

LEFT The Murchison meteorite, recovered from the town of Murchison, Australia, 1969. Next to it is held a phial of material extracted from the meteorite and which is thought to be unchanged since its formation in a star that died long before the Sun was even born. The Murchison meteorite is also rich in amino acids, fuelling speculation that the building blocks of life may have come from outer space.

There have been criticisms levelled at the Miller-Urey experiment. Some scientists have suggested that the selected blend of chemicals was not representative of those on the early Earth. They argued that oxygen may have been present – and that this highly reactive atom could then have degraded the organic compounds. Others said that volcanic activity would have created additional gases: sulphur dioxide, hydrogen sulphide, carbon dioxide and nitrogen. However, in 2008, a team from the USA and Mexico conducted a modified version of the experiment to recreate conditions close to a volcanic eruption. Their experiment produced, if anything, even more organic compounds than the original result.

Life in space

Regardless of the atmospheric mix present on the early Earth, the mixture of gases in the Miller-Urey experiment is, as Oparin noted, found today on the other planets of the solar system. And so this raises the question: could the precursors for life exist on far-away worlds? It would seem so. Analysis of the Murchison meteorite – a rock from space that landed near the town of Murchison in Victoria, Australia, in 1969 – was found to contain more than 90 different types of amino acid, confirming beyond doubt that these chemicals are being manufactured elsewhere across the gulf of space.

Such findings fuel speculation that maybe life did not begin here on Earth after all, but – like the amino acids in the Murchison meteorite – formed elsewhere, and was then transported to our planet inside comets and space rocks. This is known as the 'panspermia' theory, first put forward by Swedish chemist Svante Arrhenius in 1903. If the theory is correct then we – and all other life forms on planet Earth – are ultimately descended from aliens.

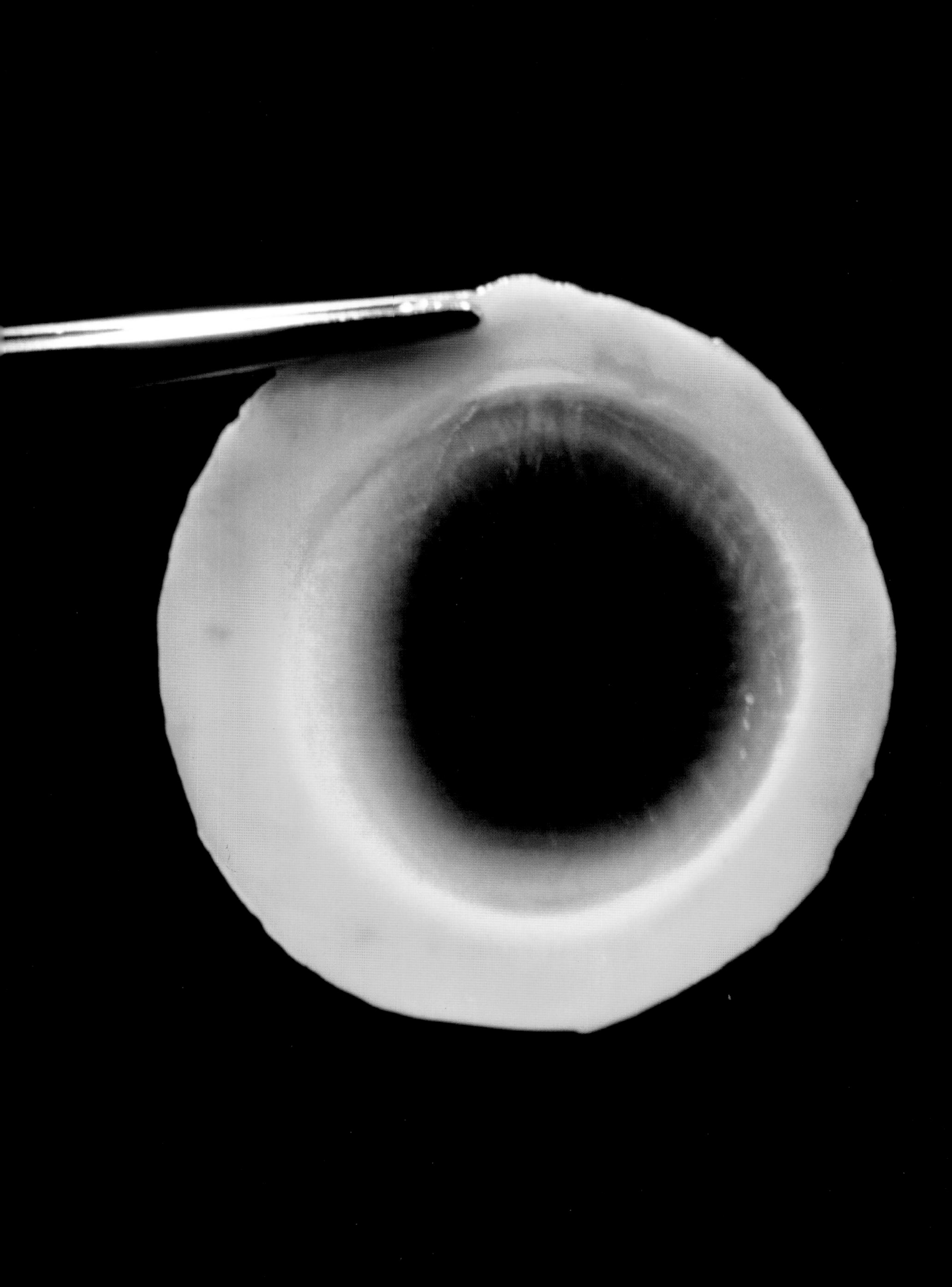

77 Transplant surgery

DEFINITION THE MEDICAL PRACTICE OF REPLACING DAMAGED OR DISEASED ORGANS AND TISSUES

DISCOVERY FIRST SUCCESSFUL TRANSPLANT BY JOSEPH MURRAY AT PETER BENT BRIGHAM HOSPITAL, MASSACHUSETTS, IN 1954

KEY BREAKTHROUGH THE TRANSPLANT WAS BETWEEN TWO IDENTICAL TWINS – THUS AVOIDING THE PROBLEMS OF ORGAN REJECTION

IMPORTANCE ORGAN TRANSPLANT SURGERY IS A LIFE-SAVING MEDICAL PROCEDURE

Transplant surgery was pioneered in the 1950s, with a kidney transplant between two twins. The twins were genetically identical – eliminating the risk of rejection by the recipient's immune system. In the 1980s, immune-suppressing drugs brought the benefits of the procedure to the masses. Today, doctors can grow transplant tissue from the recipient's own 'stem cells' – eliminating both the drugs and the need to find a suitable donor.

History is replete with accounts of medical procedures to transfer organs and tissue between one person's body and another. These run right back to the second century BC, when the Indian physician Sushruta is said to have used transplanted skin to reconstruct a patient's nose. But the authenticity of this and other ancient accounts is doubtful. Transplant surgery is a complicated field of medicine – and one that, even today, carries a high risk of failure.

The first formally documented example of a successful transplant operation between humans was a blood transfusion. It was performed in 1818 by English obstetrician James Blundell. But the treatment was not reliable until the early 20th century, with the discovery of blood groups that enabled blood to be accurately cross-matched between patients and donors.

The 20th century also saw the first experiments with tissue transplants. Austrian ophthalmologist Eduard Zirm performed the first transplant of a human cornea (the front covering of the eye) in 1905. And during the First World War, much progress was made in the development of skin-grafting techniques – repairing areas of skin damaged, typically, by burns injuries, using skin taken from another part of the patient's body.

LEFT Human cornea harvested from the eye of a deceased donor. Many transplants are rejected because tissue is attacked by the recipient's immune system. With this transplantation, however, the lack of blood flow to the cornea means there is no such danger.

But transplanting internal organs proved much harder. A replacement organ will normally have come from another human being (what doctors call an 'allotransplant'), unlike a skin graft that comes from the patient (an 'autotransplant'). This means that the transplanted organ is genetically different to the body into which it is implanted – something that the patient's immune system can often recognize. If this happens it will reject the organ as a foreign body – attacking it with antibodies. Cornea transplants are an exception to this because they are not integrated with the patient's blood supply, making it hard for the antibodies of the immune system to reach them. Organs, however, are integrated with the blood supply – and it was for this reason that all early attempts at human organ transplants failed.

Tissue types

The first successful human organ transplant operation took place in 1954, when American surgeon Joseph Murray transplanted a kidney between two identical twins. Because the twins were an exact genetic match, the transplant (referred to in this case as an 'isotransplant') ran no risk of rejection and was successful. The recipient, Richard Herrick, lived for a further eight years.

This was followed by numerous attempts to transplant a human heart – most of which failed within months of the procedure, mainly due to immune-system rejection. That began to change in the 1960s, with the development of 'tissue typing' (a way to assess the compatibility of patients and candidate donors), as well as effective immunosuppressant drugs – medications that damp down the body's immune response.

Blocking cytokines

The real breakthrough was an immunosuppressant drug called cyclosporin. It works by blocking the production of cytokines – chemical messenger molecules that are central to the functioning of the immune system. It was approved for clinical use in 1983 and greatly improved the survival chances of transplant patients. So much so that by the mid-1980s, surgeons were able to transplant the kidneys, heart, lungs, liver and pancreas. But an even more incredible possibility was just around the corner.

'Stem-cell therapy now offers the possibility of growing transplant organs and replacement tissues that are an exact genetic match to the patient – and so cannot be rejected.'

Limb transplants

In 1998, New Zealander Clint Hallam underwent transplant surgery to replace an entire hand that he had lost in a circular-saw accident. The entire operation, performed by the French surgeon Jean-Michel Dubernard, took a total of 13 hours. Although it was a clinical success and, with the help of immunosuppressant drugs, Hallam learned to use his new hand, he ultimately became unhappy with it and eventually requested that the hand be amputated – which it was in 2001. Several hand transplants have been performed since then, including a number of double procedures.

In 2005, transplant surgery stepped up to another new level with the completion of the first partial face-transplant operation. Frenchwoman Isabelle Dinoire had been mauled by her dog, losing the lower portion of her face. Surgeon Bernard Devauchelle was given permission to repair her injuries by grafting a triangle of skin taken from the face of a brain-dead donor – including the chin, lips and nose. The operation was a success, and Dinoire now has full use of her new face. Like all other transplant recipients, she must take immunosuppressants – though she has stated in interviews that one of the biggest difficulties was coming to terms with wearing a face that is not her own. The first full-face transplant was carried out by Spanish surgeons in March 2010.

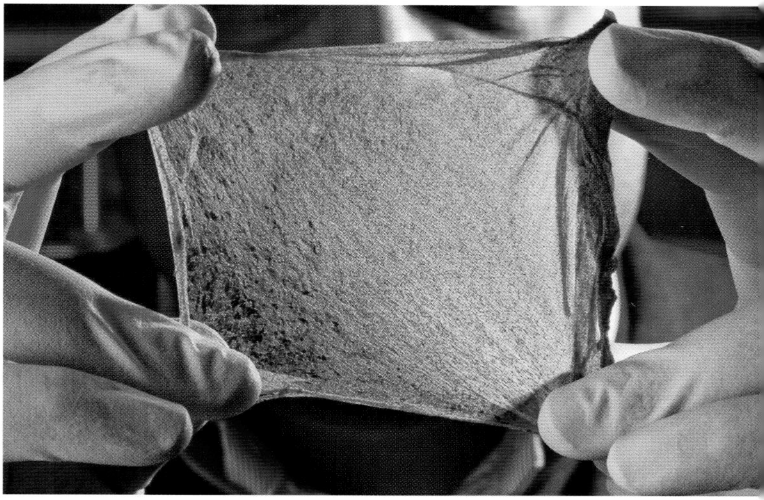

ABOVE Sample of bio-artificial tissue made by printing a mixture of living cells and a sticky polymer into a pre-prepared tissue scaffold. This technique, known as 'pressure-assisted spinning' could allow organs to be made to order, revolutionizing transplant surgery.

Bespoke organs

In the future, transplant patients may become less reliant on immunosuppressants – which at present they must take for the remainder of their lives. A new technique called 'stem-cell therapy' now offers the possibility of growing transplant organs and replacement tissues that are an exact genetic match to the patient – and so cannot be rejected. The procedure is controversial because – in some cases – it involves the creation of a cloned embryo of the patient from which 'embryonic stem cells' are then extracted – mutable cells that can be coaxed to grow into whatever tissue type is required (see page 389). Other types of stem-cell therapy use 'adult stem cells', extracted from the donor's bone marrow.

In 2008, it was announced that a 30-year-old Colombian woman had become the first patient to receive a transplant created from her own adult stem cells, when she was given a new section of windpipe to repair damage caused by the disease tuberculosis. Doctors removed a windpipe from a deceased donor and treated it with chemicals to destroy all the living cells, leaving just a scaffold made of collagen fibres. Into this they implanted the woman's stem cells, which grew to form a complete new section of windpipe that could then be surgically implanted.

Professor Martin Birchall, of the University of Bristol, one of the surgeons who performed the procedure, commented afterwards that in as little as 20 years it could be possible to manufacture almost any kind of rejection-proof transplant organ this way.

78 Lasers

DEFINITION A DEVICE THAT GENERATES MONOCHROMATIC AND COHERENT LIGHT BASED ON STIMULATED EMISSION

DISCOVERY THE PRINCIPLES OF THE LASER WERE ESTABLISHED BY AMERICAN PHYSICIST CHARLES TOWNES IN 1957

KEY BREAKTHROUGH TOWNES ADAPTED A MICROWAVE AMPLIFIER TO MAKE IT OPERATE AT VISIBLE LIGHT WAVELENGTHS

IMPORTANCE INNUMERABLE APPLICATIONS, INCLUDING FIBRE-OPTIC COMMUNICATIONS AND MEDICINE

If it was not for lasers, we would not have had music CDs, nor any movies on DVD or Blu-Ray. And high-speed, fibre-optic communication lines, responsible for high-speed internet access, would also be impossible. The good news is lasers are here to stay, having been invented in 1957 by American physicist Charles Townes.

The foundation for laser science was laid down in 1917 by Albert Einstein. He was studying a new theory of the atom, which had been put forward by his colleague, the Danish physicist Niels Bohr. In Bohr's theory, the positive charge of the atom was concentrated at the atom's centre – a region called the nucleus – while electrons orbited the outside. But the electrons did not just orbit anywhere: they only occupy certain, well-defined energy levels.

The model explains why atoms give off light only at specific wavelengths – and this has long been known to scientists working in the field of spectroscopy (see page 149). The idea was that electrons can jump down from a high-energy level to a lower-energy one, in the process giving off a photon (a particle of light) with an energy equal to the energy gap between the levels. Because the wavelength of a photon is uniquely specified by its energy, the set of characteristic gaps between an atom's energy levels specify a unique set of characteristic wavelengths at which the atom can give off light.

LEFT Laser interferometer, used to study airflow around objects placed in a wind tunnel. Half of the laser beam is allowed to pass through the moving air in the tunnel, while the other half is kept intact. When the beams are recombined they produce a visible interference pattern, the form of which depends sensitively on how the beam in the tunnel is affected by the swirling air.

To test Bohr's theory, Einstein decided to use it to try and re-derive Max Planck's law of radiation, which predicted the wavelength of the light given off by hot objects (see 'Photons', page 201). Planck's law was known to be correct, but Einstein could only extract it from Bohr's model if he made an additional assumption. He had to assume that an electron in a higher-energy state can be stimulated to drop down and emit a photon – with the characteristic wavelength determined by the gap to the energy level below – if a photon with that same wavelength happened to pass by. So, for example, an atom with an electron that is poised to emit a photon of red light can be stimulated to do so by the passage of a red-light photon of the same wavelength. Einstein called this process 'stimulated emission', and it was confirmed experimentally in the 1920s.

Microwave power

But the idea would not be applied to build lasers until much later. During the Second World War, Charles Townes had been studying microwave radar sets. His investigations of the properties of microwaves led him to suspect that these electromagnetic waves could be especially useful for investigating the properties of materials via spectroscopy.

'Transmission speeds of up to 100 billion binary digits (bits) per second have been achieved using laser-based fibre optics – thousands of times quicker than conventional copper wires.'

After the war, Townes was appointed as assistant professor at Columbia University, New York, where he began to try out his ideas. Sure enough, he was able to obtain microwave spectra from molecules – in which gaps between energy levels of electrons gave off photons of microwave radiation. But then he had a novel idea. Having read Einstein's work into stimulated emission, he realized it would be possible to use the molecules at which he was looking to build a kind of microwave generator, in which each microwave photon produced would coax atoms to emit more identical photons. He named the device a 'maser' – an acronym, standing for microwave amplification by the stimulated emission of radiation.

Masers to lasers

In 1953, Townes and his students put the idea into operation and built a maser. By 1957, Townes and his colleague Arthur Schawlow began looking at a version of the maser that could work at optical frequencies. They published their findings on what, for a time, was known as the 'optical maser'. In 1959, Columbia University graduate student Gordon Gould came up with a new name – replacing 'microwave' in the acronym with the word 'light'. Gould had coined the term 'laser'. In 1960, Theodore Maiman, at Hughes Research Laboratories, California, built the world's first working laser. Although it was only able to generate pulses of laser light, another group soon designed one that could generate a continuous beam.

Modern lasers work using a cylindrical piece of material – called the lasing medium – with mirrors at either end. The medium is 'pumped' – energy is injected into it, either electrically or by using flash tubes, to raise electrons to higher energy levels. As the electrons start to drop back down, they give off photons of light at the characteristic wavelength of the electron gaps in the lasing medium used, which then bounce up and down the cylinder,

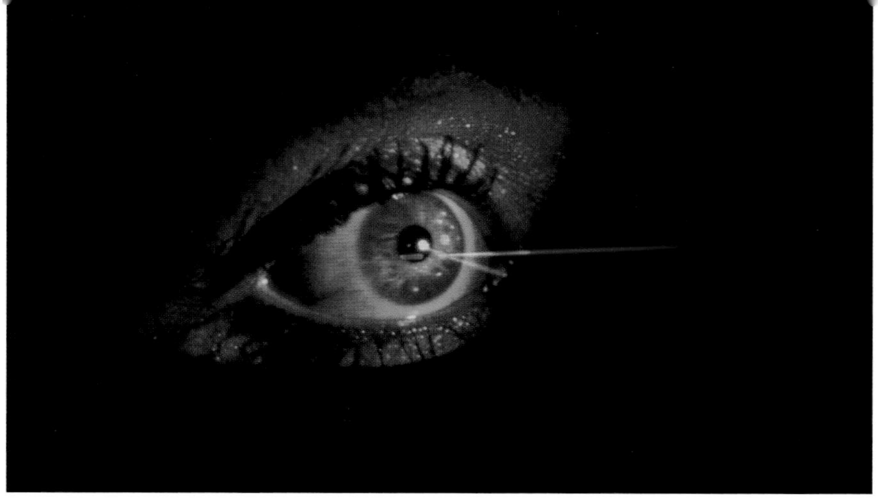

LEFT One common use of lasers is in refractive eye surgery. A laser beam is used to burn away abnormalities in the shape of the cornea, thus correcting sight defects and removing the need for corrective lenses.

stimulating other atoms to emit similar photons. One of the mirrors at the ends of the lasing medium is only partially silvered, allowing some of the light to escape – and this forms the laser's beam.

Light fantastic

Laser light is special because it is 'monochromatic' (all its photons are the same wavelength) and also 'coherent' – meaning that all its waves are vibrating in lockstep, with peaks and dips of the waveforms neatly aligned. Without such a high-quality light source there would be no optical media, such as DVDs. Lasers are an essential part of all fibre-optic communications, which encode data as pulses of laser light and send them down strands of plastic or glass. Transmission speeds of 100 billion binary digits (bits) per second have been achieved using laser-based fibre-optics – thousands of times quicker than copper wires. Lasers are also used in medicine – for example, in eye surgery and for breaking up kidney stones. And they are employed by astronomers to gauge the distortions caused by turbulence in the upper atmosphere, so they can subtract these distortions from images of the night sky.

Fire starter

Lasers could even play a role in nuclear fusion – the process of joining lightweight atomic nuclei to release energy. This process, which powers the Sun, is regarded as a potential energy source for planet Earth. The problem has been achieving the high ignition temperatures needed to start the process, typically tens of millions of degrees. Pulsed lasers could offer a solution, with the record for the highest peak power achieved by a pulsed laser standing at an enormous 1300 million million watts. That is nearly 100 times more than the average energy consumption rate of the entire planet (although it was only sustained for a fraction of a second).

The laser is one of those technologies that has changed everything – and will continue to do so. Accordingly, Charles Townes shared the 1964 Nobel Prize in Physics for his part in bringing this incredible device into existence.

79 Parallel universes

DEFINITION OTHER UNIVERSES – LIKE OURS BUT SUBTLY DIFFERENT; PREDICTED BY AN INTERPRETATION OF QUANTUM THEORY

DISCOVERY FIRST SUGGESTED FROM A SCIENTIFIC STANDPOINT BY AMERICAN PHYSICIST HUGH EVERETT IN 1957

KEY BREAKTHROUGH APPLYING QUANTUM THEORY TO MANY UNIVERSES NATURALLY EXPLAINED QUANTUM UNCERTAINTY

IMPORTANCE PARALLEL UNIVERSES SEEM TO BE CRUCIAL TO THE OPERATION OF SUPERFAST 'QUANTUM COMPUTERS'

Somewhere out there are universes where you are variously the Emperor of Japan, the winner of a reality TV show and the author of this book. Parallel universes are universes similar to our own, but where history unfolds differently. In the 1950s, scientists realized that quantum physics could be understood within a parallel universe framework. Here, particles can be in different places at the same time – just not in the same universe.

The notion of parallel universes has long been a feature of science fiction stories. But in 1957, physicist Hugh Everett at Princeton University brought the fiction to life when he put forward a new interpretation of quantum theory in which multiple universes are turned into a fundamental feature.

In the 1940s and 1950s, Richard Feynman had developed an approach to quantum theory known as 'path integrals' – where the probability of finding a quantum particle at any given point in space is calculated by adding up the probabilities of all the various routes it could take to get there. Everett's stroke of genius was to wonder whether these 'alternate histories' are not just possibilities, but actual realities. This led him to the idea of an infinite stack of parallel universes in which all possible histories are played out for real.

LEFT Artwork showing a funnel-shaped tunnel through space and time, linking two parallel universes. The idea of such a mechanism for passing between parallel worlds arose from stretching space and time in relativity theory to its limit.

Multiverses

This multitude of parallel worlds has since become known collectively as the 'multiverse'. Thus the 'many worlds interpretation' (MWI) of quantum theory was born. Weird quantum phenomena, such as 'uncertainty' – particles being in two places at the same time – can then be understood as a kind of interference between these parallel universes. Everett took a particle's wavefunction which, in previous interpretations of quantum theory, gave the probability of finding the particle at any point in space in our one universe – and broadened it to the multiverse. Stating that a particle has a 50 percent chance of detection at a particular point was equivalent to saying that it would be found at that point in 50 percent of all universes.

Until this time, physicists followed the Copenhagen interpretation of quantum theory that says a quantum particle behaves like a wave that is smeared out in space until it is measured, at which point the interference caused by the observer's meddling causes the wavefunction to collapse into a particle located at a definite point. Everett disliked the reliance on the observer in the process.

'Deutsch thinks parallel universes are the key to the operation of quantum computers – experimental computing machines that have already been built in labs around the world.'

In the MWI, the transition from waves to particles happens when interactions of the particle with its environment cause the parallel universes in the multiverse to 'peel apart', losing their quantum connection and forcing each universe to place the particle at a single definite location – a process called 'decoherence'. Quantum physics then becomes a branching process, with universes peeling away every time a quantum event takes place – to create subsets of universes in which every outcome eventually happens. Despite this neat explanation, many physicists have criticized the many worlds interpretation because it seems untestable. They say that it makes no predictions that distinguish it from other interpretations of quantum theory.

Quantum suicide

In the late 1990s, MIT physicist Max Tegmark came up with a thought experiment to illustrate how the MWI might actually be tested (though you would have to be convinced of the theory first to try it out). Called quantum suicide, it is a ghoulish twist on the Schrödinger's cat experiment (see page 241). An experimenter sits in a chair with a rifle aimed at his head. The rifle is hooked up to a radioactive source so that every second it will either fire a live round or click on an empty chamber, depending whether or not the radioactive source has spat out a particle in the preceding second.

Because the decay of the radioactive source is a quantum process, in the many worlds interpretation the state of the source in any one second of time is given by a quantum wave spread across all possible universes. Because the experimenter's final state at the end of that second is directly linked to the decay of the source, it is also described by a quantum wavefunction spanning the multiverse. When each second is up, in some universes the gun will click while in others it will fire – and the experimenter and their consciousness will either live or die, respectively. Tegmark argues that if the many worlds concept is correct then the experimenter must always find themselves in one of the universes in which they survive – because at the end of the second their consciousness will only exist in those universes.

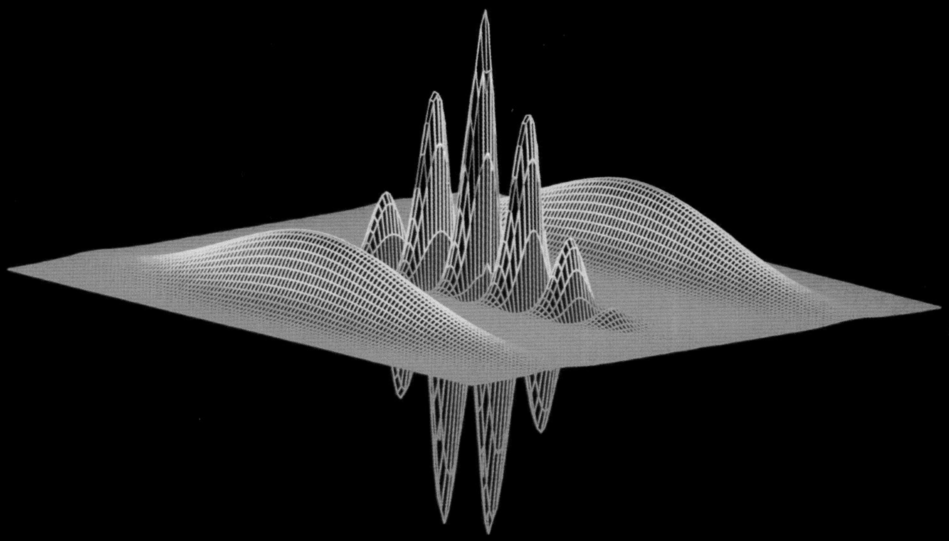

Representation of a bit and a qubit on a quantum computer. Quantum computers harness extra power from copies of themselves in parallel universes. Standard computers record data as bits (either 1 or 0), whereas quantum computers deal with qubits, which can be a mixture of 1 and 0 at the same time. Here, the two mounds are the possible 1 or 0 values of the bit. The peaks represent the merger of these two states to form a qubit.

In universes where the gun fires, their consciousness no longer exists and they do not see anything. Instead, as the seconds tick by, the experimenter sees the gun make an endless succession of clicks. But of course, only the experimenter can actually see this. Anyone else watching will see the gun click and fire at random – with ultimately unpleasant results.

Parallel computing

David Deutsch at the University of Oxford has pointed out more evidence for multiverses. He thinks parallel universes are the key to the operation of quantum computers – experimental machines that have already been built around the world. The computers that sit on our desktops today are 'classical' computers: they operate according to the laws of classical physics – storing data using the charge of electron particles, with definite 'charged' and 'not charged' states corresponding to the 1s and 0s of binary digits, or 'bits'. Deutsch realized that at the quantum level there are no definite states.

Just as particles can be in two places at the same time in quantum theory, so a bit of information can be both 1 and 0 at the same time. This is known as a quantum bit, or qubit. Performing a computation on a qubit effectively does the computation for 1 and 0 at the same time. And this is what a quantum computer does – making them potentially very fast processors. A quantum byte made of eight qubits can store 256 numbers all at the same time, which a quantum computer can process simultaneously. To operate so quickly, a quantum computer must simultaneously manipulate more bits of information than there are atoms in our one universe. Deutsch believes this is evidence that quantum computers must work by utilizing copies of themselves in other universes – supporting the many worlds interpretation of quantum theory. Otherwise, he asks, where is this vast mass of information getting processed and stored?

80 Chaos theory

DEFINITION EXTREME SENSITIVITY OF SOME MATHEMATICAL SYSTEMS, MAKING THEIR BEHAVIOUR SEEM RANDOM

DISCOVERY FIRST DEMONSTRATED BY EDWARD LORENZ IN 1963

KEY BREAKTHROUGH LORENZ'S COMPUTER MODEL OF CONVECTION IN WEATHER SYSTEMS SHOWED THAT TINY VARIATIONS IN INPUT BROUGHT VAST DIFFERENCES IN OUTPUT

IMPORTANCE CHAOS THEORY IS FOUND IN PHENOMENA RANGING FROM QUANTUM THEORY TO THE PHYSICS OF RELATIVITY

You set out for work in the morning but forget your wallet. You do not have enough money for your bus fare, you are late arriving at work and, as a result, you get the sack. It is an example of how small events can have large consequences – which is the essence of chaos theory. Chaos has a bearing on everything from quantum physics to economic models.

Chaos is the emergence of apparently random behaviour from well-defined systems of mathematical equations. It typically crops up in systems of equations that are nonlinear – that is, equations that involve squares, cubes and other higher powers. Chaos manifests itself as extreme sensitivity in the solutions of the equations to their initial conditions. So, for example, in a chaotic set of equations describing the weather you might expect to get wildly different results from starting conditions of 18°C and sunny compared with 17.5°C with scattered cloud.

The first person to realize that physical and mathematical systems can behave this way was the French mathematician Henri Poincaré. In the 1880s, he was studying solutions to Newton's law of universal gravitation (see page 157) describing three massive bodies moving in each other's mutual gravitational field. He found that – whereas for two bodies there exist well-defined closed orbits, and the bodies return periodically to their starting positions – in the three-body problem there is often no such periodicity.

Poincaré wondered about the long-term behaviour of such complex motion – realizing that reliable predictions could only be possible if the initial state of all three bodies could be measured to extremely high precision.

LEFT Flying monarch butterflies fill the air in Mexico – and who knows what effect their beating wings might have on the world's weather? In chaos theory, the butterfly effect describes the extreme sensitivity of chaotic systems, of which the weather is one. Edward Lorenz coined the phrase with his talk entitled, 'Does the flap of a butterfly's wings in Brazil set off a tornado in Texas?'

Throughout the first half of the 20th century, other mathematicians examined similar phenomena, where extreme complexity seemed to arise from seemingly innocuous maths and physics. But the theory they developed failed somehow to gain credence as a mainstream discipline, remaining an esoteric and obscure line of research until the 1960s.

'The equations of convection were so sensitive that the tiny differences in the input data were being amplified into a vast difference in the final results.'

It was the invention of computers that ultimately put chaos theory on the map. Their ability to churn out calculations to order could show anyone how dramatically the solutions to complex equations could be made to vary simply by changing their input conditions. Nowhere is this more true than in attempts to predict the weather. There is no such thing as a long-term weather forecast – we all know that predictions are only good for a few days, beyond which they become no more than random guessing. And in 1961, a meteorological researcher called Edward Lorenz got a first-hand demonstration of the reason why.

Born in 1917, in Connecticut, Lorenz first trained as a mathematician – earning his bachelor's degree from Dartmouth College in 1938 and then a master's from Harvard in 1940. During the Second World War, he put his mathematical ability to use working as a meteorologist for the United States Army. Meteorology turned out to be a subject he liked so much that after the war he returned to education to study it further, gaining a doctorate in the subject in 1948 from the Massachusetts Institute of Technology – where he then remained as a professor.

In 1963, Lorenz began a programme of research to study a phenomenon in physics known as convection. It is the process that makes hot air rise and is an essential part in the explanation of weather phenomena, such as hurricanes. Lorenz programmed the equations describing convection into a computer, and was in the process of investigating the solutions when he noticed something very weird.

Diverging solutions

If he took a print-out of the data from halfway through a run of his program and then fed this back in manually and started the program running again, he ended up with an answer that was very different to what he got when he left the program to run all the way through. He soon realized what the problem was. The computer was handling numbers to six decimal places but the print-out was automatically truncating these figures to just three. So, for instance, if the number in the computer was 0.568345 then the print-out would read 0.568.

Ordinarily, the difference between the two numbers – amounting to a few hundredths of a percent – should translate into a negligible difference between the two solutions. But the equations of convection were so sensitive that the tiny differences in the input data were being amplified into a vast difference in the final results.

Lorenz had provided the first concrete evidence for chaos theory, although it would not be named as such for another 12 years, when American mathematicians James Yorke and Tien-Yien Li used the term in an article published in 1975 in the journal *American Mathematical Monthly*. Lorenz did, however, come up with another term that captured the essence of the theory: the 'butterfly effect'. This is a metaphor to illustrate that a small change in one place can have large effects elsewhere. The equations describing the weather are so sensitive, reasoned Lorenz, that even a butterfly beating its wings can have a significant effect on the atmosphere that may ultimately alter the path of a tornado on the other side of the world.

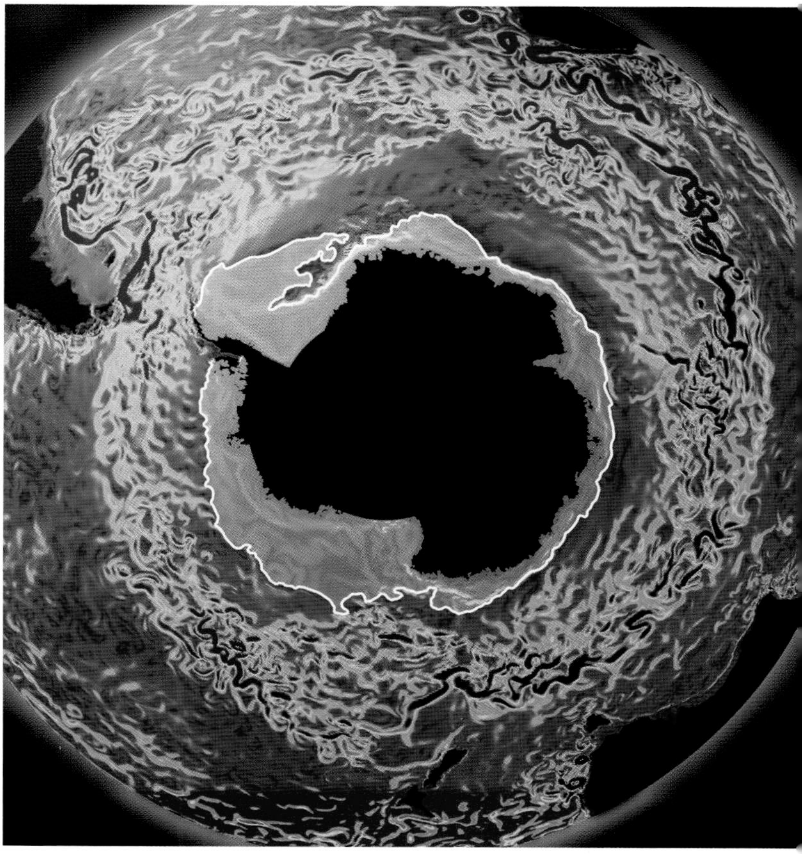

ABOVE Reconstruction of circumpolar ocean currents around Antarctica, created using satellite data, floating drifters and radio-tagged marine animals. Studies have found the currents to be chaotic – that is, their form cannot be predicted in advance.

Fractal geometry

Chaos theory gave birth to a related topic in geometry, known as 'fractals'. These are strange disjointed shapes that look the same regardless of what scale they are viewed at. The simplest fractal can be made by removing the middle third from a straight line, then repeating the process ad infinitum on all the remaining fragments. Zoom in on any fragment and you will see the same picture repeating itself.

Fractals are a representation of the dynamics of chaotic systems. Such dynamics are often analyzed using a geometrical plot known as phase portrait. A common example is the plot of velocity against position and, for well-behaved systems, the phase portrait usually gives a neat, well-ordered shape. For instance, a swinging pendulum has a position-velocity phase portrait that's just a circle. Chaotic systems, however, typically have fractal phase portraits. Indeed, when Edward Lorenz plotted a phase portrait for convection he found a fractal resembling a distorted figure '8'. Chaos is now a ubiquitous phenomenon in mathematical physics. It does not just arise in meteorology, but in turbulence, ocean currents, quantum theory, general relativity, financial models – and even the timing of a dripping tap.

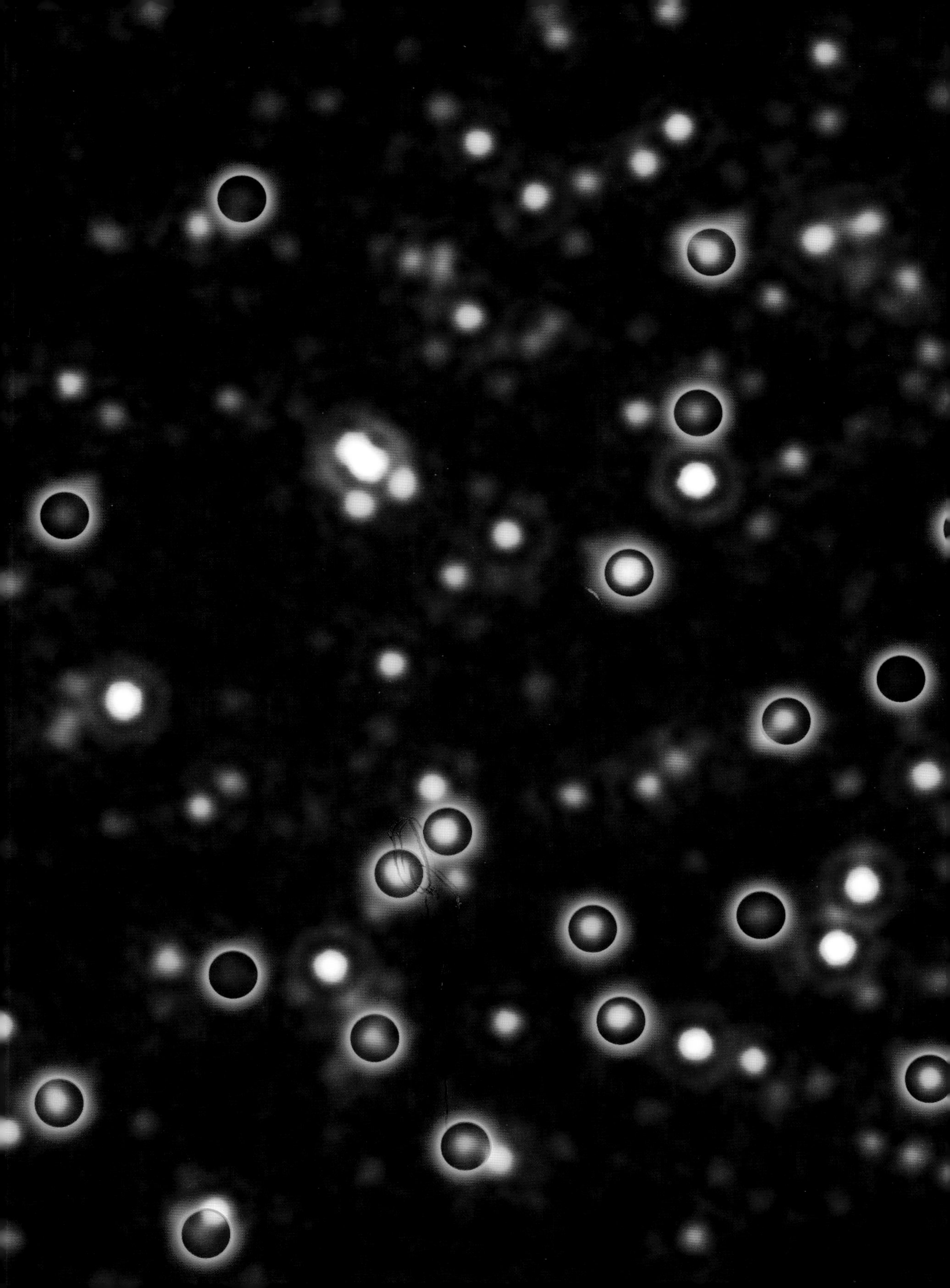

81 Quasars

DEFINITION FIERCELY BRIGHT GALAXIES LYING BILLIONS OF LIGHTYEARS ACROSS SPACE

DISCOVERY FIRST THOUGHT TO BE NEARBY STARS, THEIR TRUE NATURE WAS DEDUCED BY MAARTEN SCHMIDT IN 1963

KEY BREAKTHROUGH REALIZATION THAT PATTERNS IN THE SPECTRUM OF QUASAR LIGHT WERE CAUSED BY THEIR DISTANCE FROM EARTH

IMPORTANCE QUASARS ARE EVIDENCE THAT THE UNIVERSE IS EVOLVING AND THEY OFFER INSIGHTS INTO ITS EARLIEST MOMENTS

Quasars are cosmic record holders, being both the brightest and the most distant objects in the observable universe. A quasar can shine up to 100 times as brightly as a typical galaxy, and 1000 billion times brighter than a Sun-like star. This fierce luminosity makes them visible from enormous distances – often from billions of lightyears away.

During the 1950s, astronomers discovered hundreds of small, but bright radio sources on the night sky. But none of these mysterious objects were visible at optical wavelengths – or at least, if they were, they were very faint.

That changed in 1960, when American astronomer Allan Sandage obtained an optical image of a quasar known as 3C48 (the 48th object in the Third Cambridge Catalogue of astronomical radio sources), though, if anything, what he saw just deepened the mystery as to what a quasar was. No one had ever seen anything like this before. Sandage suspected it was a galaxy. However, rather than the fuzzy, smeared glow that normally accompanies such objects, the image revealed a tight point of light – more like a star.

Sandage was further baffled when he looked at 3C48's spectrum: a diagram made by splitting the light from the object into its constituent colours and then plotting the brightness of each colour. Common astronomical objects have characteristic spectra that are determined by their temperature and chemical composition and easily identified. But 3C48's spectrum looked unlike anything he had ever seen. The puzzle was solved in 1963 by Dutch astronomer Maarten Schmidt. He was using the powerful Hale telescope at the Mount Palomar Observatory, California, to study another of the

LEFT Quasars – shown here in orange – are fiercely bright galaxies at the edge of the observable universe. Quasars are powered by material releasing its energy as it falls into a supermassive black hole. Each of these black holes can weigh several billion solar masses.

quasars, called 3C273, when he noticed that some of the features in its spectrum resembled those produced by hydrogen gas. Only the features were not where they should be – they were shifted greatly towards the red end of the spectrum.

Galaxy redshift

Astronomers already knew that the light from distant galaxies gets redshifted as the galaxies are swept away from one another by the expansion of the universe, and that each galaxy's degree of redshift increases with its distance. But 3C273's redshift was so great that it had to be a whopping two billion lightyears away – so far away that its light, travelling at 300,000km/s (186,411 miles/s), must take two billion years to reach us.

Because 3C273 and the other mysterious radio sources looked star-like – or stellar – they were classified as 'quasi-stellar' objects – later shortened to just 'quasars'. The word was first used in an article that appeared in the journal *Physics Today* in 1964, by astrophysicist Hong-Yee Chiu.

For quasars to be visible at such distances they must be extremely bright. So what exactly was powering them? Soon after the first quasars were discovered, it was suggested that they could be hypothetical objects known as white holes: the opposite of a black hole. In the 1930s, Albert Einstein and his colleague Nathan Rosen had shown that the mathematical equations governing a black hole are symmetrical so that the distance from the hole, r, can quite validly be replaced by $-r$. This led Einstein and Rosen to speculate that both matter and energy can fall into a black hole but instead of being destroyed at $r = 0$ they could pass straight through and spew out on the other side – where r is negative.

'Because 3C273 and the other mysterious radio sources looked star-like – or stellar – they were classified as "quasi-stellar" objects, later shortened to just "quasars".'

During the 1970s, however, the theory of white holes fell from favour when a more plausible explanation emerged. Rather than resorting to such exotic objects, it seemed that quasar luminosity could be explained instead by black holes – objects in space for which there was now mounting evidence.

Quasar power

The black holes in question are enormous. It is now believed that each quasar is powered by a central 'supermassive' black hole – weighing hundreds of millions of times the mass of our Sun. The quasar's power comes from matter falling into the black hole – which gets squashed violently, emitting a dying scream of energy as it is heated to terrific temperatures. Such is the efficiency of this process that the hole only needs to swallow about ten stars a year to account for the quasar's enormous energy output. Each quasar is now known to be surrounded by a galaxy similar to our own Milky Way, and this is thought to be the source of the stars on which the quasar feeds.

While the surrounding galaxy spans roughly 100,000 lightyears, studies have shown that the quasar itself is usually just a few light-days across – about the same size as our solar system. Many quasars also eject material from their poles in two spectacular 'jets', reaching millions of lightyears into intergalactic space.

Images returned from the Earth-orbiting Hubble Space Telescope in the 1990s confirmed that supermassive black holes lie at the centre of most galaxies. Hubble's images showed stars near the centres of galaxies circling so fast, and on orbits so tight, that the central object cannot be anything but a black hole.

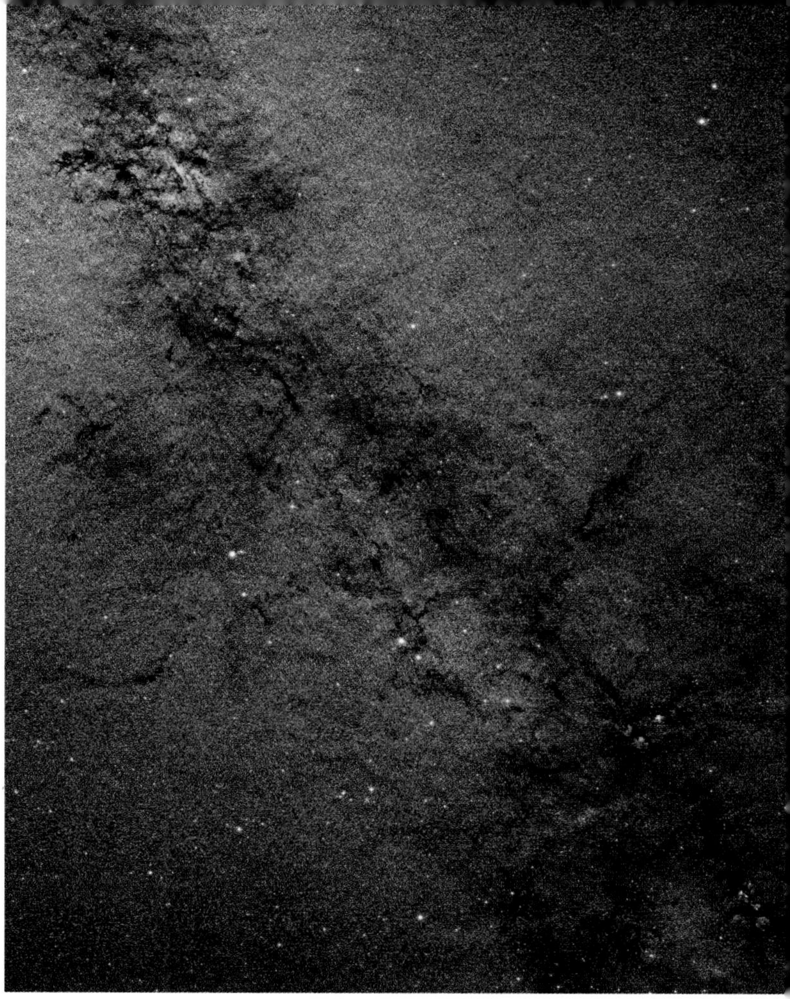

ABOVE The view towards the centre of our Milky Way: each pinprick of light is a distant star, while the black marks across the centre are caused by clouds of dust in the galaxy's mid-plane. The Milky Way is thought to have been a quasar in its youth. If so, it will still have a supermassive black hole in its core today.

Cosmic evolution

The supermassive, black-hole theory also explains another seemingly odd feature – namely, that quasars only seem to have existed when the universe was young. We know this because all the quasars that have been discovered so far are at a terrific distance from Earth and, because of look-back time (objects a lightyear away are seen as they were a year earlier), our observations of distant objects mean we are seeing them as they were billions of years ago.

This had been realized in the 1960s. And at the time, it was a vital piece of evidence for the Big Bang theory, supporting the view that the universe is evolving with time rather than remaining static – which was the thrust of the rival steady-state theory for the universe.

So where did the quasars all go? In the supermassive black-hole theory they simply burnt themselves out – running out of fuel to feed the ravenous black holes in their cores, at which point they reverted to being ordinary galaxies.

Today, there are some 200,000 known quasars and they reveal to astronomers what the universe was like billions of years ago. Even our own Milky Way is thought to contain a supermassive black hole – and probably started its life as one of these fiery beacons located at the edge of the primordial cosmos.

82 Higgs boson

DEFINITION A SUBATOMIC PARTICLE BELIEVED TO HAVE GIVEN MASS TO ALL OTHER PARTICLES OF MATTER

DISCOVERY THE PARTICLE WAS PROPOSED THEORETICALLY IN 1964 BY PETER HIGGS AT EDINBURGH UNIVERSITY

KEY BREAKTHROUGH REALIZATION THAT PARTICLE MASSES CAN BE EXPLAINED VIA THEIR INTERACTIONS WITH THE HIGGS BOSON

IMPORTANCE IT IS THE ONLY PARTICLE IN THE ACCEPTED STANDARD MODEL OF PARTICLE PHYSICS THAT IS YET TO BE DETECTED

It is not hard to see why it has been called the God particle. The Higgs boson is a hypothetical subatomic particle that, according to many physicists, is what gave all other particles of matter in the universe their mass. Put forward by Peter Higgs, it is the only element in the standard of model particle physics that is, as yet, undetected experimentally. The world's biggest and most powerful particle accelerators are working to rectify this.

Just as photons are the particles of the electromagnetic field, the Higgs boson is a particle associated with a field of energy known as the Higgs field. It was proposed as a way to explain the varied masses of subatomic particles. Peter Higgs was motivated by a branch of particle physics that sought to unify quantum electrodynamics (the quantum theory of the electromagnetic field, see page 301) with the weak nuclear force operating in atomic nuclei and which is responsible for certain types of radioactive decay (see page 185).

The photons that carry the electromagnetic force are known to be massless. But for the theory to work, the particles responsible for the weak force – which would later be called the W and Z – would need to be some of the most massive individual particles that are known, weighing more than entire atoms of iron. How could such wildly different masses for the force-carrying particles be incorporated into a single, unified mathematical theory?

LEFT Simulation of how scientists might detect a Higgs boson particle. Although itself short-lived, the Higgs boson decays into characteristic particle patterns – the detection of which infer its existence. Here, the Higgs boson has decayed into four muon particles, each of which is made up of two quarks.

Higgs mechanism

The 'Higgs mechanism' provided the answer. Peter Higgs constructed a theory whereby the interaction between particles of matter and a field of energy pervading space is what determines the particles' masses. The different masses of different particle species arise through different interaction strengths with the field. The Higgs field is a bit like a sea of sticky treacle through which all particles attempt to move. The motion of the particles that get snagged in the treacle becomes slow and sluggish, just like a particle with mass – these are the heavy particles. On the other hand, those particles that slip through the treacle, as if coated with nonstick Teflon, have low mass.

If that all sounds a bit contrived, then think again – a similar phenomenon has been observed in crystals. The nuclei of the atoms in a crystal form a positively charged rigid lattice. A negatively charged electron moving through it feels an electrical attraction to the lattice, which slows it down in just the same way as the Higgs field. Electrons in crystals have been seen to effectively increase their mass by a factor of 40 because of this effect.

'The Higgs field is a bit like a sea of sticky treacle through which all particles move. The motion of the particles that get snagged in the treacle become slow and sluggish, just like a particle with mass.'

In 1968, a working unified theory of the weak force and electromagnetism – called the electroweak theory – was put forward jointly by physicists Steven Weinberg, Abdus Salam and Sheldon Glashow. The W and Z particles that it predicted to carry the weak interaction were duly detected in particle-accelerator experiments and, in 1979, Glashow, Weinberg and Salam received the 1979 Nobel Prize in Physics for their work.

As anticipated, their theory required the Higgs mechanism to explain the different particle masses – but how could this aspect of the theory be tested? Peter Higgs knew that if the Higgs mechanism is correct then it should be possible to detect particles of the Higgs field.

Bosons and spin

The theory predicted that Higgs particles should take the form of 'bosons'. This is a specialist technical term taken from quantum theory, referring to a particle's spin – which is broadly analogous to spin in the everyday world but with a few quantum peculiarities. Each subatomic particle species has a fixed quantum spin, which can either be an integer (such as 0, 1, 2) or a half-integer (such as 1/2, 3/2, 5/2). Particles with half-integer spin are called fermions, while those with integer spin are known as bosons. The Higgs particle is predicted to have spin 0, and is therefore classed as a boson.

The electroweak theory also predicted that the Higgs boson should be very massive – around 140 times the mass of a proton. The usual way to detect subatomic particles is to manufacture them inside machines called particle accelerators, which whiz beams of particles around at close to light speed and then crash them into one another. The energy of the collision is converted into the mass of new particles (through Einstein's famous equation linking mass and energy, $E = mc^2$). These new particles then spew out from the collision site and can be picked up using particle detectors. However, this method can only be used to make particles with masses equal to or less than the collision energy. And it is only recently that particle accelerators have been built that can crank up their collision energy high enough to, theoretically, manufacture the hefty Higgs particle. The situation is complicated further by the fact that the Higgs boson cannot be seen or detected directly – but it should decay rapidly into characteristic patterns of other particles that can be picked up by particle detectors.

Higgs hunters and Higgs sceptics

Foremost among the projects to track down the particle is the Large Hadron Collider (LHC) at the CERN particle physics laboratory. After a fault was discovered when the accelerator was first switched on in 2008, it was shut down for an entire year while repairs were made. But by March 2010, operators had turned the collision energy up to half of the LHC's designed maximum. As yet, though, the elusive Higgs boson is still to show its face. Indeed, it is the only subatomic particle in the standard model of particle physics for which we do not have a single shred of evidence.

A minority of physicists are even suggesting that the particle may never be found. They believe that other effects could be responsible for the generation of mass – such as extra dimensions of space and time or even supersymmetry, which supposes that each boson particle has a hidden fermion partner, and vice versa. Other experts wonder if a Higgs-like particle could arise as a kind of composite made by bolting together other more familiar particles in nature. Whatever the outcome, the next few years promise to be a very exciting time indeed for experimental particle physics.

LEFT Professor Peter Higgs inspecting the ATLAS experiment, part of the Large Hadron Collider. ATLAS is one of two experiments that will try and detect the elusive particle. The Higgs boson was proposed by Peter Higgs in 1964.

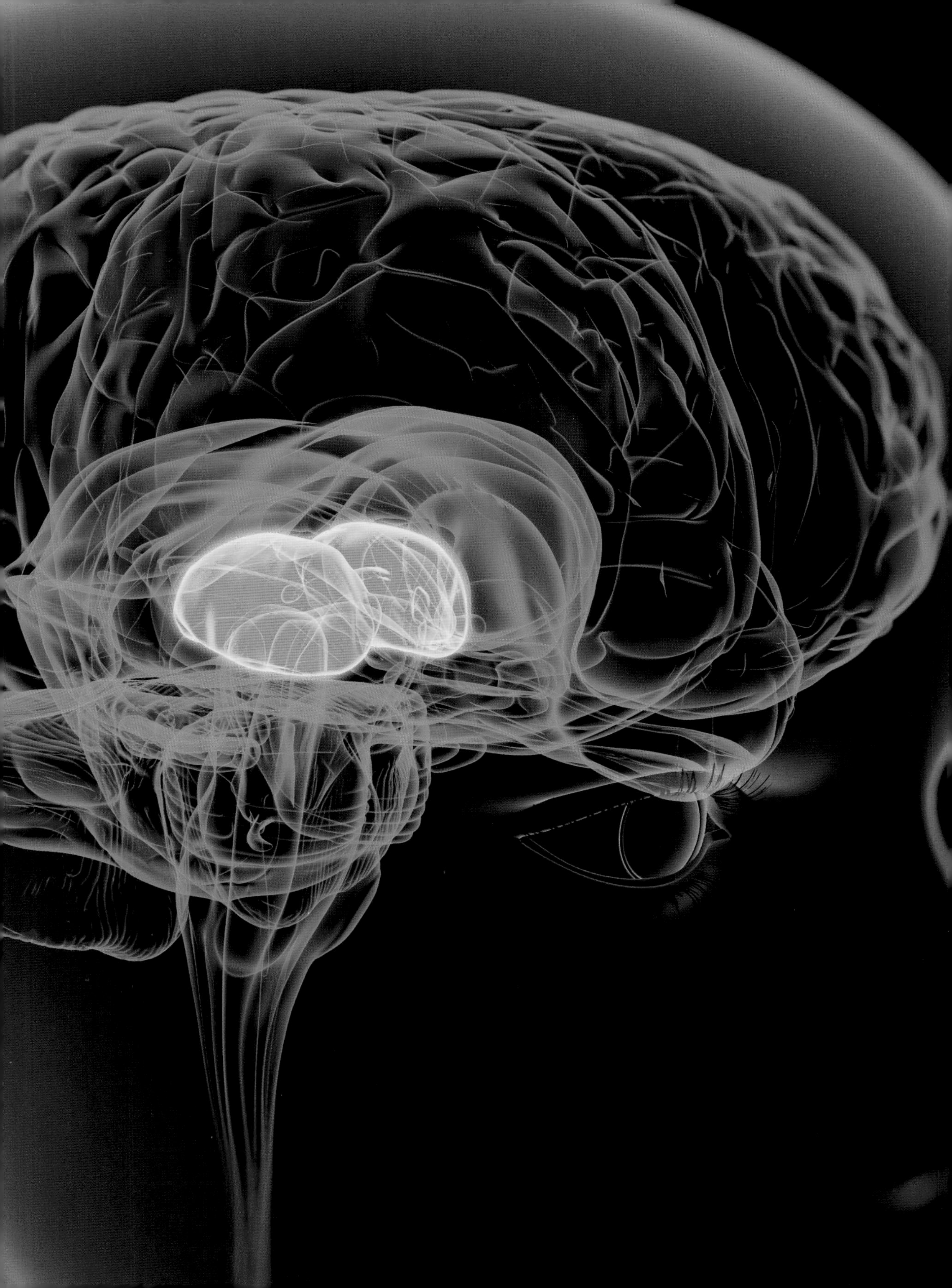

83 Cognitive psychology

DEFINITION A THEORY OF PSYCHOLOGY THAT PUTS THE EMPHASIS ON UNDERSTANDING THE THOUGHT PROCESSES INSIDE THE BRAIN

DISCOVERY FORMALLY ESTABLISHED BY AMERICAN PSYCHOLOGIST ULRIC NEISSER IN HIS 1967 BOOK OF THE SAME NAME

KEY BREAKTHROUGH APPLIED IDEAS FROM THE FIELD OF COMPUTING BY TREATING THE BRAIN AS AN INFORMATION PROCESSOR

IMPORTANCE THE MOST SCIENTIFIC FORM OF PSYCHOLOGY, IT HAS LED TO ALTERNATIVE METHODS OF PSYCHOTHERAPY

Cognitive psychology is a scientific theory of psychology that replaced Freud's psychoanalysis and the ideas of behaviourism. The theory focuses on thought processes – 'cognition' – by which it tries to understand the brain by treating it as a computer. Cognitive psychologists study the specifications of this computer and the software that each of us has running on it.

Psychology is about what goes on inside the mind. Perhaps one of the first psychologists was philosopher René Descartes, who mused over whether the mind and the body are made from the same stuff. He argued that they are not – but now we know he was wrong (otherwise, for example, anaesthetics injected into the body would not subdue the mind during surgery). One of the first self-proclaimed psychologists was Wilhelm Wundt. In 1879, he established at the University of Leipzig the world's first laboratory carrying out psychological experiments, thereby establishing psychology as a science.

Fall of Freud

In the early 20th century, two schools of psychological thought emerged – and promptly clashed. On one side was psychoanalysis, developed by Sigmund Freud. His central idea was that the mind is in a constant state of turmoil between the rational 'ego', the moralistic 'superego' and the mind's depraved dark side, called the 'id'. This battle for supremacy, Freud argued, occurs in the subconscious – though occasionally the id gets one past the ego, which is when we accidentally drop a 'Freudian slip' into conversation. Freud's ideas are criticized for being unscientific. He was a psychiatrist rather than a psychologist, and his research was based on patient anecdotes rather than scientific research. Nevertheless, he remains the father of psychology.

LEFT The human brain: the orange region is the thalamus, believed to help transfer signals between different brain areas. The relationship between these processes and a person's behaviour is examined through cognitive psychology.

In opposition to psychoanalysis was the behaviourist movement. In this view, all that mattered about a person's psyche was what could be measured and quantified – in other words, their behaviour. Behaviourist psychology boiled down to predicting observed behaviours with little regard for the thoughts, feelings and other internal processes of the mind.

Reward and punishment

Behaviourism began in 1913 with the work of the American psychologist John B. Watson. He argued that Ivan Pavlov's dog research – in which Pavlov famously trained dogs to salivate at the sound of a bell, by ringing it every time they were fed – could be applied to people. This became known as 'operant conditioning' and stated that learning is achieved through a regime of reward and punishment – with rewards emerging as the more effective of the two.

'Cognitive psychology was manifestly scientific in its approach, underlining that all reasoning is the product of processes in the brain, and the brain is simply a physical system operating according to scientific laws of biology, chemistry and physics.'

The scientific foundations of behaviourism led to its triumph over psychoanalysis, and it held sway until the 1950s. However, even this theory's days were numbered. Behaviourism put the emphasis on 'nurture' – that is, learning through our past experiences. But with the increasing understanding of genetics, it was clear that not all behaviours are nurtured. Many are given to us instead by 'nature' – that is, inherited from our parents. This was something for which behaviourism did not account.

This realization, accompanied by broadsides delivered against behaviourism by intellectual heavyweights, such as American philosopher Noam Chomsky, saw the theory replaced by new ideas. These new ideas took the emphasis off behaviour and focused on the thought processes inside the mind.

Computer consciousness

It was no coincidence that this shake-up of psychological reasoning took place shortly after the invention of computers. Alan Turing, who developed the theory underpinning modern computers, also investigated the idea of a computer that could mimic the learning ability of the human brain.

In some ways the revolution in psychology turned Turing's idea on its head, arguing that thinking boils down to information-processing inside the brain. In this view, the brain is like a computer, while the thought processes and personality are the software running on it. It was a radical departure from old ways of thinking, where actions were driven by desires and instincts – now they were the product of experiences, beliefs and reasoning.

The new psychology

In 1967, these ideas were pulled together into a book by the American psychologist Ulric Neisser. Its title finally gave a name to the revolution that

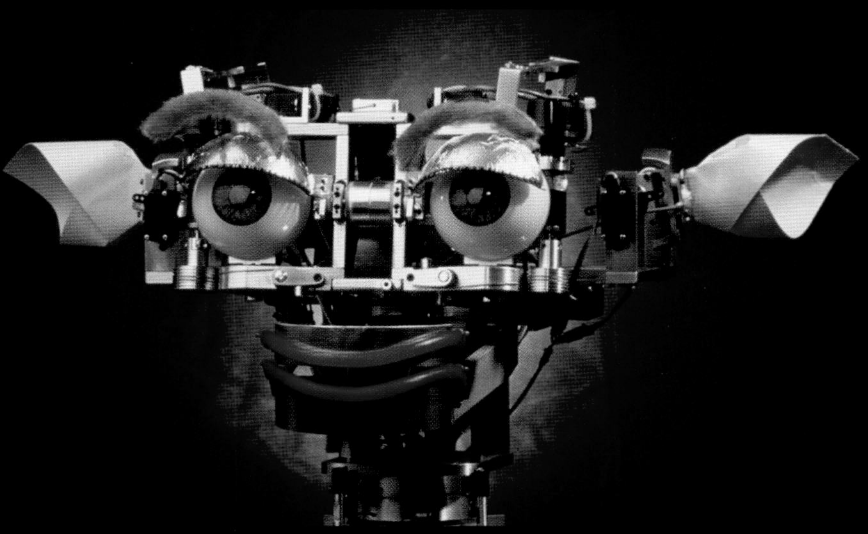

LEFT Kismet is a robot that can be programmed to adopt certain mood states, such as happiness, and express them through its humanoid features. Since cognitive psychology describes the mind in terms of data processing, this has led to a crossover between the discipline and research into artificial intelligence.

had been gathering momentum over the past decade – 'cognitive psychology', meaning psychology driven by thought processes, or 'cognition'. Neisser's book was instrumental in defining this field and made a compelling case to the psychology community. By the 1970s, cognitive psychology had replaced behaviourism as the primary model for understanding the mind.

Cognitive psychology was manifestly scientific in its approach, underlining that all reasoning is the product of processes in the brain, and the brain is simply a physical system operating according to scientific laws of biology, chemistry and physics. Indeed, most researchers agreed that it was the most scientific interpretation of psychology so far.

Cognitive science

The new picture of psychology was certainly more able to integrate with other scientific disciplines. Today, cognitive psychologists work alongside neuroscientists, philosophers, linguists and artificial intelligence computer programmers, with all of these scientific disciplines productively feeding into one another to form the broader field of 'cognitive science'.

Cognitive psychology has been applied clinically to great effect, with the development of cognitive behavioural therapy (CBT) as a treatment for mental-health issues, including depression, anxiety, eating disorders and addictions. CBT is a form of counselling that works by helping patients recognize the thought processes that are causing the problem and then helping them learn how to change those thought processes – in a way, reprogramming the defective part of the brain's software. CBT is considered to be one of the most effective treatments for depression and anxiety, assuming all other physical causes have been eliminated.

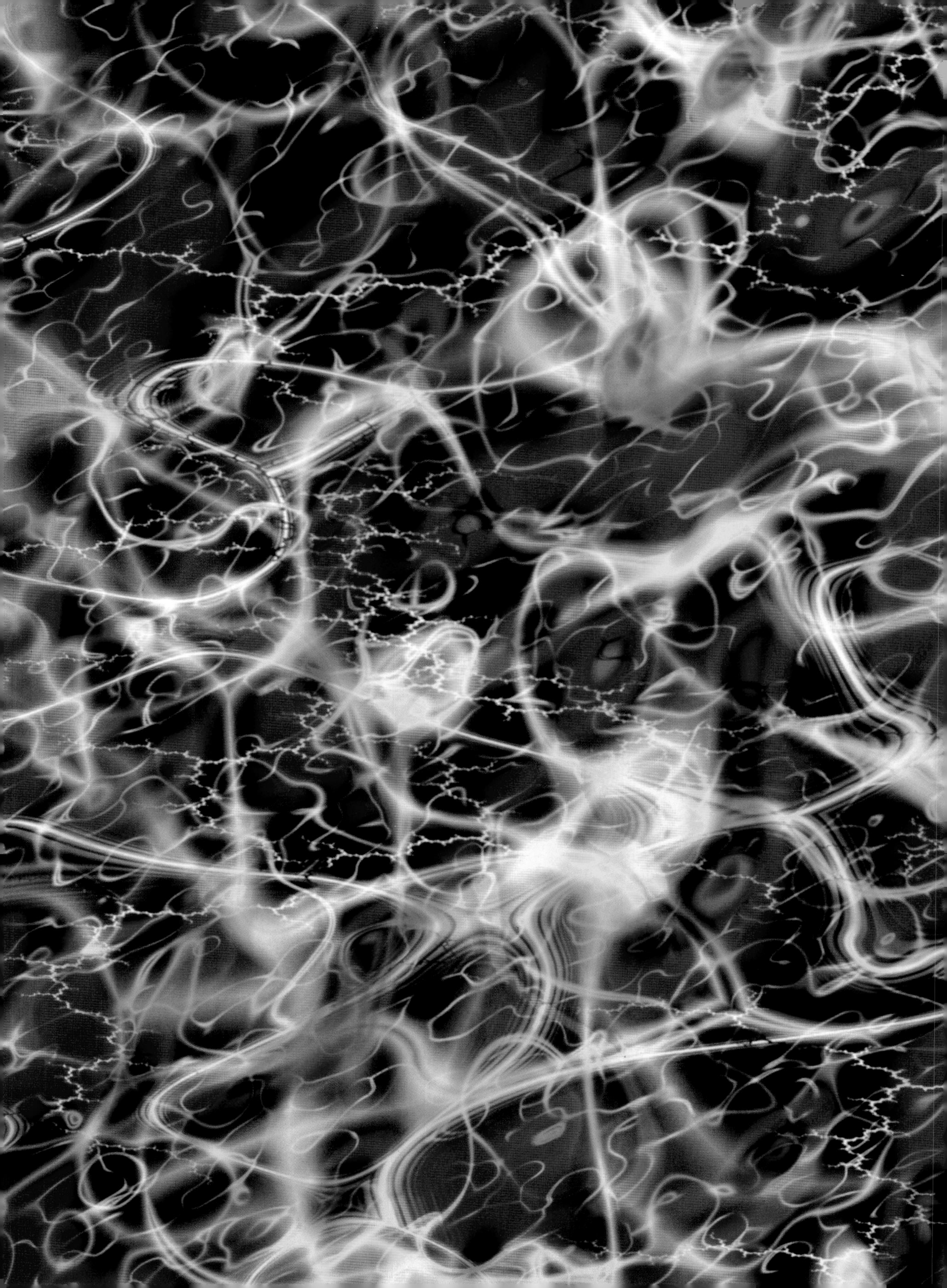

84 String theory

DEFINITION A THEORY OF PARTICLE PHYSICS IN WHICH POINT PARTICLES ARE REPLACED BY VIBRATING 'STRINGS' OF ENERGY

DISCOVERY IT AROSE FROM AN ATTEMPT TO EXPLAIN THE STRONG NUCLEAR FORCE BY GABRIELE VENEZIANO IN 1968

KEY BREAKTHROUGH PHYSICISTS REALIZED THAT VENEZIANO'S MATHS COULD BE INTERPRETED AS DESCRIBING VIBRATING STRINGS

IMPORTANCE STRING THEORY IS THE MOST PROMISING CANDIDATE FOR A UNIFIED THEORY OF THE FUNDAMENTAL FORCES OF NATURE

It sounds like a radical proposition – that the universe is made of string. But that is the idea behind the most promising theory we have to consolidate the fundamental forces of nature (gravity, electromagnetism and so on). String theory states that particles are made of tiny, one-dimensional loops of energy and that to accommodate them, space and time must have ten dimensions.

When it was published, Einstein's general theory of relativity was a revolutionary theory of the gravitational force. At the time, only one other force of nature was known – electromagnetism – and James Clerk Maxwell had delivered the definitive theory on that back in 1861. Electromagnetism had been formulated by unifying electricity and magnetism. Soon, Einstein and others became convinced that the next step for theoretical physics would be the discovery of a theory to unify general relativity and electromagnetism.

Unified field theory

Einstein referred to such a model as a 'unified field theory'. One candidate for this theory was offered in 1919 by German mathematician Theodor Kaluza. He added an extra dimension to the four-dimensional spacetime of relativity to formulate an equation describing five-dimensional spacetime that seemed to account for the behaviour of both electromagnetism and gravity. In 1926, Swedish physicist Oskar Klein devised a follow-up scheme explaining why we do not see the extra dimension, stating that it is curled up, or 'compactified' – in the same way that the two-dimensional surface of a rolled-up paper tube looks more one-dimensional the tighter it is rolled. The Kaluza–Klein theory was plagued by problems and eventually dropped. Einstein continued to search for the correct unified field theory, to no avail.

LEFT Superstrings are tiny, one-dimensional strands of energy from which subatomic particles, such as quarks and electrons, could be constructed. Superstring theory seeks to unify all the forces of nature into a single law of physics. Physicists are now seeking evident to support the theory.

Strength of strings

In the 1930s, two further forces of nature – in addition to gravity and electromagnetism – were discovered to complicate the situation. These were the strong and weak nuclear forces that represented interactions within the nuclei of atoms. It seemed with the addition of these two forces that a unified description of fundamental physics was further away than ever. But, in 1968, an attempt to better understand the strong force inside atomic nuclei led to perhaps the most promising unified theory that we have today. A 26-year-old Italian physicist called Gabriele Veneziano developed a theory of the strong interaction based on an approach to quantum physics developed in 1941 by Werner Heisenberg, known as S-matrix theory. A number of other physicists noticed that Veneziano's mathematics could be interpreted as describing vibrations as tiny, one-dimensional lengths of energy – or 'strings'.

And so 'string theory' was born – particles were represented as 'notes' played on a string. And like the Kaluza–Klein theory, strings required extra dimensions of space – ten or 26, depending on which version of the theory you used (though most now work with ten). The strings themselves were minute – so small that, as physicist Brian Greene put it, if an atom were as big as our solar system then a string would be the size of a tree. Veneziano's theory of the strong interaction was soon eclipsed by quantum chromodynamics (see page 303). However, interest in the theory lived on.

'String theory and its derivatives represent the most successful attempts to construct a unified picture of physics.'

Incompatible partners?

But what of the hunt for a unification theory? Because the weak and strong nuclear forces only existed in atomic nuclei, they were manifestly quantum entities. Physicists concluded that any theory embracing all four forces must therefore be quantum in nature. But there was a problem – Einstein's general relativity seemed to be utterly incompatible with quantum theory as it stood.

Attempts to quantize general relativity resulted in 'divergences', when the predicted values of physical quantities would blow up to infinity. Worse still, these divergences could not be removed by the process of renormalization used to fix the theory of quantum electrodynamics. Some physicists wondered whether divergences were arising from modelling particles as zero-dimensional points in space – when, in reality, they must have some degree of physical extent. And attention thus turned to string theory as a potential solution. In the 1980s, American physicist Edward Witten proved that replacing zero-dimensional particles with one-dimensional strings did indeed lead to a workable quantum theory of gravity with no divergences. The force of gravity would be mediated by the exchange of particles, called 'gravitons' – which, in turn, are a particular vibration mode of the strings. Similarly, the particles carrying the other forces – electromagnetism, weak and strong – correspond to different string states again, so forming a unified theory of all four interactions.

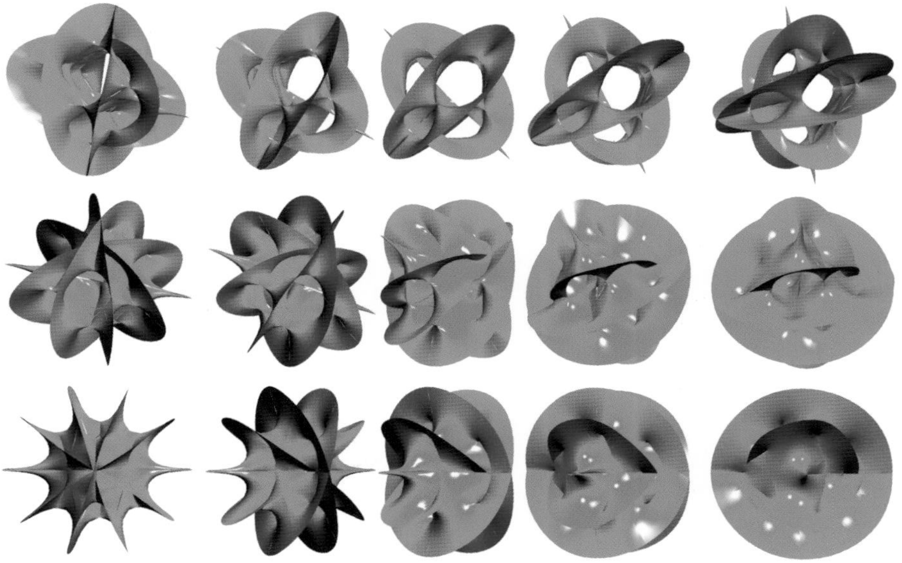

LEFT Selection of Calabi-Yau manifolds. These are complex configurations of space that possess six extra dimensions and so satisfy the requirements of string theory. These dimensions cannot be seen because they are curled up.

One significant problem with string theory as it stood was that it was not just one theory but several, depending how you specified the parameters. In the mid-1990s, Witten carefully pulled together these sub-theories into one all-encompassing model, and named it M-theory. Whereas strings were one-dimensional objects in ten dimensions, M-theory dealt with two-dimensional 'membranes' in a total of 11 dimensions. Strings could then be regarded as neat slices cut through these membranes (lowering the dimensionality by one), with the orientation of each slice defining the particular string-theory model at which you are looking.

Hard evidence

String theory, and its derivatives, represent the most successful attempts to construct a unified picture of physics. However, as yet there is no compelling evidence for it. Indeed, the theory is untestable at low energy. A particle accelerator 100,000 billion times more powerful than the Large Hadron Collider (LHC) would be needed to put string theory to the test. However, experiments are now being run at the LHC to investigate the possible existence of the extra dimensions required by string models. Researchers are looking out for tiny black holes created in particle collisions: the idea is that graviton particles could leak out of the hidden dimensions, boosting the force of gravity locally and so causing the black holes to form.

If the LHC is successful in finding evidence for these extra dimensions – which are curled up far too tightly for us to see ordinarily – it would be an incredible discovery in itself. But it would also give us the first evidence that physics is, in fact, unified, and that matter on the tiniest scales might just be made out of string.

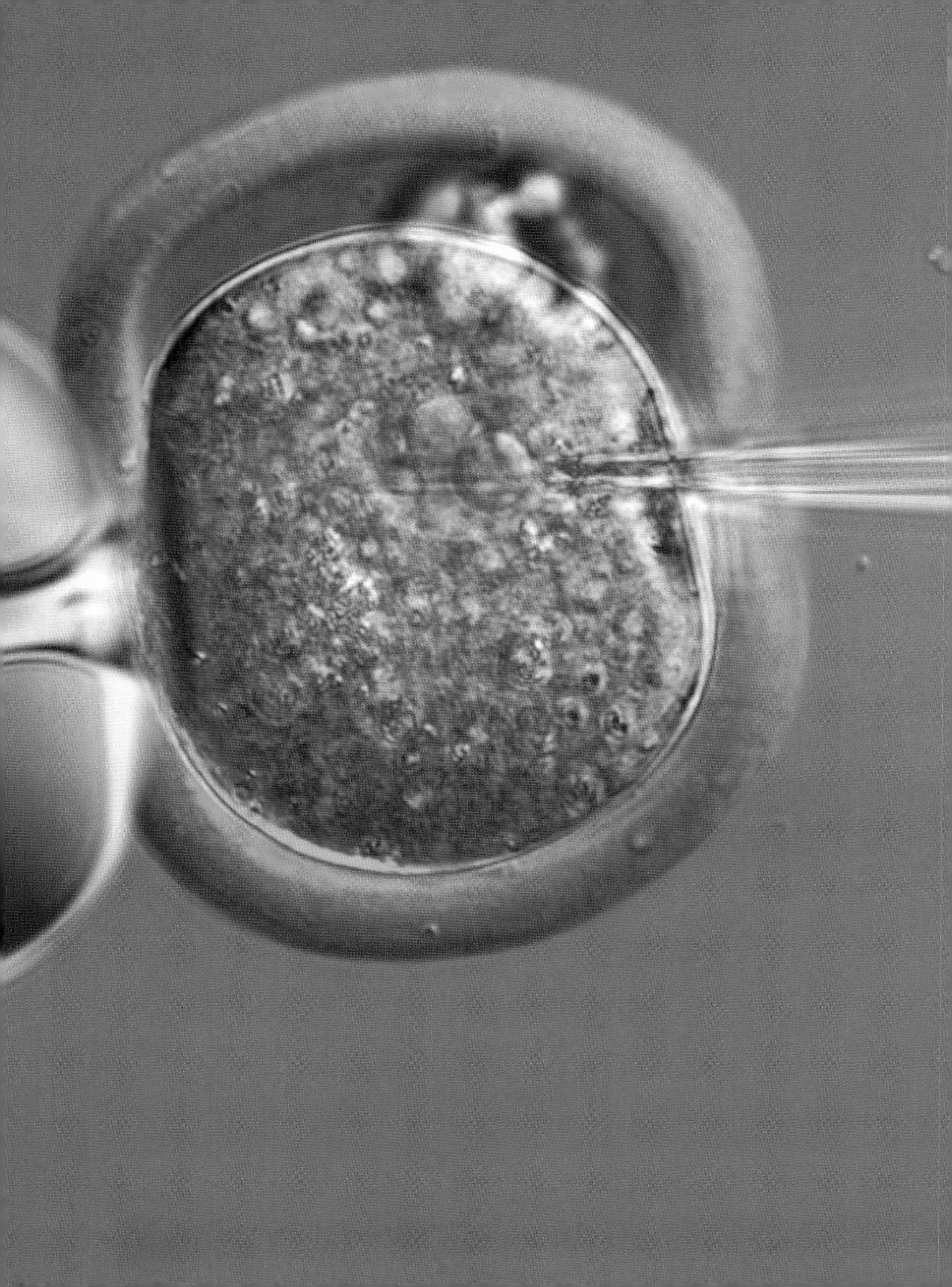

85 Genetic engineering

DEFINITION MANIPULATING THE GENES OF LIVING ORGANISMS IN WAYS THAT DO NOT OCCUR NATURALLY

DISCOVERY IN 1972, PAUL BERG WAS THE FIRST TO SPLICE TOGETHER GENETIC MATERIAL FROM TWO DIFFERENT LIFE FORMS

KEY BREAKTHROUGH USING ENDONUCLEASE ENZYMES TO CUT AND REJOIN LENGTHS OF DNA

IMPORTANCE BY MODIFYING DNA, GENETIC ENGINEERS CAN MAKE CUSTOMIZED ORGANISMS, SUCH AS DISEASE-RESISTANT CROPS

Genetic engineers use chemical techniques to chop and change the structure of DNA into which genes are written. This enables them to alter the characteristics of organisms for various uses: to create disease-resistant crops, new drugs and to prevent and cure illness. And yet concerns over the safety of tinkering with nature leave the field mired in controversy.

Pure science leads to engineering – the application of scientific knowledge to produce new technology. It happened with the physics of matter, electricity and nuclear reactions. And in the 1970s it happened with genetics.

In 1944, physician Oswald Avery had proven that genes – the biological data that we inherit from our parents – are written into a molecule at the heart of cells called DNA. Then, in the 1950s, James Watson and Francis Crick revealed the structure of DNA and the chemical behaviour that enables it to replicate and manufacture proteins – the principal constituents of cells. Throughout the 1960s, understanding continued to grow both of the genetic code and of the almost unbelievably long lengths of DNA upon which it is written – a single strand of DNA inside a human cell is almost a metre long.

Recombinant DNA

By the early 1970s, biologists knew enough to start manipulating DNA. They developed processes that could cut a DNA strand and then rejoin it with a new piece of DNA – thus modifying the genetic code of the host organism. DNA with genetic material from other organisms spliced into it is called 'recombinant DNA'. It is to genetic engineering what the wire and the battery are to electronics – absolutely fundamental. Recombinant DNA was

LEFT Transgenesis is the injection of new genetic material (DNA) into a living cell. In this case, the process is being conducted on the female reproductive cell of a mouse.

first made in 1972 by American biologist Paul Berg: he cut lengths of DNA from the tumour-causing monkey virus SV40 and spliced them together with DNA from a lambda phage virus – a virus that infects bacteria in order to reproduce itself.

Cut and snip

The key technology that made all this possible was the development of endonucleases – special kinds of enzymes or proteins that stimulate biochemical reactions. An endonuclease is essentially a pair of chemical scissors that causes a chemical reaction to cut a strand of DNA at a predetermined point. The most basic kinds simply cleave the double helix to leave a square end. But other kinds make a Z-shaped cut, also breaking some of the bonds between nucleotide bases (see page 305). This leaves the cut end in a 'sticky' state, with one or more unpaired bases ready to attach to new bases – or, indeed, an entirely new strand of DNA that has had its own end cut similarly. And this is how lengths of DNA are joined together.

'Taking the genes that stop an Antarctic fish from freezing in cold polar waters, and then putting it in the genetic code of a potato, gives you a new strain of potato that is frost-resistant.'

Once the genetic engineers have selected the required pieces of DNA, the next job is to replicate them many times over. This is done using a chemical process called a polymerase chain reaction (PCR). It involves first heating the sample to force the two strands in the double helix to peel apart. The sample is allowed to cool slightly under the action of enzymes: this encourages new nucleotide bases to attach themselves along the length of the two single strands and form two new double helices. The process is then repeated many times.

Paul Berg's hybrid virus never went any further than this stage – it simply existed as modified DNA. But it was not long before scientists were implanting their genetically engineered sequences back into living creatures. The first such 'transgenic' organism was created in 1973, when researchers modified the genetic sequence of an *E. Coli* bacterium to make it resistant to antibiotics. And the following year came the first genetically engineered mammal – a mouse with viral DNA slotted in amongst its own.

Gene machines

Modified sequences of DNA can be inserted back into organisms in a number of different ways. The simplest method is to simply inject the new genes directly into the cell nucleus, using 'micro-injection', a technique using a needle one two-hundredth of a millimetre across guided by a microscope. As the cell then divides naturally, the new DNA is spread throughout the organism. Another option is to splice the modified genetic code into the DNA of a virus. Viruses work by infecting cells, hijacking their DNA-copying machinery to make new viruses. Scientists can commandeer this process, so the viruses force the cells to copy not their own DNA but the genetically engineered substitute that they have been given.

Saving lives

In 1978, one of the first practical uses for genetic engineering came to light. By taking the section of the human genetic sequence that codes for the blood hormone insulin, and introducing it into the DNA of *E. coli* bacteria, the bacteria were made to grow their own human insulin – which could then be harvested to treat insulin-deficient diabetes patients. Up until this point, researchers extracted medicinal insulin from animals and purified it as best they could.

ABOVE Genetically modified corn kernels (maize) sprouting in a Petri dish. Genetic modification improves the characteristics of commercial crops, such as making them resistant to weed killer.

Genetic engineering – or 'genetic modification', as it has become known – has also found applications in food production. For example, taking the genes that stop an Antarctic fish from freezing in cold polar waters, and then putting them into the genetic code of a potato, gives you a brand new strain of potato that is frost-resistant.

Research in this area has also yielded food crops that have extended shelf life, that are resistant to herbicides and pest attack – even foods that are enriched with elevated nutritional content.

'Igor, it lives . . .'

Despite its benefits, however, the Frankenstein metaphor for biotechnology-gone-wrong has loomed large over genetic modification (GM) technology, leading to public concern over its safety. The extent of the backlash was such that some supermarkets removed genetically modified foods from their shelves. Others simply refrain from publicizing the genetic origin of produce, although in many countries there are regulations to prevent this.

Proponents argue that GM technology is entirely safe, and that research into it follows a strict protocol that has been adhered to by scientists since as far back as 1975. And, of course, indirect genetic modification has already been going on for thousands of years through traditional agricultural cross-breeding of both plants and animals – with no apparent ill-effects.

One reason why the furore over genetically modified tomatoes seems to have died down is because the science of genetics has moved on, taking the focus of public concern with it. Rather than being content to simply alter the DNA of existing organisms, researchers can now apply advanced GM techniques to genetic sequences that originate in the biologist's imagination and notebook, rather than to naturally occurring life forms. This research field is known as synthetic biology (see page 405) and has been described as 'genetic engineering on steroids', because its aims have never been loftier: to create new types of life where before there were none.

86 Hawking radiation

DEFINITION RADIATION EMITTED BY QUANTUM MECHANICAL PROCESSES ON THE EVENT HORIZON OF A BLACK HOLE

DISCOVERY THEORETICAL PROOF WAS FIRST PUBLISHED BY STEPHEN HAWKING IN 1974

KEY BREAKTHROUGH SHOWED HOW VIRTUAL PARTICLE PAIRS DO NOT HAVE TO RECOMBINE AND VANISH IN A STRONG GRAVITY FIELD

IMPORTANCE HAWKING RADIATION IS A PRIME TEST-BED FOR CANDIDATE QUANTUM THEORIES OF GRAVITY

As Stephen Hawking himself famously put it: 'Black Holes ain't so black'. In 1974, he applied quantum physics to general relativity to show that black holes are not just one-way matter disposal units – but can give off particles as well. This is known as Hawking radiation, and may well be our first vista of the undiscovered country of quantum gravity.

Black holes are objects whose gravity is so intense that not even beams of light can escape. Anything crossing the outer surface of a black hole – its event horizon – is doomed to be crushed out of existence at the hole's core. By gobbling up everything that comes its way, a black hole gets bigger.

Or does it? In 1974, English physicist Stephen Hawking published landmark research while based at the University of Cambridge. He showed that black holes do actually give off a small amount of radiation. The effect has become known as Hawking radiation or, because it can cause the black hole's mass to decrease, Hawking evaporation. If he is right, it overturns the picture of black holes as one-way mouths that devour everything – and could even mean that it is possible for a black hole to vanish into nothing.

LEFT Simulation of a black hole in the galaxy MCG-6-30-15 expelling radiation. Black holes are usually portrayed as ravenous mouths that swallow matter and energy but, in 1974, Stephen Hawking proved otherwise. The effect became known as Hawking radiation.

Hawking's derivation made use of an idea from quantum theory, called the uncertainty principle. This was put forward in 1927 by German physicist Werner Heisenberg and states that it is impossible to know everything about a quantum particle at the same time. For example, if I measure a particle's speed then I lose accuracy in my knowledge of its position. If I measure its energy then the precise time at which I make the observation becomes less well determined. In the late 1920s, physicists realized that this means space must be filled with pairs of virtual particles winking in and out of existence. In any given period of time, a particle pair (one particle of matter, one of antimatter – see page 261) can be created, so long as the energy and the length of time for which it exists satisfies Heisenberg's principle. The smaller the energy of the pair the longer the particles can exist for, before they recombine and vanish – as if they were never there in the first place.

'There could exist tiny primordial black holes – created by the terrific pressures and densities shortly after the Big Bang – that are giving out terrific amounts of Hawking radiation and could even be evaporating away to nothing in the sky today.'

Virtual particles fill empty space everywhere, and are responsible for important processes in physics, such as the forces between molecules and the interaction between two magnets. Hawking wondered what would happen when this process inevitably takes place in the space just outside a black hole. What if, he supposed, a virtual pair appeared just outside the hole's event horizon so that, before the particles can recombine, one of them is swallowed up by the black hole while the other escapes.

Hawking's calculations suggested that there should indeed be just such a flux of orphaned particles drifting away from the black hole. And this is Hawking radiation. What is more, he found the particles that fall in must have negative energy – and thus negative mass by Einstein's $E = mc^2$ formula. The cumulative effect of these negative-mass particles raining down on the black hole's horizon decreases its mass over time and the hole really does 'evaporate'.

Fire in the hole

Hawking found that a black hole radiates like any other hot body – with the amount of radiation given off determined by its temperature. He showed what this temperature should be, depending on the black hole's mass. Counter-intuitively, though, small black holes have higher temperatures than larger ones. A black hole weighing about the same as our Sun would have a tiny Hawking temperature, 60 billionths of a degree Celsius above absolute zero. That is less than the temperature of the cosmic microwave background radiation – the relic radiation pervading space from the Big Bang – and so a solar-mass black hole would still gain energy (and hence mass) from its environment even after accounting for Hawking radiation.

The microwave background has a temperature of 2.7°C (37°F) above zero. A black hole with this temperature – which would be in equilibrium with its surroundings – would weigh about the same as the Moon. Lighter black holes have higher temperatures and give out more energy than they receive.

Some astrophysicists suggest that there could exist tiny primordial black holes, created by the terrific pressures and densities after the Big Bang, giving out terrific amounts of Hawking radiation and evaporating away to nothing. This final stage of Hawking radiation would most likely manifest itself as a violent flash of high-energy gamma radiation. NASA's Fermi

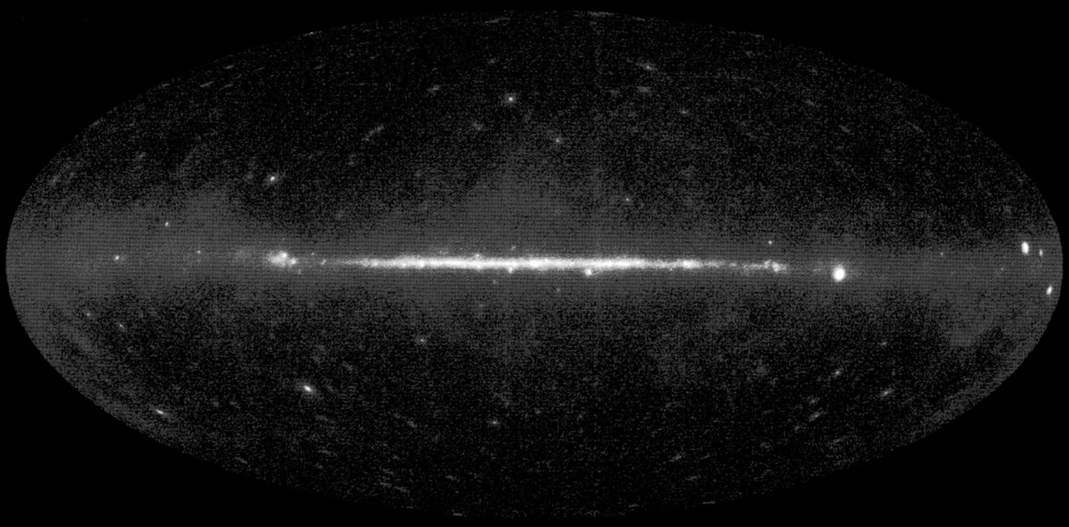

Gamma-ray Space telescope, launched in 2008, is currently searching for these flashes. Other researchers are trying to find evidence for Hawking radiation by studying analogues of black holes in laboratories here on Earth. In September 2010, a team from Edinburgh and Milan fired laser light into a piece of glass, radically altering the speed that light can pass through it. The effect of this was to set up a kind of horizon, much like the event horizon in a black hole, from which light cannot escape.

ABOVE Gamma-ray emissions from our galaxy, imaged by the Fermi Gamma-ray Space Telescope. Hawking radiation predicts that when a black hole is starved for long enough, it explodes and creates a flash of high-energy gamma rays.

Or so they thought. The team subsequently observed odd flashes of light that seemed to defy the ban – appearing on the forbidden side of the horizon. The flashes did not seem to be stray laser light. The laser was operating at a wavelength between 600 and 700 billionths of a metre, whereas the flashes were seen between 850–900 and at around 300 billionths of a metre – fitting the predicted wavelength of Hawking radiation.

It is an intriguing result, and one that many believe bodes well for the eventual detection of the effect from a real black hole. If and when that happens, it may well be the first piece of experimental evidence in the hunt for a quantum theory of gravity – a model of the gravitational force that naturally embraces quantum effects.

Hawking's analysis has been criticized by some because he jammed together quantum physics with general relativity (a theory that is overtly at odds with quantum rules). Observations of a real evaporating black hole, and how its behaviour differs from general relativity's predictions, will offer a big hint to physicists trying to deduce the correct quantum gravity theory – a puzzle that has been baffling them for the last 80 years.

87 Death of the dinosaurs

DEFINITION THE EVENT THAT WIPED OUT THE DINOSAURS, AND 70 PERCENT OF ALL OTHER LIFE ON EARTH, 65 MILLION YEARS AGO

DISCOVERY THE NATURE OF THE EVENT WAS REVEALED IN A SCIENTIFIC PAPER BY LUIS AND WALTER ALVAREZ IN 1980

KEY BREAKTHROUGH FROM LEVELS OF EXTRATERRESTRIAL IRIDIUM, THEY CONCLUDED THAT AN ASTEROID WAS RESPONSIBLE

IMPORTANCE THE DISCOVERY HIGHLIGHTED THE DANGER POSED BY COSMIC IMPACTS, AND OUR NEED TO DEFEND AGAINST THEM

One day, 65 million years ago, the dinosaurs vanished from planet Earth, taking many other species with them. For years, the reason why was a mystery. Then, in 1980, Luis and Walter Alvarez found huge quantities of the extraterrestrial mineral iridium in the layer of the fossil record corresponding to the dinosaurs' demise. The implication was clear: the dinosaurs had been wiped out by an asteroid impact.

It had long been known that large and terrifying creatures once walked the Earth. Dinosaur fossils were unearthed by ancient civilizations, which, no doubt, served as the source for legends of mythical beasts, such as dragons.

Scientists began to study these fossils during the late 1600s in England. And in 1842, the English palaeontologist Richard Owen coined a name for them – 'dinosaurs', a portmanteau of the Greek words for 'terrible' and 'lizard'. Dinosaur remains were recovered from a broad swathe of the fossil record. The depth from which a fossil is recovered indicates the era in which it lived, and the dinosaurs appear to have inhabited an era spanning some 160 million years – from 225 million years ago until 65 million years ago, at which point the dinosaur fossils abruptly stop. For some reason, the dinosaurs had died out. And they were not alone – if the fossil evidence was anything to go by, an estimated 70 percent of species on the planet had gone with them.

LEFT The Chicxulub crater in Mexico is thought to have been formed by the asteroid that wiped out the dinosaurs 65 million years ago. The crater rim is shown by the green, yellow and red rings, and measures 180km (112 miles) across. The white line through the centre is the coastline, with land located in the lower half.

Events such as this – where large numbers of species suddenly disappear from the fossil record – are known as 'mass extinctions'. A total of five big ones are known, the earliest of which occurred about 450 million years ago. Mass extinctions usually form a transition between geological ages. The largest mass extinction wiped out 96 percent of all marine species. Known as the Permian–Triassic extinction, it took place about 251 million years ago. The specific event that killed the dinosaurs is known as the Cretaceous–Tertiary, or K–T extinction, because it separates the Cretaceous and Tertiary periods in the fossil record (however, geologists recently voted to rename the lower Tertiary rock layer as the 'Palaeogene', meaning that the shorthand K– Pg is likely to start replacing K–T in scientific literature). The K–T layer is clearly visible in rock formations as a 1-cm (0.4-inch) thick layer of dark clay separating the predominantly light-coloured Cretaceous rock below and the darker-coloured Tertiary-period rock above.

'A re-examination of rock bed surveys from the 1970s showed up a crater dating to exactly the era when the K–T layer was laid down and the dinosaurs died out. The crater was gigantic, spanning approximately 180km (112 miles) – so big that the asteroid causing it must have been at least 10km (6.2 miles) across.'

Numerous causes have been put forward by scientists for the Big Five mass extinctions – including naturally occurring climate change, volcanoes and even events in space, such as the explosion of a nearby star as a supernova. But there was particular interest in what caused the death of the dinosaurs. What could have obliterated this mighty race of animals from the planet? Whatever it was cleared the way for the rise of the mammals – that is, us. That's the good news. The bad news was that the same force, whatever it was, could occur again and unseat human beings in a single geological heartbeat.

Alvarez and son

In the 1970s, geologist Walter Alvarez was examining rock strata in Italy with his father, the American physicist Luis Alvarez. Walter showed Luis the K–T layer and explained that this was the point in the Earth's geological timeline where the dinosaurs became extinct – and yet no one really knew why. Luis Alvarez believed that the clue might be locked inside the K–T layer of clay, and so he took away a piece for analysis.

The first thing Luis Alvarez wanted to do was figure out how quickly the clay layer had been deposited. Did it represent millions of years of geological evolution, or much less? It struck him that one way to find out would be to measure the amount of the rare element iridium contained within the layer of clay. Iridium is largely absent from the Earth's surface (it is an iron-loving element that would have sunk to the Earth's core during the planet's formation), but it does arrive slowly in the form of micrometeorites from space, that scatter the element lightly over the planet's surface. Alvarez knew the rate at which iridium usually arrives from space and therefore by measuring how much of the mineral was in the K–T layer, he could determine how long it had taken to form.

But when Luis did this he got a shock. There was so much iridium in this thin layer of clay, that there was no way it could possibly have been put there by tiny, grain-size micrometeorites. The Alvarezes knew then that the only thing that could have brought this much iridium to Earth so quickly was the impact of a large asteroid. The theory triggered a heated debate amongst geologists, but another piece of evidence would soon arise to silence almost all the dissenters.

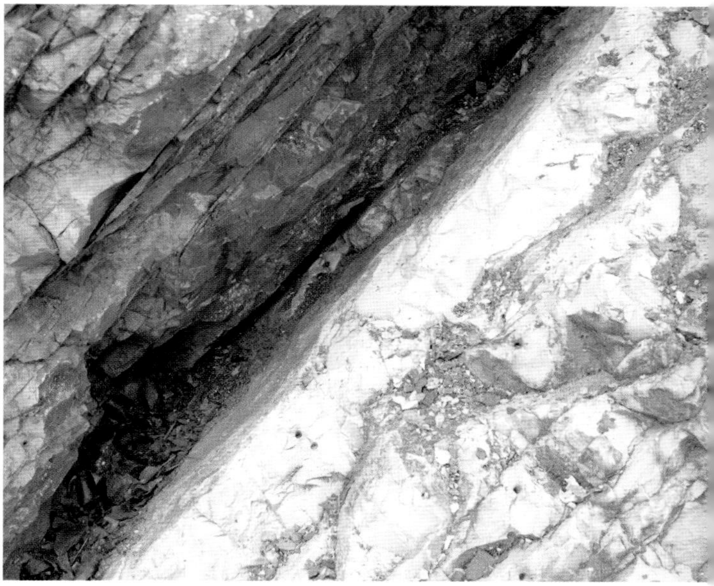

ABOVE The infamous K–T boundary in the fossil record, dividing fossils from the Cretaceous (K) period (shown as white sandstone) and those from the Tertiary (T) period (darker layer above). The boundary is a thin layer of dark clay, rich in the extraterrestrial element iridium, suggesting something from space caused the dinosaurs' demise.

Smoking gun

In 1990, a re-examination of rock-bed surveys from the 1970s showed up a crater dating to exactly the era when the K–T layer was laid down and the dinosaurs died out. The crater was gigantic, spanning approximately 180km (112 miles) – so big that the asteroid causing it must have been at least 10km (6.2 miles) across. The impact of an object this size would have exploded with the force equivalent to a billion Hiroshima bombs, unleashing firestorms and throwing clouds of ash and soot into the atmosphere that would have blocked out the Sun for months, meaning death for plants – closely followed by animal life.

The crater was off the coast of Mexico's Yucatán Peninsula, centred on the town of Chicxulub, meaning that large tidal waves would also have swept around the planet following the impact. Evidence for these 'megatsunamis' has since been found, along with other supporting evidence for the Alvarez hypothesis, including soot and heat-melted glass fragments inside the clay of the K–T layer. In March 2010, a panel of scientists writing in the prestigious research journal *Science* came to the unanimous conclusion that the Chicxulub impact was the cause of the extinction that killed the dinosaurs.

Space guard

The implication for humans was clear: watch the skies. Astronomers are now scanning the skies for other potentially hazardous comets and asteroids. Meanwhile, space scientists are devising schemes by which we could deflect any unwanted visitors on a collision course. Their strategies involve attaching rocket engines to the asteroids, and even painting them to make them more reflective – so that they can be pushed away by sunlight.

The European Space Agency is designing a spacecraft to directly test one plan, to deflect an asteroid using a kinetic impactor – effectively, a big cannon ball. Scientists hope the mission, called Don Quijote, will fly by 2015. We can only hope that the lessons we learn from Don Quijote, and other projects like it, will help humanity avoid the fate of their reptilian predecessors all those millions of years ago.

88 Scanning tunnelling microscope

DEFINITION A DEVICE THAT TAKES ADVANTAGE OF QUANTUM PHYSICS TO CREATE IMAGES OF OBJECTS AS SMALL AS ATOMS

DISCOVERY FIRST BUILT BY PHYSICISTS GERD BINNIG AND HEINRICH ROHRER AT IBM IN SWITZERLAND, 1981

KEY BREAKTHROUGH DRAGGING A QUANTUM PROBE OVER A SURFACE TO MEASURE OBJECTS AS SMALL AS 0.01 NANOMETRES ACROSS

IMPORTANCE SCANNING TUNNELLING MICROSCOPES ARE THE MOST POWERFUL MICROSCOPES KNOWN TO SCIENCE

Anyone who has looked down an optical microscope at school will have been amazed by the sight of single-celled creatures measuring just a thousandth of a millimetre across. In 1981, Gerd Binnig and Heinrich Rohrer, two researchers working at IBM in Switzerland, built a new device that could make out details 100,000 times smaller.

The first electron microscopes were built in the 1930s by German physicist Ernst Ruska and electrical engineer Max Knoll. The devices worked on the principle of wave-particle duality (see page 237), which meant that particles – like electrons – behave like waves, and can therefore be focused and used to create images. Because electrons pack more punch than photons of light, their equivalent wavelength is much smaller. And this meant that illuminating an object with an electron beam reveals a far greater level of detail than using a beam of light – typically better by hundreds of times.

The trouble was that whereas light beams can be easily focused and manipulated using lenses and mirrors, electron beams tended to pass straight through. Electrons were, however, electrically charged – and Ruska and Knoll exploited this to control their beams using electric and magnetic fields.

They developed two types of electron microscope. A transmission electron microscope (TEM) works by passing a beam of electrons through a wafer-thin sample of material to produce an image of the material's surface. Ruska and Knoll first demonstrated this technique in 1931. Later, in 1935, they put forward the scanning electron microscope (SEM), which works by sweeping an electron beam back and forth across the target sample and

LEFT A scanning tunnelling microscope image of a scanning tunnelling microscope probe. The needle at the probe's tip is just a few atoms wide. It is electrically charged and dragged over the surface of the sample under examination. Variations in the current flowing from the needle then reveal the tiniest lumps and bumps in the sample.

RIGHT European hornet, as revealed by a scanning tunnelling microscope. For the STM technique to work, the object must be able to conduct an electric current, and so this hornet has first been 'sputter coated' with a layer of gold.

measuring the electrons that get bounced back from it. The image resolution of these techniques was astonishing, able to show up surface details as small as a nanometre (a billionth of a metre) in size. This was fine-scaled enough to produce images of virus particles and the structure of cells – as well as terrifying images of house dust mites and other bugs in extremely high detail.

In 1981, two physicists at IBM in Zurich figured out that it was possible to do even better. German Gerd Binnig and Swiss Heinrich Rohrer invented a device called the scanning tunnelling microscope (STM). As the name suggests, the operation of the STM is based on a phenomenon called 'tunnelling' – which arises in quantum physics.

Tunnelling happens because of the indeterminate nature of quantum physics. It is impossible to say that a quantum particle is at a particular location in space at a given time – all you can do is specify the probability of the particle being found in that location, or indeed any other.

Quantum tunnelling

In ordinary classical physics, two electric charges of the same polarity that are brought close to each other will experience a force of repulsion pushing them apart. But in the indeterminate quantum world 'close to each other' loses much of its meaning, and this has the effect of lowering the force required to push the particles together. And this is quantum tunnelling.

This phenomenon is partly why the Sun shines. The Sun is powered by nuclear fusion reactions in its core, in which hydrogen nuclei combine and fuse together, releasing energy in the process. Getting the nuclei to combine in order for fusion to begin means slamming them together very hard to overcome their mutual electrical repulsion (hydrogen nuclei have positive charge and therefore repel one another). The Sun does this by virtue of its temperature, which makes the nuclei cavort energetically and inevitably crash together. However, ordinarily the temperature in the Sun's core would be insufficient to drive two nuclei together hard enough. Tunnelling, however, greatly reduces the force and thus the temperature required, allowing fusion to take place.

Tunnelling current

STMs utilize tunnelling to operate in a very different way to conventional electron microscopes. Rather than firing out a beam of electrons, a metal probe is dragged across the surface of the target sample. The probe is extremely sharp – just one atom across at its tip. It does not actually make contact with the sample, but hovers about two atomic diameters above it. A voltage is then applied between the tip of the probe and the sample. Because of tunnelling, this sets up a tiny current of electrons that flows between the sample and the tip.

'STM images can reveal fine surface features 100 times smaller than those visible in SEM images – measuring just 0.01 nanometres, or one hundred-millionth of a millimetre across.'

The magnitude of the current can be measured by the instruments in the microscope. If, as the tip is dragged across the surface of the sample, it gets closer to the surface of the sample, the current gets stronger; while if the tip gets further away, the current gets weaker. Sensing these changes in current allows an actuator on the microscope to adjust the tip's height in order to keep the electron current constant. These minuscule variations in height as the tip moves are logged in a computer. When the tip has scanned back and forth across the whole of the sample, the result is a super-detailed relief map of its surface.

Ultimate microscope

STM images can reveal fine surface features 100 times smaller than those visible in SEM images – measuring just 0.01 nanometres, or one hundred-millionth of a millimetre across. Amazingly, this is good enough to reveal images of individual atoms – or rather, images showing the atoms' locations. They show up in these pictures as upturned cones.

More recently, scientists have even turned the STM into a tool for manipulating matter at the atomic scale. They have found that it is possible to make atoms stick to an STM probe, allowing them to be moved around at will. In 1989, scientists at IBM famously exploited the technique to write the letters 'IBM' in atoms of the element xenon.

The scanning tunnelling microscope has opened up an incredible new window through which we are able to both view and interact with the subatomic world. Binnig and Rohrer won the 1986 Nobel Prize in Physics for their invention.

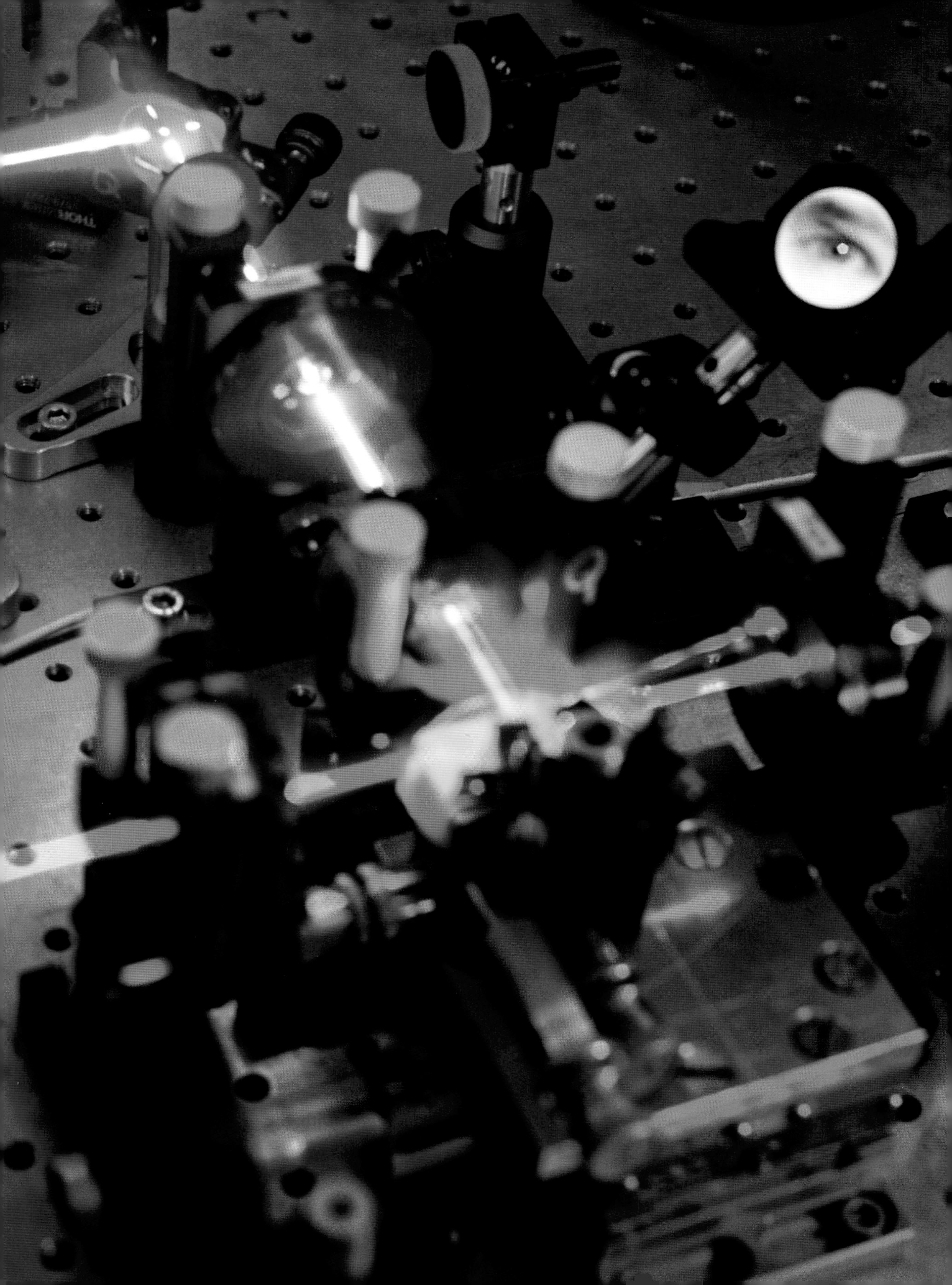

89 Quantum entanglement

DEFINITION THE WAY PAIRS OF QUANTUM PARTICLES CAN REMAIN LINKED DESPITE BEING SEPARATED BY VAST DISTANCES

DISCOVERY THE EXISTENCE OF ENTANGLEMENT WAS CONFIRMED EXPERIMENTALLY BY FRENCH PHYSICIST ALAIN ASPECT IN 1982

KEY BREAKTHROUGH ASPECT TESTED A MATHEMATICAL RULE FROM QUANTUM THEORY, PROVING ENTANGLEMENT IS CORRECT

IMPORTANCE ALLOWED NEW APPLICATIONS OF QUANTUM PHYSICS, SUCH AS QUANTUM COMMUNICATION AND TELEPORTATION

Quantum physics enables pairs of subatomic particles to be linked – so that if you move them lightyears apart and wiggle one, then the other wiggles back instantaneously. The effect is called 'quantum entanglement' and was first put forward by Einstein, ironically to prove that quantum theory was wrong. In 1982, quantum entanglement was confirmed as a reality.

Einstein famously loathed quantum theory. He believed that theories in physics should make hard-and-fast predictions about the world – they should be 'deterministic' theories. But quantum theory was anything but. It dealt with hazy probabilities and wavefunctions that gave the likelihood of finding subatomic particles either here or there, with no determinism at all.

Proven absurdity

In 1935, Einstein – together with two colleagues at the Institute for Advanced Study in Princeton, Boris Podolsky and Nathan Rosen – constructed a thought experiment to highlight what they viewed as the absurdity of quantum physics and its probabilistic take on reality. The dominant view of quantum theory at the time was the Copenhagen interpretation, stating that particles remain in an non-deterministic hazy state until they are measured, at which point their wavefunctions collapse and they become classical or 'nonquantum' entities, with well-defined states and positions.

Einstein, Podolsky and Rosen imagined an electron and a positron (see page 261) that had been created as a matter-antimatter pair from the decay of another subatomic particle, called a pion. The researchers considered the 'quantum spin' of the particles – a subtle property, broadly analogous to

LEFT The eye of an observer is reflected in a mirror in quantum cryptography apparatus. Quantum cryptography is based on the principle of entanglement, a property of a pair of particles in which a change to one has an instantaneous effect on the other, no matter how far apart they are.

classical spin but with a few quirks. Because quantum spin obeys a standard conservation law (see page 93) then the total quantum spin of the electron and positron must add up to the total spin of the original pion – which is known to be zero. Electrons and positrons can have a spin of either +1/2 or −1/2 and, so to add up to zero, one of them must take the value +1/2 while the other takes the value −1/2.

The clever bit

But which particle takes which value? Until a measurement is made, it is impossible to tell – each particle has a 50 percent probability of being in one spin state or the other. And as soon as one particle is measured, the other must instantly snap into the opposite spin state to satisfy the conservation law. But then came the clever bit. Einstein and colleagues imagined taking the electron and positron far apart – perhaps to opposite sides of the universe – before a measurement was made. Then, measuring one of the particles instantly fixes the state of the other even though it is many lightyears away, contradicting normal notions of causality.

'Aspect's experiment had proven that subatomic particles really do maintain a spooky quantum mechanical link regardless of how far apart they are taken.'

Einstein famously derided this 'spooky action at a distance', which meant that the state of the distant particle was effectively known without being measured – contradicting the Copenhagen interpretation. The thought experiment became known as the EPR paradox, after the initials of the three scientists.

Einstein used the result to advocate an alternative interpretation of quantum theory, known as 'hidden variable theory'. This says that the indeterminism of a quantum system is an illusion. In this view, the system behaves deterministically but the values of the deterministic quantities are not observable – they are hidden from view.

In the quantum theory view of the EPR paradox, the spin states of the two particles do not exist until a measurement is made. In hidden variable theory, however, the spin states are there all along – you just need to make a measurement in order to see them. It is a subtle difference. Imagine the two particles are empty boxes, and the two different spin states are represented by a blue ball and a red ball. In hidden variable theory, each box definitely contains one ball or the other – and so it is no surprise that looking inside one and finding, say, the red ball tells you straight away that the blue ball is in the other. However, in true quantum theory neither ball exists at all as a definite entity inside either box until you actually open one and take a look.

Bell's inequalities

So how do you tell which is correct – proper quantum theory, or Einstein's hidden variables? The difference between the two interpretations was so subtle that for years no one knew. Then, in 1964, Irish physicist John Bell worked out a way. He was able to prove that if hidden variables theory was correct, then certain mathematical relationships between observable properties of quantum systems must hold true. These relationships were known as Bell's inequalities.

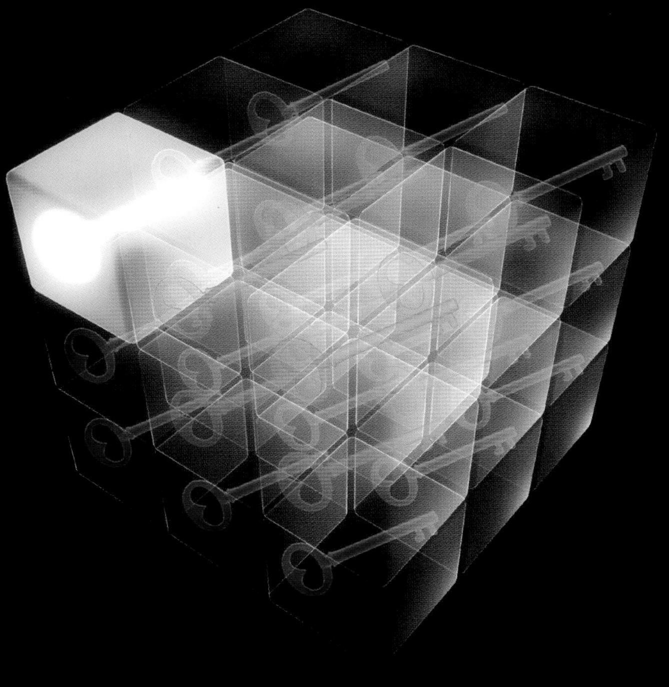

ABOVE Representation of the concept of quantum encryption. This field of technology exploits the link between entangled photons to encode messages in an unbreakable form of encryption. In 2007, quantum encryption was used to secure electronic votes cast in the Swiss elections.

In 1982, a research team led by French physicist Alain Aspect carried out a ground-breaking experiment to test Bell's inequalities. The results were clear beyond doubt – the inequalities were violated, the idea of hidden variables is wrong and the standard interpretation of quantum theory is correct. Aspect's experiment had proven that subatomic proton particles really do maintain a spooky quantum mechanical link regardless of how far apart they are taken. This property of some pairs of particles, whereby their quantum wavefunctions remain linked over great distances, is now generally known as quantum entanglement.

New physics

Entanglement is central to the more modern picture of quantum theory that replaces the original Copenhagen interpretation. It takes the emphasis off the measurement-making observer and says that quantum systems make the transition from quantum to classical behaviour not through collapse of the wavefunction when they are measured, but when they inevitably interact with their surroundings. This interaction is known as 'decoherence'. Maintaining a quantum state is a bit like building a house of cards – all it takes is a gentle gust of wind from the surroundings to bring the whole thing tumbling down.

Physicists have utilized this new view of the quantum world to achieve some quite amazing results. In so-called 'quantum communication' experiments, entangled channels are used to transmit information with absolute security against eavesdroppers (because anyone trying to listen in will, by definition, cause the quantum system to decohere, thereby revealing their presence). In even more incredible research, scientists have used entangled states to teleport matter from one place to another. In May 2010, scientists in China succeeded in teleporting the states of photons of light over a record distance of 16km (10 miles). Despite this, most physicists agree that teleporting larger objects, such as people, will be a considerably more complex task – if it is even possible at all.

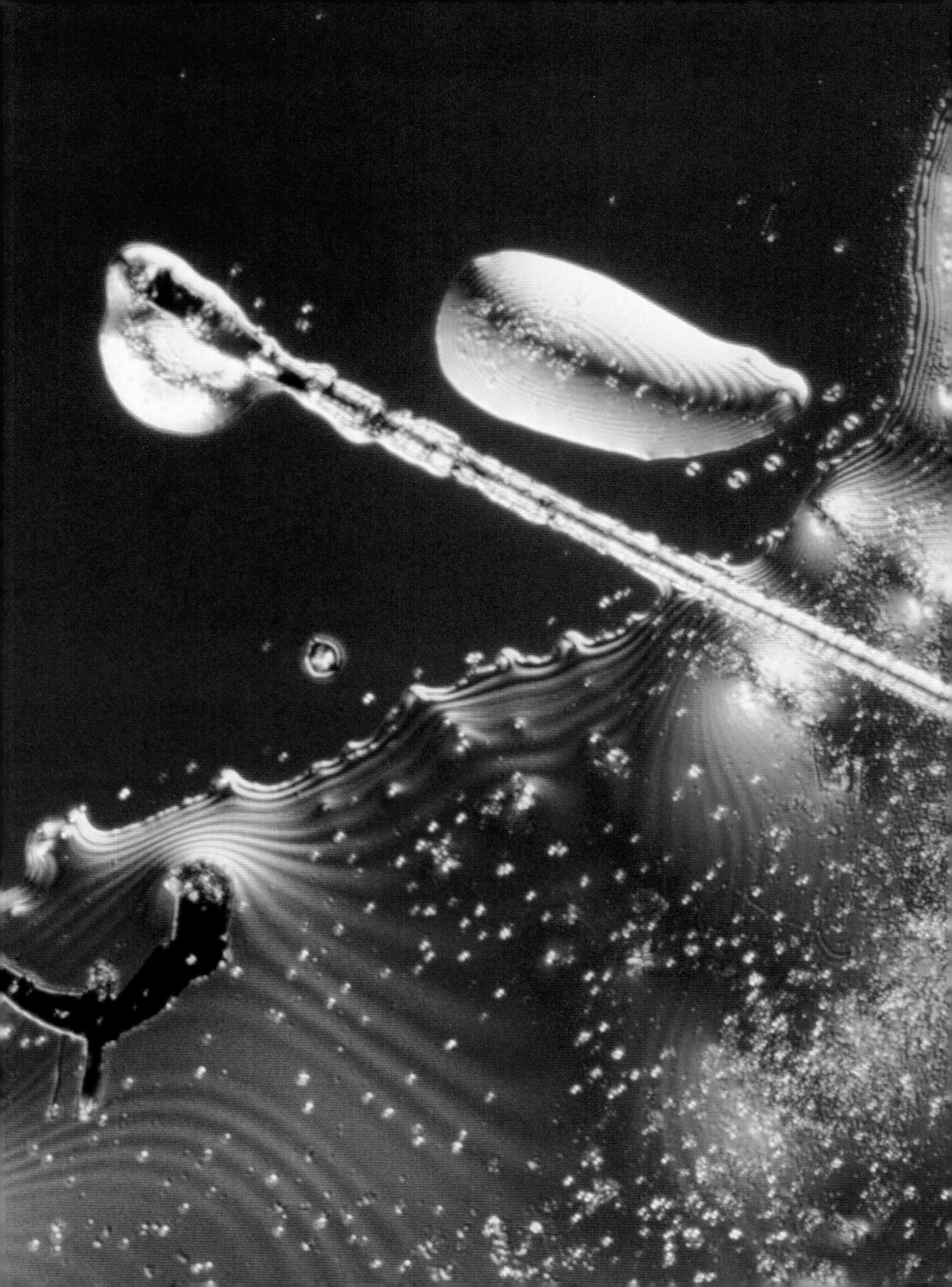

90 Fullerenes

DEFINITION NEW KINDS OF CARBON MOLECULES MADE BY BOLTING TOGETHER LARGE NUMBERS OF CARBON ATOMS

DISCOVERY THE FIRST FULLERENE – BUCKMINSTERFULLERENE OR 'BUCKYBALL' – WAS DISCOVERED IN 1985

KEY BREAKTHROUGH ACCIDENTAL CREATION OF BUCKYBALLS IN AN EXPERIMENT TO RECREATE CARBON COMPOUNDS AROUND A STAR

IMPORTANCE FULLERENES OFFER EXCEPTIONAL ELECTRICAL, MECHANICAL AND THERMAL PROPERTIES

Carbon molecules resembling soccer balls or long tubes of chicken wire, fullerenes are a revolutionary new form of carbon. They include carbon nanotubes: tiny tubes of carbon molecules that form a super-rigid material with 20 times the tensile strength of steel for a fraction of the weight of aluminium – as well as astonishing electrical and thermal properties. Fullerenes are the engineering materials of the future.

Carbon is a part of over 90 percent of the chemicals known to science, and is the primary chemical element upon which life on Earth is based. Carbon has long been known to take three different forms, or 'allotropes' – essentially different molecular arrangements of carbon atoms.

The allotrope we are most familiar with is probably ordinary soot, which is also known as amorphous carbon. Here carbon atoms bond together randomly, to form a disordered molecular pattern with no crystal structure. Another form is graphite, where the carbon atoms bond rigidly to form sheets, making a material with desirable electrical and mechanical properties. Finally, there is diamond, where the atoms of carbon have been arranged into a highly ordered crystal lattice. Diamond is formed by compressing carbon under extremely high pressure (roughly 50–60 times Earth's atmospheric pressure). Diamond is the hardest naturally occurring substance known.

LEFT Crystals of buckyballs, derived from pure carbon. Buckyballs were the first type of fullerene and were made by joining 60 atoms of carbon. The amazing engineering properties of fullerenes have led to various applications, from body armour to antibacterial drugs.

Red giant carbon

In 1985, an international team of researchers discovered a new allotrope of carbon. English chemist Harold Kroto was using microwave spectroscopy (see page 149) to study the different kinds of carbon molecules that occur in space, near to very large stars known as red giants. These are old stars that have begun advanced nuclear reactions, enabling them to generate their own carbon. Kroto was interested to know its type.

Kroto was in touch with two American chemists – Richard Smalley and Robert Curl – who were carrying out studies of various chemicals in their laboratory, using a powerful laser device that could vaporize the chemicals into a high-temperature, gaseous plasma. Kroto, Curl and Smalley used the apparatus to try to recreate the carbon chemistry in the searing environment close to a red giant – by using the laser to vaporize a sample of graphite.

The chunks of carbon vaporized from the graphite were analyzed using a mass spectrometer. This device filters atoms and molecules according to their mass – giving the number of particles of different mass in any sample that is fed in. It works by electrically charging each particle and then firing them at constant speed through a magnetic field. Charged particles moving through a magnetic field are deflected into a circular path, with the radius of the circle increasing with the particle's mass – so allowing the number of particles of different masses to be deduced.

New revelations

But when the researchers did this for the carbon produced by their laser apparatus, they were in for a surprise. The graph produced by the spectrometer, showing the number of molecules versus mass, had an enormous peak at 60 times the mass of a carbon atom. They had been expecting a much more even distribution. Further research revealed that the only stable carbon molecule with 60 atoms would have to be spherical – a ball. Indeed, a ball with 60 vertices would need to be made of 12 regular pentagons and 20 regular hexagons – exactly the configuration of a soccer ball.

'Carbon nanotubes form a type of fullerene material with 20 times the tensile strength of steel and yet weighing just half as much as aluminium.'

Because these spherical molecules reminded them of the geodesic architectural designs of American engineer Richard Buckminster Fuller, the team named their new carbon-60 molecule 'buckminsterfullerene' – or buckyballs for short. They published their discovery just weeks later in the research journal *Science*. Since then, other spherical carbon molecules have been discovered, with 70, 76 and 84 atoms, forming a family of molecules known by the shortened name of 'fullerenes'.

Carbon nanotubes

Buckyballs have enabled the synthesis of over 1000 new chemicals, leading to many new patents. But perhaps most exciting of all, research into buckyballs has led directly to the discovery of another related allotrope: carbon nanotubes. These are molecules that resemble sheets of graphite rolled up into tubes just a few billionths of a metre in diameter. Carbon nanotubes form a type of fullerene material with 20 times the tensile strength of steel and yet weighing just half as much as aluminium.

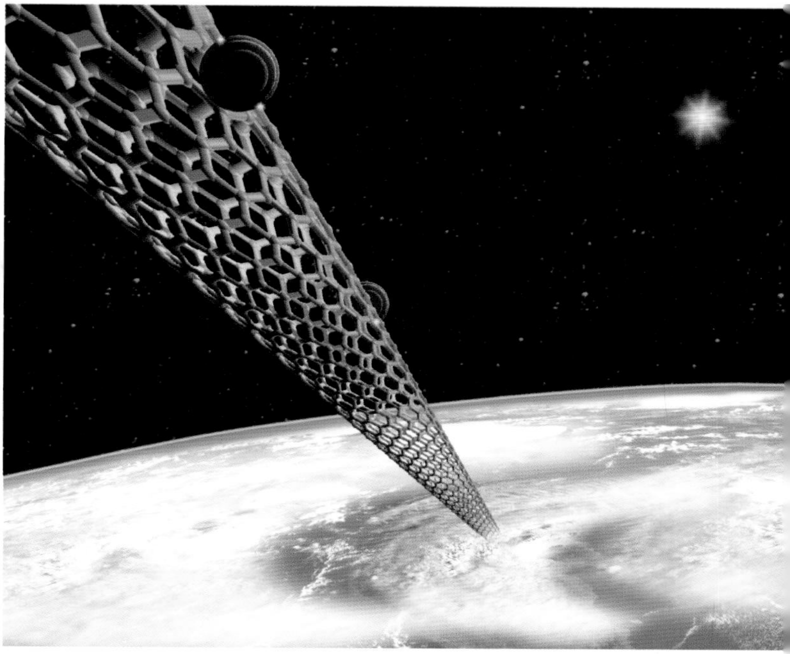

Fullerenes also possess astonishing conductivity of both heat and electricity. And adding potassium to the chemical mix even transforms them into superconductors (see page 293). This special quality, coupled with their extreme resilience, makes them excellent filaments for X-ray machines, high-intensity lamps and as materials for the probe tips of scanning tunnelling microscopes.

The longest continuous nanotubes that have been engineered so far measure just a few centimetres, but new techniques are improving on this, and nanotubes are widely touted as a building material of the future. Other applications for these hardwearing materials include body armour, super-thin batteries and scaffolds to encourage the growth of new bone tissue.

ABOVE The carbon nanotube space elevator could, hypothetically, offer a new method for travelling into space. It employs an orbiting platform from which payloads are winched up on fullerene cables. Traditional steel cables would break under their own weight, but fullerenes are both lighter and stronger.

Space elevator

Carbon nanotubes could even have a part to play in space research. In 1959, Russian engineer Yuri Artsutanov put forward the idea of a space elevator – a platform in Earth orbit from which payloads could be winched up into space from the Earth below. The concept was popularized by the English science-fiction author Arthur C. Clarke, in his 1979 novel *The Fountains of Paradise*. Such a technology has the potential to be vastly cheaper than launching payloads into space using rockets and better for the environment.

The trouble is that there is no known material strong enough to serve as the cable for such a device. Steel cable will break under its own weight as soon as the length reaches around 30km (18.6 miles). On the other hand a strong, lightweight cable made from carbon nanotubes could support itself even if it was thousands of kilometres in length.

Kroto, Curl and Smalley received the 1996 Nobel Prize in Chemistry for their discovery of fullerenes – a truly 21st-century material, and a hotbed for research in chemicals and materials science. With new applications emerging in fields ranging from flat-screen displays to bicycles, it seems likely we have only just begun to unlock their true potential.

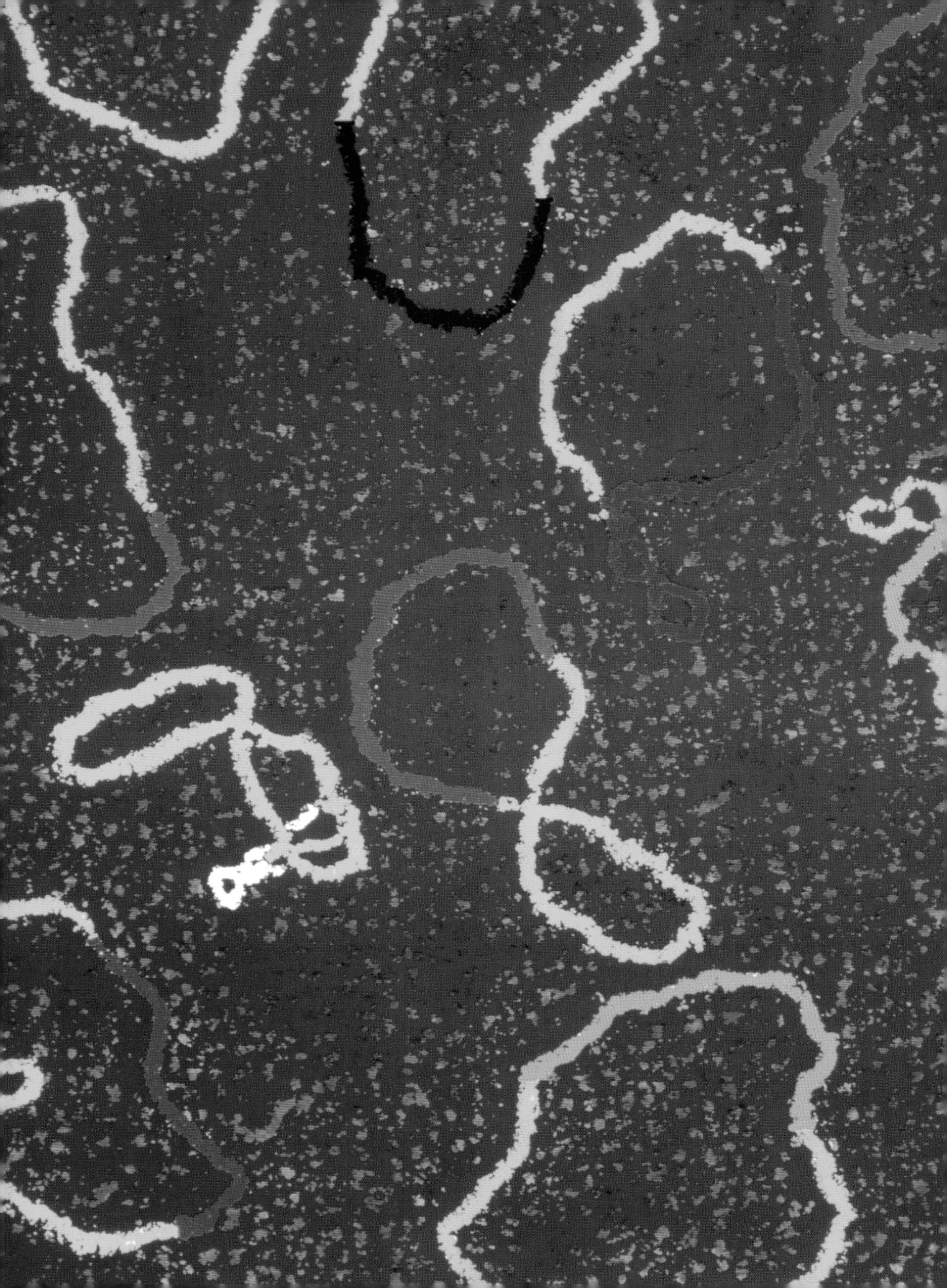

91 Gene therapy

DEFINITION THE APPLICATION OF GENETIC ENGINEERING TO HUMAN PATIENTS IN ORDER TO TREAT DISEASE

DISCOVERY FIRST CLINICAL TRIAL IN 1990 BY WILLIAM ANDERSON, MICHAEL BLAESE AND KENNETH CULVER

KEY BREAKTHROUGH TREATING AN IMMUNE SYSTEM DISORDER BY REPLACING DEFECTIVE GENES IN WHITE BLOOD CELLS

IMPORTANCE OFFERS HOPE OF A CURE FOR DEADLY AND DEBILITATING ILLNESSES, SUCH CANCER, AIDS AND HUNTINGTON'S

Some illnesses, such as cystic fibrosis and cancer, are caused by faulty genes. Gene therapy treats these conditions using genetic engineering to replace the bad genes with healthy ones. However, the complexity of the human genetic code and the tragic deaths of some patients during clinical trials, mean that gene therapies have been slow to gain approval.

Right at the start of the genetic-engineering revolution, scientists saw the potential benefits for medicine. Genetic engineering is all about altering the genetic make-up of organisms by tweaking DNA to alter their physical traits and characteristics. In 1972, American geneticists Theodore Friedmann and Richard Roblin suggested in the journal *Science* that the same techniques could be used to treat human genetic diseases by removing the defective DNA from patients and splicing in healthy copies instead.

Potential pitfalls

Many ailments, such as sickle-cell anaemia and cystic fibrosis, are caused by one defective gene. Scientists had hoped that it would be a relatively simple task to deploy genetic therapies to treat these conditions. However, altering even a single gene safely has proven extremely difficult. The human genetic code is three billion base pairs long, thus requiring pinpoint accuracy to cut and paste genes at the right place. It is not simply a case of tweaking the odd pair of nucleotide bases – the average human gene is 3000 base pairs long.

That is why it took almost two decades from Friedmann and Roblin's plan for gene therapy to progress from the germ of an idea to the first clinical trial in a human patient. This trial took place in 1990, at the US National

LEFT Loops of DNA plasmids (shown in yellow), within cells of the bacteria *E. coli.* Lengths of alternative DNA have been spliced into the plasmids (shown in blue, red and black). Plasmids modified in this way have been used in human gene therapy to transport new DNA into a patient's cells.

Institutes for Health (NIH) Clinical Center in Maryland. The patient was a four-year-old girl whose body – owing to a genetic defect – was unable to produce the enzyme adenosine deaminase (ADA) that is essential for the successful functioning of the immune system. As a result, the girl had become exceptionally prone to infection and developed a condition known as severe combined immunodeficiency (SCID), or 'bubble boy syndrome', because sufferers have to live in a sealed sterile cocoon in order to survive.

Full steam ahead

Doctors at the NIH were given clearance by the US Food and Drug Administration (FDA) to carry out a clinical trial of a gene-therapy technique developed over the preceding years by scientists William French Anderson, Michael Blaese and Kenneth Culver. The procedure involves extracting white blood cells from the patient's body – these are the cells that are meant to fight infection in a healthy person – and then inserting into them the missing genes needed to make ADA. Finally, the corrected white blood cells are reinjected back into the patient's bloodstream.

BELOW Deformed red blood cells in a sufferer of sickle-cell anaemia – a hereditary disease in which red blood cells assume an elongated shape that stops them working. The condition is caused by one defective gene, making it potentially treatable by gene therapy.

The 1990 trial of the technique was a partial success. The girl's ADA levels did improve, but the corrected gene was not passed on to new immune cells – meaning she has had to continue the therapy, as well as taking an ADA-replacing drug. Importantly, though, there were no obvious side-effects.

One of the principal difficulties with gene therapy is passing the therapeutic genetic material into the patient's genetic code. Simply injecting it into their bloodstream is not enough. Genes, specifically the lengths of DNA onto which they are written, need to be physically slotted into the patient's own DNA – which is locked inside the nucleus of every one of their cells.

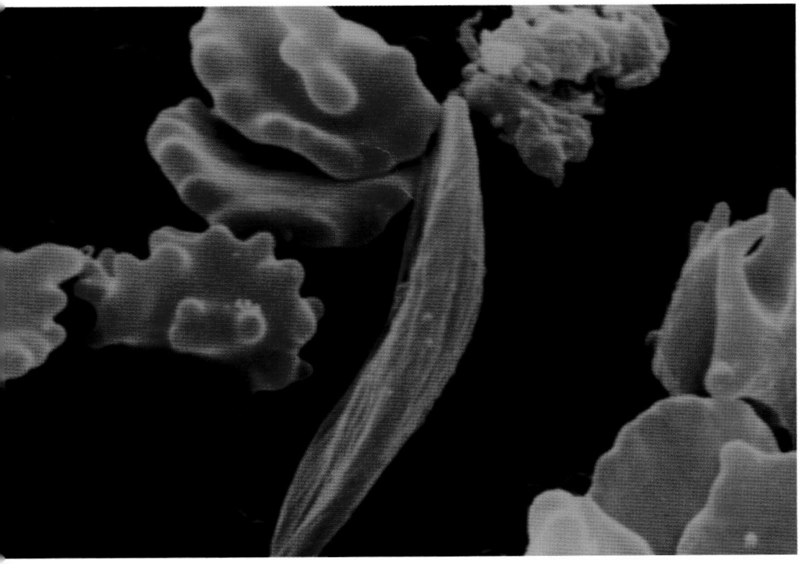

Viral vectors

Methods for achieving the insertion of new DNA are known as 'vectors'. Viruses are a commonly used vector. These infectious agents are normally associated with the spread of disease: they hijack the DNA-replicating machinery inside cells, and use it to make copies of themselves, which burst forth, destroying the cell. But by cleverly engineering out the destructive elements of a virus and implanting the therapeutic genes in amongst the viral DNA, it is possible to reprogram the virus to deliver new, modified genes right where they are needed. This was the method used in the 1990 clinical trial to modify the defective white blood cells.

DNA rejection

Gene therapy is not without its risks. If the new genes are put in the wrong place they can cause genetic mutations leading to potentially cancerous tumours. Introducing new genes can also trigger a dangerous immune response, with the new genes being rejected as foreign – in much the same way that transplanted organs can sometimes be rejected.

In 1999, 18-year-old Jesse Gelsinger died from just such a complication during a clinical trial of a gene therapy for liver disease. He was lacking an enzyme for processing ammonia – a natural by-product of protein breakdown in the body – allowing ammonia to accumulate and reach life-threatening levels. A virus carrying a gene to correct the condition was injected directly into him, but his immune system attacked it. He died four days later from organ failure. Tragic cases such as Gelsinger's are a reminder that rearranging the DNA of a living, breathing patient carries an inherent risk. And this is why research into gene therapy has been so painstaking.

'By cleverly engineering out the destructive elements of a virus and implanting the therapeutic genes in amongst the viral DNA, it is possible to reprogram the virus to deliver new, modified genes right where they are needed.'

But progress is being made. In 2006, researchers in Milan developed a technique for masking new genes – effectively making them invisible to immune cells and so lessening the risk of an immune response. In other research, scientists are developing therapies to combat cancers, AIDS, and even diseases that are not caused by genetic abnormalities. Cells can be altered to make them more responsive to radiotherapy and chemotherapy, and by programming infected cells to self-destruct.

Gene silencing

One promising research field is 'RNA interference', or RNAi. This works by blocking the action of ribonucleic acid (RNA) within cells, which act as a chemical messenger to carry genetic information to the sites where it is used to build proteins. By stopping particular genes from making proteins it is possible to prevent the HIV virus from reproducing, and to silence the genes responsible for some brain diseases, such as Huntington's, which had previously been untreatable. The American researchers who pioneered RNAi were awarded the 2006 Nobel Prize in Physiology or Medicine.

In the future, gene therapy could offer cures for other common ailments, such as depression and even baldness. But there may be a long wait for these treatments to materialize, while medical researchers – not wanting to repeat the tragedies of past clinical trials or neglect the sensitive ethical issues surrounding gene therapy – pick their way through the code of human life slowly but surely.

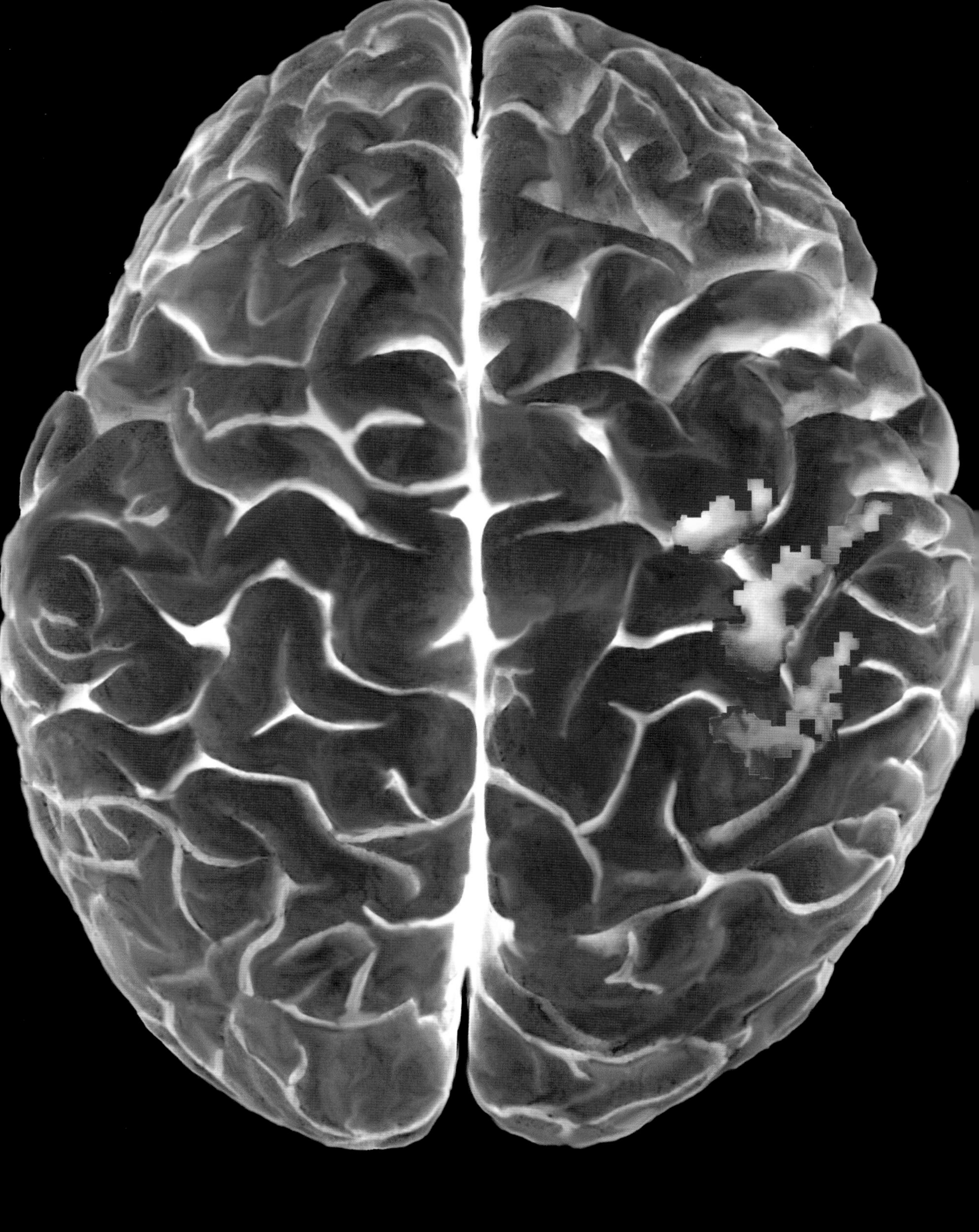

92 Functional MRI scanners

DEFINITION A WAY OF PRODUCING IMAGES OF ACTIVITY IN THE BRAIN, USING MAGNETIC FIELDS AND RADIO WAVES

DISCOVERY THE PRINCIPLE UPON WHICH FMRI SCANS ARE BASED WAS DISCOVERED BY JAPANESE RESEARCHER SEIJI OGAWA IN 1990

KEY BREAKTHROUGH OGAWA DISCOVERED HOW ACTIVE BRAIN AREAS HAVE DIFFERENT MAGNETIC PROPERTIES TO INACTIVE AREAS

IMPORTANCE IT IS A NON-INVASIVE TECHNIQUE FOR DIAGNOSING BRAIN DISEASE WITHOUT THE HARMFUL RADIATION OF AN X-RAY

Functional magnetic resonance imaging, or fMRI, is a way to look inside someone's head and tell which parts of the brain are functioning normally, and which are not. It allows doctors to diagnose brain illnesses and brain surgeons to plan operations with pinpoint accuracy. The technology of fMRI can also reveal what people are thinking, leading to its use in lie detection.

For millennia, doctors had only been able to examine patients externally. Then, in the early 20th century, the development of medical X-ray machines allowed physicians to see inside patients without cutting them open. But there was still a problem. X-rays are a harmful form of radiation and continued exposure leads to cancer and genetic mutations.

MRI springboard

In the 1970s, a new scanning technique became available, known as MRI, or magnetic resonance imaging. MRI exposed the patient to magnetic fields and radio waves, but no harmful X-rays. It applies a very strong magnetic field to the human body – around seven teslas (by comparison, the magnetic field of the Earth is about three ten-thousandths of a tesla, while the field of a fridge magnet measures around five thousandths of a tesla).

Generating magnetic fields this large takes a very big magnet, and most medical MRI machines take up a whole room. The enormous field strength also means that no one is allowed to bring metal objects near an MRI machine. Even a tiny nickel earring can become a lethal projectile in such strong magnetic fields. This is also the reason why anyone with metallic prosthetic implants, such as a pacemaker, is ineligible for an MRI scan.

LEFT This fMRI scan of a human brain shows brain activity while the patient moves their left hand (red/brown area). The technology of fMRI can also reveal higher thought processes, such as telling the difference between feelings of love and lust.

On the pulse

The magnetic field of the MRI scanner causes proton particles inside water molecules in the patient's body to snap into alignment with the field. This happens because charged subatomic particles are themselves rather like tiny magnets – they align with the field in the same way a compass needle aligns itself with the magnetic field of the Earth. Next, the patient is subjected to a pulse of radio waves. Some of these radio waves are absorbed by protons inside the patient's body, knocking them out of alignment. The effect is only momentary, though, and a very short time later the protons spring back, re-releasing a burst of electromagnetic energy as they go. This pulse of energy is detected by the MRI machine and used to form an image.

Different tissue types have different magnetic properties and this causes their particles to snap back into line with the magnetic field at different rates, which is how the resulting image is able to discern between different tissues, such as fat, muscle and, perhaps, a tumour. MRI is used for non-invasive imaging of all areas of the human body. The resolution (the level of detail visible) of the images is comparable with that revealed by an X-ray, but the contrast between different tissue types is normally better, with the added benefit that the MRI is less damaging than X-rays.

'The images produced by an fMRI scan are accurate to around a millimetre. That means if surgery is required to treat a condition, a scan allows neurosurgeons to plan their procedures with pinpoint accuracy.'

Functioning regions

The early 1990s brought an even more important discovery in MRI research, when it became clear that the technique could be used to map brain activity. This is known as fMRI, or functional magnetic resonance imaging, because it highlights 'functioning' regions of the brain, rather than different tissue types.

The key breakthrough that enabled fMRI was made in 1990 by the Japanese scientist Seiji Ogawa. Ogawa discovered an effect in the brain called blood-oxygen-level dependence, or BOLD for short. Brain activity requires oxygen and this is provided by the blood – red blood cells carry molecules of oxygen from the lungs around the rest of the body. When the oxygen is used up, the blood becomes de-oxygenated. Ogawa discovered that the brain overcompensates for this de-oxygenation by massively increasing the blood flow to active areas, thus creating an excess of oxygenated blood in those regions.

Magnetic blood

Crucially, oxygenated and de-oxygenated blood have different magnetic properties. Oxygenated blood is called 'diamagnetic', which means that when it is exposed to an external magnetic field it tends to set up an opposing magnetic field inside itself and thus repels the external magnetic field. On the other hand, de-oxygenated blood is 'paramagnetic', meaning that it creates a magnetic field that attracts any external field applied to it. An fMRI scan can detect these magnetic differences and tell which regions of the brain are active. The scan can be used to diagnose brain diseases by comparing the observed brain functionality with that of known healthy brains. The images produced by an fMRI scan are accurate to around a millimetre. That means that if surgery is required to treat a condition, a scan allows neurosurgeons to plan their procedures with pinpoint accuracy.

One drawback of MRI imaging is that while an X-ray is taken instantaneously, an MRI scan can take up to one hour to complete. The patient is laid on a table that is then slid into a cavity in the heart of the machine, which in some cases can be just 50cm (20 inches) high, making it a somewhat claustrophobic way to spend 60 minutes. In addition, the patient must remain perfectly still for the entire duration.

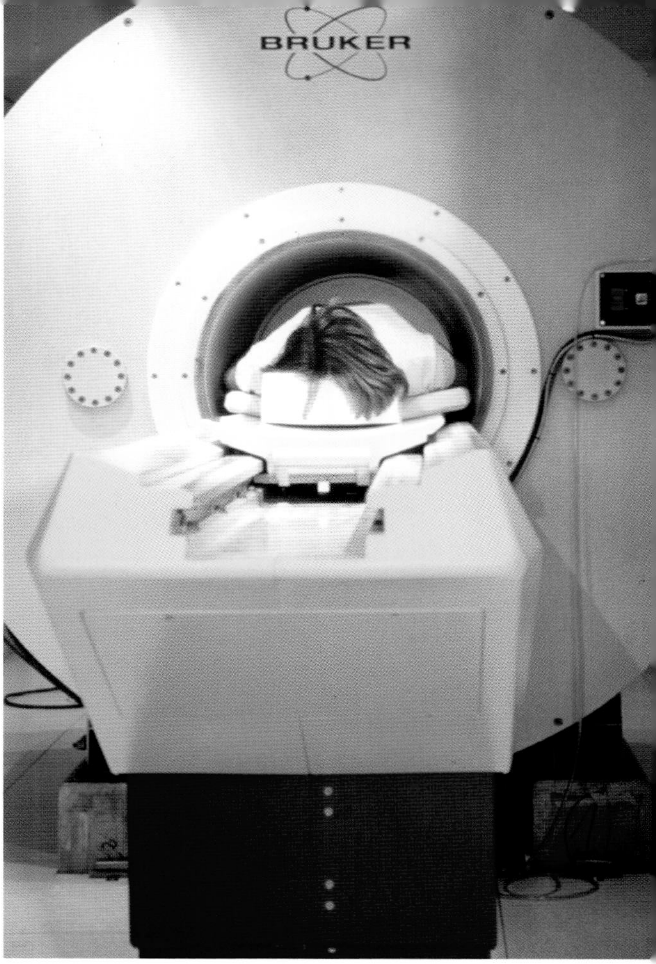

ABOVE Body scan inside an fMRI machine. It is a claustrophobic experience for the patient because the cavity in the scanner is very small. The machine is also noisy as the magnetic field moves back and forth. Patients may listen to music as a distraction or take sedative drugs.

Mind-reading

Intriguingly, fMRI brain scanning is now being applied outside of medicine as a way to, quite literally, figure out what people are thinking. The key is that different types of thoughts take place in different areas of the brain. For example, in one study carried out by researchers at Stanford University, a group of volunteers were scanned while being shown a range of products that they had the option to buy. The nucleus accumbens – a brain region associated with pleasure – became active when the products were viewed. But when the price was shown, the medial prefrontal cortex – associated with balancing gains against losses – lit up. Being able to view brain activity while individuals perform tasks, or answer questions, opens the door to new research applications in psychology, marketing and lie detection.

A liar can often be rumbled on an fMRI scan just by looking at the volume of activity going on in his or her brain. Whereas telling the truth simply involves retrieving and recounting information from memory, lying is harder because the liar needs to continually check that the new elements of their story are consistent with what they have already said.

Despite ethical concerns, other researchers are applying fMRI in marketing by finding out what it is in the brain that makes us willing to part with our money when shopping. It seems that fMRI, a technique that revolutionized medical imaging 20 years ago, is now seeding a new revolution by giving scientists the incredible ability to read other people's minds.

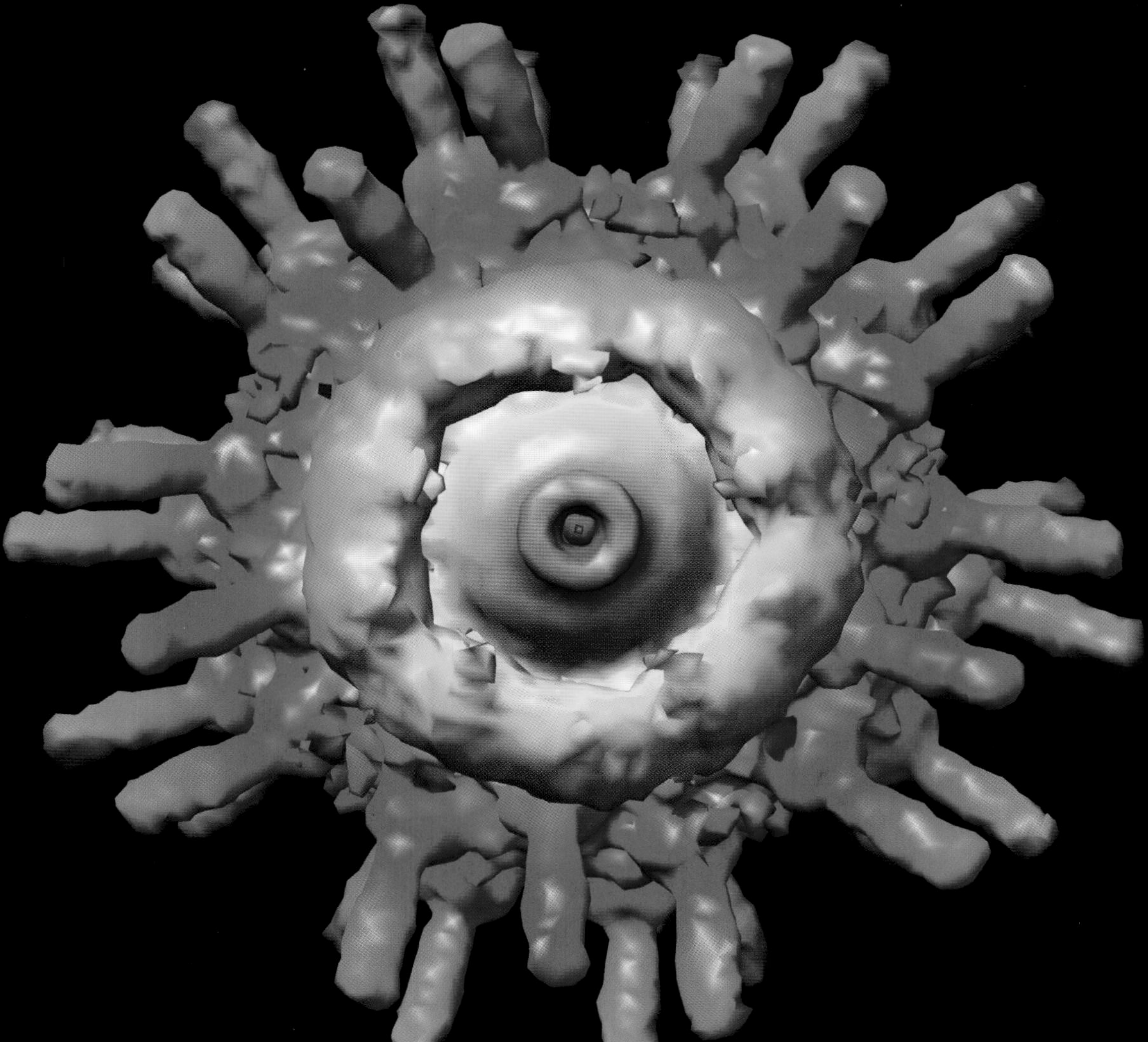

93 Evidence-based medicine

DEFINITION THE APPLICATION OF STATISTICAL AND COMPUTERIZED SYSTEMS TO OPTIMIZE DECISION-MAKING IN CLINICAL MEDICINE

DISCOVERY BECAME A REALITY IN 1993 WITH THE ESTABLISHMENT OF THE COCHRANE LIBRARIES OF CLINICAL TRIALS DATA

KEY BREAKTHROUGH THE LIBRARIES WERE SET UP TO IMPLEMENT RECOMMENDATIONS MADE BY DOCTOR ARCHIE COCHRANE IN 1971

IMPORTANCE EVIDENCE-BASED MEDICINE ENSURES HEALTHCARE RESOURCES ARE SPENT ON THE MOST EFFECTIVE TREATMENTS

Most of us would expect modern medical care to be based on solid scientific evidence, and it generally is. However, evidence-based medicine takes this a step further, demanding rigorous recourse to statistical data or detailed computer models to guide the course of decision-making in both healthcare policy and in the treatments prescribed for individual patients.

In 1747, a Scottish naval surgeon called James Lind carried out an experiment that marked a turning point in medical history. At the time, the disease scurvy was wreaking havoc in the British Royal Navy by killing high numbers of sailors on long journeys. To see if he could come up with a treatment, Lind took a group of 12 serving sailors who were suffering from the disease. He divided them up into six groups of two and gave each group a different daily supplement to their rations – such as vinegar, cider and garlic. One group was given citrus fruits and after just six days of taking the ration supplements, this group of sailors had fully recovered from the scurvy.

Lind had carried out the first-ever clinical trial – a controlled test of potential treatments to see which ones work and which do not. Modern clinical trials often rely on hundreds or thousands of volunteers. This ensures the test is based upon a large-enough sample so that positive results can clearly show themselves, and that any side effects can also be caught and noted.

LEFT Model of a bacteriophage virus created using Chimera software. The program sifts through electron microscopy data to construct its molecular models. This gathering and interpretation of data is a type of evidence-based medicine known as bioinformatics.

Placebo factor

Modern clinical trials make use of what are called placebos. These are dummy medications and contain no active ingredients. The trials are designed in such a way that whoever receives the real drug or the placebo is determined randomly. The randomization is 'double-blinded', which means that neither the patients nor the doctors administering the trial are told who has a placebo and who has the drug. The idea is to look for treatments that perform better than the placebo without introducing dangerous side-effects.

Clinical trials form part of an essential element of modern medical practice called 'evidence-based medicine'. Most of us would probably – and quite rightly – expect all medicine to be based on evidence, and indeed it is. Evidence-based medicine, however, involves applying the lessons learned in both clinical trials and in clinical practice systematically and comprehensively.

Help with difficult decisions

It is all about optimized decision-making. Global corporations use this approach when faced with decisions that are too complex or based on too many variables for the correct decision to be obvious. Instead, they employ risk assessors and operations researchers to mathematically model elements of their business (such as the pricing of stocks and shares) so that they can calculate the best possible decision (that is, whether to buy or sell). Evidence-based medicine essentially means applying the same approach in medical decision-making – both at the policy level and in the treatment of individual patients.

'Evidence-based medicine involves applying the lessons learned in both clinical trials and in clinical practice systematically and comprehensively.'

For example, in healthcare policy, imagine a set of guidelines are to be drafted to advise general practitioners on how they should deal with cases of a particular illness. The traditional method would be for a panel of medical experts to convene and pool their knowledge, arriving at a set of guidelines on diagnosing and treating the condition. The guidelines could be based upon the knowledge, experience and anecdotes of less than ten people.

In evidence-based medicine, on the other hand, the approach is more systematic. There would be a lengthy phase of data-gathering and statistical analysis, so that the guidelines are based on the strength of indisputable facts rather than reported ideas and subjective opinions. Insights from practising healthcare professionals remain important, but a much wider spectrum of individuals would be canvassed.

Another example might be containing the outbreak of a deadly flu epidemic. Rather than basing the response on suggestions and ideas, authorities adopting an evidence-based approach would consult mathematical epidemic models that take into account the best information about how quickly the virus spreads, the time since the first case was reported and a multitude of other factors to arrive at the best course of action. The same principles apply in an evidence-based approach to individual healthcare, and this is where clinical trials come in. In 1971, Scottish doctor Archie Cochrane published a book entitled *Effectiveness and Efficiency: Random Reflections on Health Services*

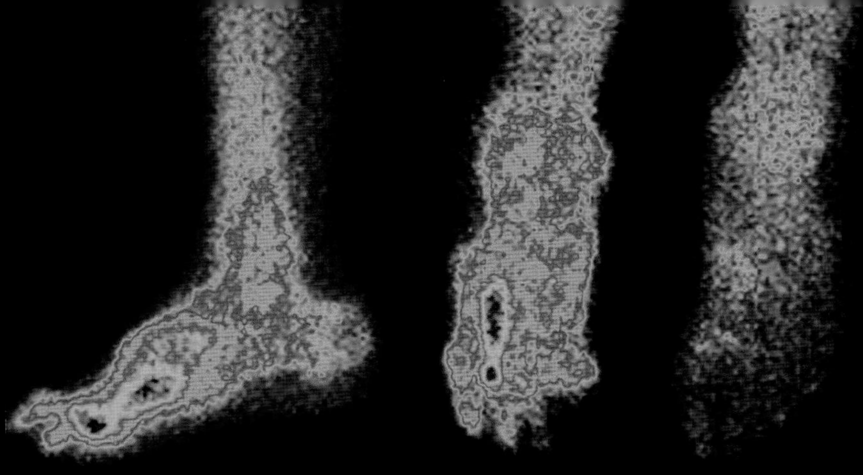

LEFT Gamma-ray scan of a foot ulcer – a common complication in diabetes. The scan works by injecting a radioactive isotope into the blood, then monitoring radiation emissions to gauge the blood flow around the infection. The yellow and black areas reveal a reduced blood supply. Scans are essential in gathering medical data and planning surgery.

in which he called for the establishment of a regularly updated database of systematic reviews of clinical trials results.

Data mining

Cochrane's motivation was simple – just as a small panel of medical experts putting together a set of guidelines is limited by the experience and knowledge of those individuals, so a doctor's ability to diagnose a patient and prescribe the correct treatment is constrained by his own clinical experience. What if, wondered Cochrane, all doctors had access to a universal database where they could look up concise and up-to-date summaries and evaluations of new medical research? This would potentially give every practising doctor access to the findings and experiences of every other doctor.

Archie Cochrane died in 1988. But in 1993, the system that he had envisaged was set up as the Cochrane Library, a collection of medical databases administered by the Cochrane Collaboration – a non-profit-making organization run by volunteers in over 90 countries.

Doctor.com

Other scientists have taken evidence-based medicine even further by using software that can run an automated diagnosis – cross-matching symptoms with a library of medical data, as well as mathematical algorithms to model the behaviour of certain diseases. This has led to web portals where sufferers of chronic illnesses, such as diabetes, can receive advice on unusual symptoms. Scientists also apply computational techniques in the front line of medical research to gather and process data – a field known as bioinformatics.

Evidence-based medicine is especially effective in nationalized healthcare systems, where limited financial resources make it essential that resources are allocated efficiently. Its analytical approach will be pivotal in taking public healthcare forwards through the 21st century.

94 Extrasolar planets

DEFINITION AN EXTRASOLAR PLANET, OR EXOPLANET, IS A PLANET ORBITING A STAR OUTSIDE OUR SOLAR SYSTEM

DISCOVERY FIRST CONFIRMED BY MICHEL MAYOR AND DIDIER QUELOZ, OF THE UNIVERSITY OF GENEVA, IN 1995

KEY BREAKTHROUGH MEASURED TINY MOVEMENTS OF DISTANT STARS CAUSED BY THE GRAVITY OF PLANETS ORBITING AROUND THEM

IMPORTANCE EXTRASOLAR PLANETS HAVE INTENSIFIED INTEREST IN THE POSSIBILITY OF EXTRATERRESTRIAL LIFE

Our solar system is not the only place in the Milky Way galaxy where there are planets. Astronomers have found hundreds of new bodies – known as 'extrasolar planets' – orbiting around stars many lightyears away. The hunt is now on for extrasolar planets that resemble the Earth. On these worlds life could be thriving.

Do other stars have planets circling around them – or is our solar system unique? Until the mid-1990s, no one knew. Some astronomers even said that planets orbiting other stars are impossible to see because their low-brightness would render them invisible against the glare from their parent stars. Since then, however, nearly 500 of these 'extrasolar planets' have been discovered.

False alarms

The first claimed detection of an extrasolar planet was in 1855 when Captain W. S. Jacob, of the Madras Observatory in India, thought he had seen a planet around the star 70 Ophiuchi. But Jacob was mistaken. More false alarms followed in the 1950s and 1960s. An extrasolar planet was found in 1992, but it was orbiting a pulsar – the superdense remnant of a star that had long since blown itself apart in a supernova explosion. This world was very different to any of the planets of our solar system.

In 1995, Didier Queloz and Michel Mayor, at Haute-Provence Observatory, France, finally pulled the needle from the cosmic haystack – finding a planet in orbit around a Sun-like star. They had realized that even though the planets cannot be seen directly (the sceptics were right about that much), it might still be possible to detect them *indirectly* by the effect they have on

LEFT Artwork of the extrasolar planet OGLE-2005-BLG-390Lb. The planet was detected by the European Southern Observatory in 2005 and is located 3300 lightyears from Earth. It is five times Earth's mass and has an Earth-like rocky core and a thin atmosphere.

their parent star. As a planet orbits, its gravity pulls the star this way and that, giving it a prominent 'wobble' that can either be measured directly, or detected by looking for shifts in the frequency of the star's light.

51 Pegasi b

When Queloz and Mayor applied their method to the Sun-like star 51 Pegasi, they saw just what they were looking for. Their data showed the presence of a Jupiter-like planet orbiting at less than an eighth the distance that Mercury is from the Sun. The planet circles at a terrific speed – completing one orbit every 4.2 days. Astronomers nicknamed the new world Bellerophon, after the hero of Greek mythology who rode the winged horse Pegasus. It was later given its official designation of 51 Pegasi b.

'By late 2010, 490 extrasolar planets had been found. The closest to Earth is just 10.5 lightyears away, orbiting the star Epsilon Eridani – so close that powerful telescopes may soon be able to image it directly.'

The discovery was first announced in the science journal *Nature* on 5 October 1995, and was confirmed on 12 October by American astronomers Paul Butler and Geoffrey Marcy. Butler and Marcy have since become prolific planet-hunters in their own right, personally chalking up 120 of the first 200 extrasolar planets to be found.

But 51 Pegasi b threw up a new puzzle. It was closer to its host star than Mercury is to our Sun, yet it was colossal – half the mass of Jupiter. This was totally at odds with the theory of planet formation, which says there is simply not enough gaseous material this close to a star to make planets so large. Many of these so-called 'hot Jupiters' have now been found orbiting other stars. And they have forced astronomers to rethink their ideas on how these planets are made. The leading idea is a theory known as 'orbital migration'. It states that hot Jupiters began far out from their home star and then fell inwards as friction with clouds of gas and dust surrounding the star eroded their orbits.

In transit

Many more extrasolar planets have now been detected using Queloz and Mayor's wobble method. But it is not the only way to find planets. When designing the Kepler space mission, launched in 2009 for the specific purpose of finding extrasolar planets, NASA scientists adopted a different approach. Rather than wobbles, Kepler looks for 'transits' – the dips in brightness caused as planets move across the face of their star and block out a small portion of its star's light.

Astronomers prefer the transit technique because it allows them to directly measure the size, density and temperature of a planet. By studying such a large number of stars, Kepler provides statistics on how common Earth-like planets are around stars similar to our sun. The transit technique also means astronomers can study the planet's atmosphere using the technique of spectroscopy. This is where astronomers take starlight that has passed through the planet's atmosphere and break it up into a spectrum. The spectrum will have dips and peaks in brightness at different colours caused by the planet's atmosphere absorbing and emitting light, respectively, and these variations can be matched to the patterns of dips and peaks produced by different chemical elements – thus revealing the planet's composition.

By late 2010, a total of 490 extrasolar planets had been found. The closest example to Earth is positioned just 10.5 lightyears away, orbiting the star Epsilon Eridani – so close that powerful telescopes on Earth may soon be able to image it directly. Most amazing of all was the discovery in 2007 of a planet five times the mass of the Earth orbiting the star Gliese 581. Nicknamed 'super-Earth', Gliese 581's planet lies within the habitable zone around its star, where the range of orbital radii means the surface temperature of the planet permits liquid water to exist. And this has fuelled speculation about the possibility of life there.

SIM Lite

The search for Earth-like worlds is now the focus of NASA's extrasolar planet research. By 2020, it plans to launch its Space Interferometry Mission Lite Astrometric Observatory (SIM Lite), a supersensitive telescope that will scan the skies for distant worlds and analyze their atmospheres for tell-tale chemical signatures of life.

Because it is so difficult to take a picture of a planet as small as Earth in orbit around a distant star, SIM Lite uses the technique of optical interferometry to combine the light gathered by multiple telescopes to perform the work of a single, much larger telescope. Whereas other astronomers have found Jupiter-like extrasolar planets by measuring the wobbles of their stars, Sim Lite's interferometer will detect the tiny stellar wobbles caused by Earth-like alien worlds. Mission controllers believe the technique will reveal planets as small as one Earth mass which, if orbiting in the habitable zone, may well bear life.

ABOVE Preparation of the Kepler space telescope before its launch in 2009. Kepler has scanned more then 100,000 Sun-like stars in the Milky Way.

95 Fermat's last theorem

DEFINITION IF n IS A WHOLE NUMBER GREATER THAN 2, THEN THERE ARE NO WHOLE NUMBERS x, y AND z SATISFYING $x^n + y^n = z^n$

DISCOVERY CONJECTURED BY PIERRE DE FERMAT IN 1637; PROOF PUBLISHED BY ENGLISH MATHEMATICIAN ANDREW WILES IN 1995

KEY BREAKTHROUGH WILES PROVED THE THEOREM USING THE ABSTRUSE MATHEMATICS OF ELLIPTIC CURVES

IMPORTANCE WILES'S PROOF WAS THE CULMINATION OF MORE THAN 300 YEARS OF MATHEMATICAL RESEARCH

Fermat's last theorem is so simple that it can be understood by a high school mathematics student. And yet proving it was to be a task so unimaginably complex that it would take the best mathematical brains on the planet almost 360 years. The theorem was first stated, without proof, in 1637 – but not proven until 1995.

Many people learned about Pythagoras's theorem while at school. Named after the Greek mathematician who lived in the sixth century BC, it is a mathematical law linking the lengths of the sides of a right-angled triangle. It essentially states that the square of the longest side (z) is equal to the sum of the squares of the two shorter sides (x and y), giving $x^2 + y^2 = z^2$.

There are certain cases where it is possible to have right-angled triangles with side lengths that take whole-number values. For example, if the two shortest sides are given by $x = 3$ and $y = 4$, then Pythagoras's theorem tells you that the longest side is given by $z = 5$. Combinations of whole numbers such as this are known as Pythagorean triples.

Diophantine development

In the third century AD, the mathematician Diophantus of Alexandria studied mathematical equations of a form similar to Pythagoras's theorem. But rather than dealing exclusively with the sum of squares, these so-called Diophantine equations involved the sum of general powers, denoted by the letter n, giving the equation $x^n + y^n = z^n$. Diophantus published his work in a book called *Arithmetica*, which was finally translated into Latin (the scholarly language of Europe) in 1621.

LEFT Polynomiographic illustration of a polynomial algebraic expression in a single variable, in this case $4x^2 + 3x - 7$. An example of a polynomial in *more* than one variable is $x^n + y^n = z^n$, where n is a whole number. This equation is the basis of Fermat's last theorem.

One person who owned a copy of this translation was French mathematician Pierre de Fermat. He was born in 1601, in Beaumont-de-Lomagne, to the family of a prosperous merchant. Fermat studied first at the University of Toulouse before moving to the University of Orleans, where he trained in law. Indeed, Fermat became a lawyer by trade and his mathematical work remained an amateur pastime. He even refrained from publishing his findings as professional mathematicians do, preferring to communicate his ideas in letters to friends. This, inevitably, led to priority disputes with other mathematicians about who had been the first to make various discoveries.

Doodle clues

Fermat obtained important results in optics, probability theory and did much of the groundwork upon which calculus would later be built. But it was his research in number theory for which he is remembered best, proving numerous theorems regarding the properties of numbers and their series. And by far the most famous of these theorems was found jotted in the margin of his copy of *Arithmetica.*

'It would take the world's mathematicians over 300 years to come up with a watertight proof for this simple mathematical principle – which has now become known as Fermat's last theorem.'

The particular passage of the book in which Fermat noted down his ideas concerned quadratic Diophantine equations – those with $n = 2$, of the form of Pythagoras's theorem. Fermat had scribbled a message in the margin to say that while there are whole-number solutions to the quadratic Diophantine equations – namely, the Pythagorean triples – there are no such whole-number solutions when n is greater than 2. He added that he had come up with a 'marvellous proof' of this theorem, but that this was too long to fit in the margin of the page.

Stated formally, the theorem said that there are no combinations of three whole numbers – x, y and z – such that $x^n + y^n = z^n$, where n is a whole number greater than 2. It would take the world's mathematicians over 300 years to come up with a watertight proof for this simple mathematical principle – which has now become known as Fermat's last theorem. The first hint at a proof emerged in 1984, when German mathematician Gerhard Frey suggested that the secret could lie in the mathematics of elliptic curves – smooth geometric forms that are governed by cubic equations in two-dimensional space. Frey suspected that if Fermat's equation had a solution then a complicated theorem about the properties of elliptic curves – known as the Taniyama-Shimura conjecture – would have to be violated.

Frey's instincts turned out to be correct. In 1986, American mathematician Ken Ribet proved exactly that – the Taniyama-Shimura conjecture implies Fermat's last theorem. Now all that was needed was a rigorous proof of the conjecture – demonstrating beyond doubt that it cannot be violated, and so proving Fermat's last theorem to be correct. Enter Cambridge University mathematician Andrew Wiles. An expert in number theory and elliptic

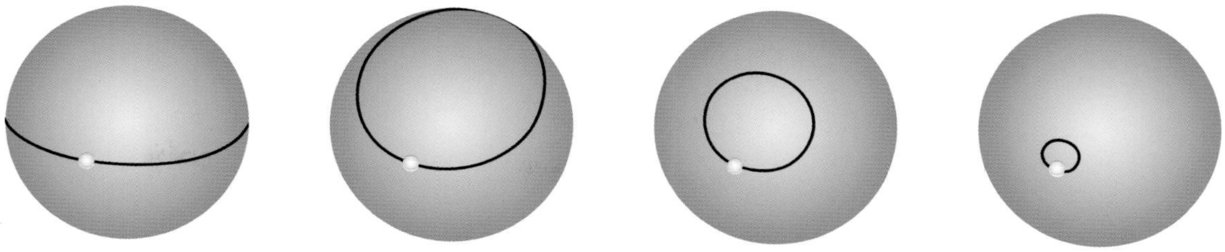

ABOVE The Poincaré conjecture: stretch an elastic band around the two-dimensional surface of a sphere and, by shifting the band over the sphere, you can shrink it down to a point. But does the same hold true for the analogue of a sphere with a three-dimensional surface? In 1904, Henri Poincaré stated that it does, but it was not proven until 2003.

curves, he was the perfect man for the job. Even so, the task before him was so formidable that it took six years of arduous brain work to produce a proof of the conjecture. And even then, in 1993, as other mathematicians were checking his workings prior to their publication, a fatal mistake emerged. It took Wiles a further year, working with colleague Richard Taylor, to finally fix the error.

At last, in 1995, 358 years after Pierre de Fermat had first written down the theorem, Wiles's proof appeared in the research journal *Annals of Mathematics*. The complex mathematical techniques used in the proof did not exist in Fermat's day, a fact that has led many scientists and historians to doubt that Fermat really possessed the proof to which he alluded in his margin note.

Millennium problems

Fermat's last theorem is not the only long-standing problem in mathematics to be recently cracked. In 2003, Russian mathematician Grigori Perelman proved the Poincaré conjecture – an idea put forward without proof 99 years earlier in 1904 by French mathematician Henri Poincaré.

If you stretch an elastic band around a tennis ball, then you know that by sliding back the band carefully over the ball's surface it can be shrunk down to a point. On the other hand, stretch a rubber band around the outside of a ring doughnut and it is impossible to continuously shrink the band to a point because the hole in the middle gets in the way. Mathematicians say that the tennis ball is 'simply connected' while the doughnut is not. Poincaré speculated that the same might be true in three dimensions – for shapes analogous to balls and doughnuts, but with three-dimensional surfaces.

Perelman proved the conjecture using complicated notions from geometry. His solution made him eligible for a grand $1 million prize that had been offered for a proof of the conjecture by the Clay Mathematics Institute, Massachusetts. Perelman graciously declined the prize, claiming he was not motivated by money. He also turned down the Fields Medal for the same discovery, the mathematical equivalent of a Nobel Prize. The Poincaré conjecture was one of seven mathematical puzzles, known as the Millennium problems, for which the Clay Mathematics Institute is offering the cash prize. As yet, it is the only one to be solved.

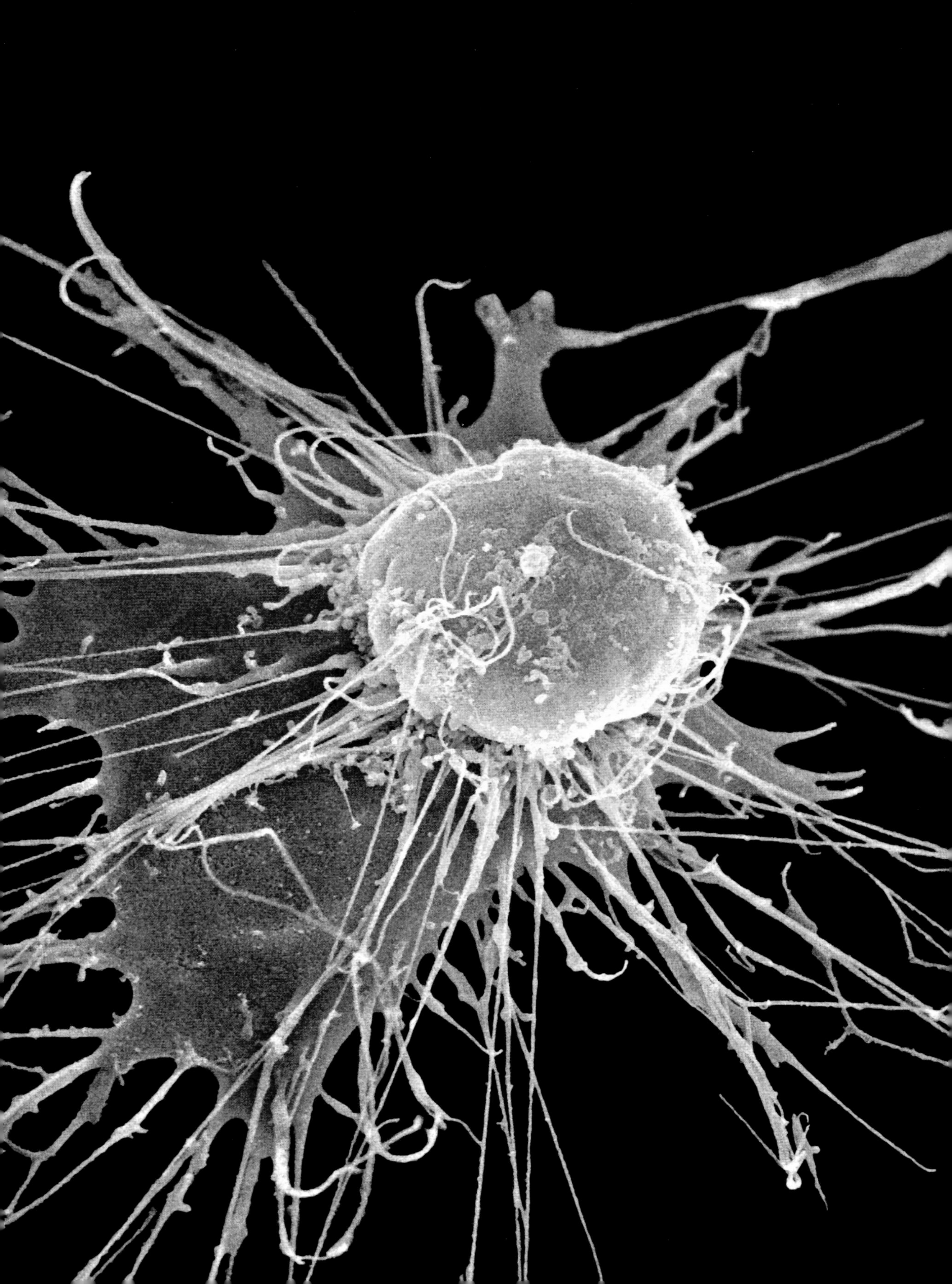

96 Cloning

DEFINITION USING GENETIC TECHNIQUES TO CREATE IDENTICAL COPIES OF ORGANISMS FROM ADULT CELLS

DISCOVERY 1996 BY IAN WILMUT AT THE ROSLIN INSTITUTE, UNIVERSITY OF EDINBURGH

KEY BREAKTHROUGH WILMUT'S TEAM PRODUCED DOLLY THE SHEEP, THE FIRST MAMMAL TO BE CLONED FROM AN ADULT CELL

IMPORTANCE CLONED HUMAN STEM CELLS HAVE THE POTENTIAL TO TREAT NEURODEGENERATIVE ILLNESSES, SUCH AS ALZHEIMER'S

In the final years of the 20th century, a research team in Scotland produced Dolly the sheep – the world's first mammal to be cloned from an adult animal cell. Since then, the procedure has been applied to mice, cows, horses and other animals. How long until the first human is cloned – indeed, could this already have been carried out?

Cloning is the production of genetically identical copies of organisms. When species reproduce sexually, the offspring have a mix of DNA from both parents – which is why we resemble our mothers and fathers but we are not direct copies. A clone, on the other hand, has just one parent to which it is an exact genetic match.

Cloning has long been taking place in the natural world. Organisms such as bacteria, as well as some plants and fungi, reproduce asexually – that is, offspring are exact copies of a single parent. In some species that reproduce sexually, a phenomenon has been observed known as parthenogenesis, where a female egg can develop into an embryo without fertilization by male sperm. This has been seen in insects, reptiles and some shark species. Even cloning by human hands is not really a new trick – any gardener who has taken a cutting from a plant and made it grow has created a clone of that original plant.

Scientists took up the cloning trail in 1952 when the English biologist John Gurdon cloned a frog. The first cloned mammals followed several decades later, when mice were cloned by Russian scientists in 1986. These animals, however, were made from embryonic cells rather than a fully grown animal.

LEFT Human stem cell – a mutable kind of cell, sometimes harvested from an embryo, which can be stimulated to grow into any other cell type. Stem cells taken from cloned embryos could enable surgeons to grow transplant organs that are an exact genetic match to their recipient, meaning there is no chance of rejection.

Using SCNT

The event that really captured the world's attention was the birth of Dolly the sheep – created by Ian Wilmut and colleagues at the Roslin Institute of the University of Edinburgh in 1996. Here, finally, was a large mammal that had been cloned from an adult cell. So how was it done?

Dolly was created by a technique known as somatic cell nuclear transfer (SCNT). The process starts with a mother animal from which an egg cell is removed. The nucleus of the cell – which contains the mother's genetic data – is extracted and discarded. Then, a donor cell is taken from the animal that is to be cloned. This cell is taken from the body of the animal – a 'somatic' cell, as opposed to a reproductive sperm or egg cell. The somatic cell's nucleus is extracted and implanted into the vacant egg cell. The next step is to stimulate the composite cell with an electric shock, which makes it start to divide. When the cell has grown into a 'blastocyst' – a pre-embryonic state of about 100 cells – it is reimplanted into the mother animal where it gestates and, assuming no complications result, is brought to full term.

'Reproductive human cloning means repeating the process used to create Dolly the sheep in a human surrogate mother, to produce a live cloned human being that is genetically identical to the cell donor.'

Only complications do result. In the case of Dolly, it took nearly 300 attempts to produce a living cloned lamb. And the problems did not stop there. After five years of life, Dolly developed age-related illnesses and was eventually put to sleep at the age of six, suffering from lung cancer and arthritis. The normal life expectancy for sheep of her breed is 11–12 years. Since then, other researchers have cloned dogs, cats, horses and even Spanish fighting bulls. The natural question was: how long before we clone humans?

Ethical dilemmas

In fact, we can clone humans right now. The procedure is no different to that in other mammals. The real question is: should we? And to answer that, human cloning is usually divided into two distinct categories – reproductive cloning and therapeutic cloning.

Reproductive human cloning means repeating the process used to create Dolly in a human surrogate mother to produce a live cloned human that is genetically identical to the cell donor. The clone would apear identical and have many similar behavioural traits to the donor. This form of cloning attracts strong opposition and is outlawed in at least 50 nations.

The purpose of therapeutic cloning, on the other hand, is to create cloned cells to treat degenerative diseases. These cells – known as embryonic stem cells – are extracted at the embryo stage of the clone's development, before they have differentiated into specific types of human tissue. The stem cells can then be tweaked to grow into the tissue type required by the patient. So, for example, they could be used to grow brain tissue – neurons – to replace neurons that have been damaged by a neurodegenerative condition, such as Alzheimer's disease. Because the cells are cloned, they are an exact genetic match, so there is no chance of rejection by the patient's immune system, as can be the case with transplant surgery. Therapeutic cloning is allowed in some countries, such as the UK and USA, and is strictly regulated. Depending who you talk to, it is either a medical miracle that promises life-saving treatments or it is the most flagrant example of scientists playing God.

ABOVE Dolly the sheep, born 5 July, 1996, was the first mammal cloned from an adult cell. She is pictured here aged eight months and lived until February 2003. Her remains are on display at the Museum of Scotland, Edinburgh.

A third kind of human cloning has also been mooted. 'Replacement cloning' involves growing back-up bodies for patients into which their brains could be transplanted if their original bodies fail – either through disease or old age. However, the difficulty of reconnecting a brain to the spinal cord of the new body (the same reason why there's currently no treatment for patients paralyzed by quadriplegia) places this firmly in the realms of science fiction – at least for the time being.

Fantasy and fiction

Science fiction and cloning are certainly no strangers. In Michael Crichton's novel *Jurassic Park* dinosaur DNA is extracted from fossils and implanted into modern-day day reptiles via SCNT, to bring the dinosaurs back from the dead. Soviet scientists have been trying for some years to recreate this process with DNA extracted from woolly mammoth remains found encased in Arctic ice. So far, however, they have been unsuccessful – DNA, it seems, is just too fragile to survive the ravages of time.

Other scientists have suggested that cloning could be a way to prevent species from becoming extinct in the first place. But conservationists generally view this with disdain – arguing that it diverts attention from the real issues of habitat preservation and species protection. Many other potential applications have also been put forward – including the cloning of high-yield organisms for agriculture, champion racehorses and even replacing beloved pets that have died with cloned replicas.

The possibilities are incredible, even if the reality may seem far-fetched. But as the boundaries between science and science fiction continue to blur, the fierce ethical and moral debate surrounding cloning and its applications looks certain to rumble on.

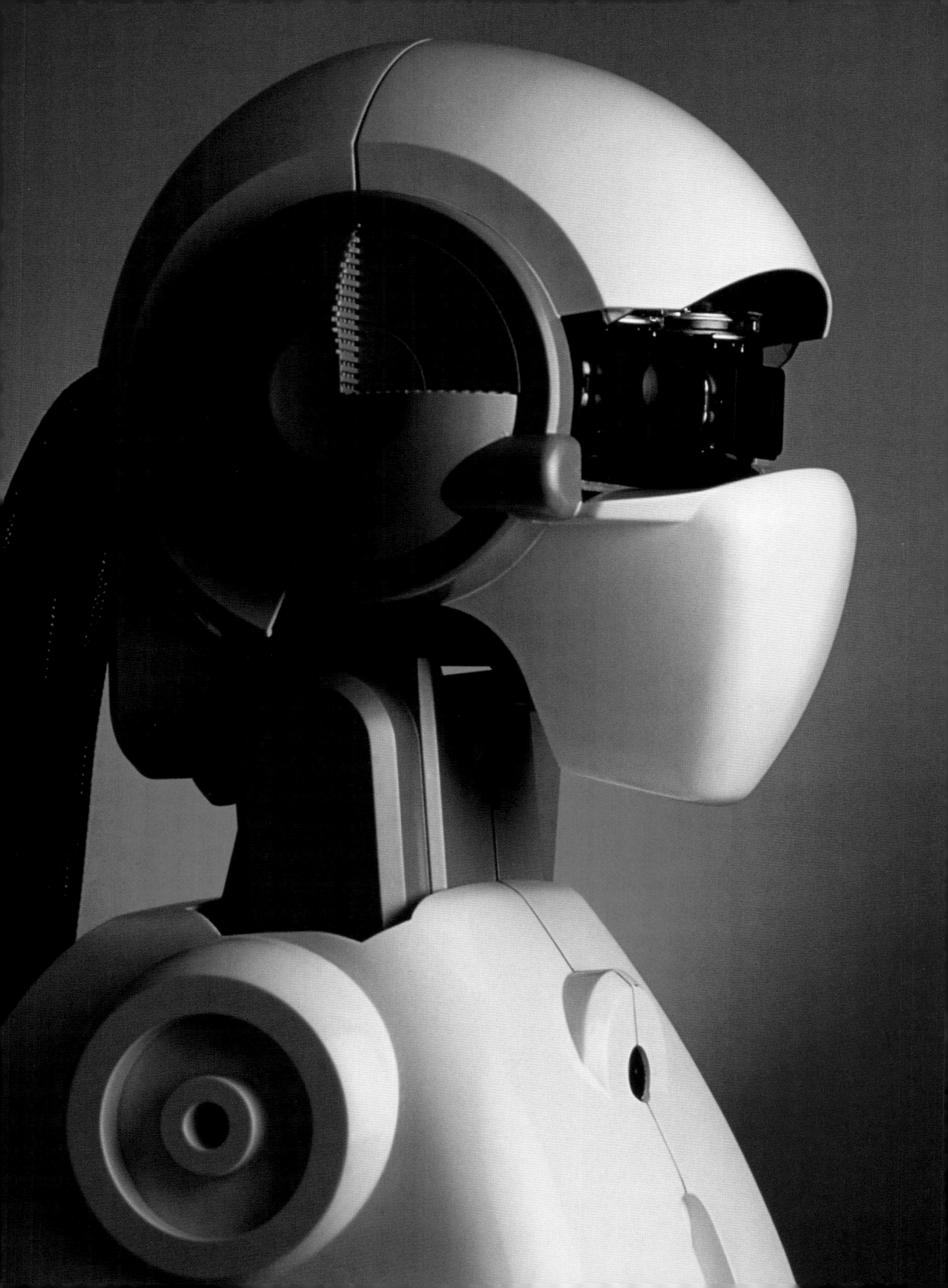

97 Deep Blue beats Kasparov

DEFINITION DEEP BLUE WAS A CHESS-PLAYING SUPERCOMPUTER DEVELOPED IN THE 1990S BY RESEARCHERS AT IBM

DISCOVERY IN 1997, DEEP BLUE WON A SIX-GAME MATCH AGAINST REIGNING WORLD CHESS CHAMPION GARRY KASPAROV

KEY BREAKTHROUGH DEEP BLUE OUTPERFORMED KASPAROV'S BRAIN

IMPORTANCE THE VICTORY OF A CHESS COMPUTER OVER A GRANDMASTER SYMBOLIZED THE RISE OF MACHINE INTELLIGENCE

It is a story that has enjoyed many an outing in science fiction – what will happen when computers ultimately outsmart the human brain? In 1997, the reigning world chess champion Garry Kasparov got a special preview of this scenario when he was beaten by the IBM chess computer Deep Blue. This triumph of a machine over a human genius was a milestone in artificial intelligence research.

History is littered with tales of inventors on quests to build machines that can think and act for themselves. The Ancient Greeks and Egyptians tried to design automatons – robotic slaves that could do their masters' bidding. And scholars in the Middle Ages are believed to have constructed their own crude prototypes.

These were primitive contraptions. At the heart of any robot that can replicate human behaviour needs to be a device. which can reproduce human thought processes. That breakthrough had to wait until the 20th century and the invention of the modern digital computer. Indeed, English mathematician Alan Turing – the inventor of the modern computer (see page 277) – was quick to realize that his invention could enable scientists to build their own artificial brains, and even began musing over his own designs.

Tragically, Turing took his own life in the early 1950s – just a few short years before the first scientific conference on such thinking machines was held at Dartmouth College, New Hampshire, in 1956. It was there that a young American computer scientist called John McCarthy coined the term 'artificial intelligence', or 'AI', to describe these devices.

LEFT Humanoid robot built as part of the Erato project in Tokyo, Japan. The project builds robots that are able to evolve their intelligence as they study their environment. Artificial intelligence came of age in 1997 with the victory of IBM's chess computer over the world's greatest human chess player, Garry Kasparov.

AI research

Initially, the AI field was buoyed up by high hopes. Research into computer science was progressing in leaps and bounds, and it was reasonable to assume that these incredible machines – which could solve complex mathematical problems in next to no time – would soon mimic other facets of the human mind. But viewing human thought processes as mathematical equations was a mistake. While aspects of logical thought, deduction and reasoning may well be intrinsically mathematical, other brain functions are inherently fuzzy in nature – such as creativity, intuition and learning. Solving these problems soon proved much harder than anticipated and the artificial intelligence growth curve flattened off.

Much of the problem was that no one really knew how the brain works, and thus what exactly it was that researchers were trying to model. Overcoming this obstacle was helped by the development of cognitive psychology, which tried to model the basic workings of the brain as a computational process. The merger of computer science and cognitive psychology gave rise to 'cognitive science', an area of research that is still thriving today.

Are you neat or scruffy?

There soon emerged a broad division of AI research into two camps – loosely branded 'neat' and 'scruffy'. Researchers adopting the neat perspective concentrated on applying logic and established programming methods to try to carefully engineer intelligent behaviour. On the other hand, the scruffies were more concerned with hacking together odds and ends of hardware and software and tweaking it until the requisite behaviour simply emerged.

Neural networks were one kind of AI system to come out of the scruffy approach. These are software creations that mimic the way that neurons in the human brain process information. No surprise then that they are extremely good at quintessentially human skills, such as recognizing faces or learning from experience. On the other hand, the neat approach led to one of the highest profile triumphs of artificial intelligence – the first victory of a chess-playing computer over a human grandmaster. In fact, Alan Turing himself had written the first program for a chess computer in the 1950s, but no computer at the time was sophisticated enough to run it.

'Deep Blue was capable of evaluating 200 million different board positions every second and holding the moves of some 700,000 games against grandmasters in its working memory.'

In the 1980s, computer scientists at Carnegie Mellon University developed a chess-playing computer called Deep Thought. It could beat other chess computers, but when in 1989 its creator,

Feng-hsiung Hsu, pitted it against world chess champion Garry Kasparov it was roundly defeated. Hsu was still a student during his work on Deep Thought. But following graduation, computer firm IBM recruited him, along with other computer scientists, to continue the research. Their new program was initially called Deep Thought 2, but renamed Deep Blue after IBM's corporate nickname 'Big Blue'.

LEFT Audience members look on as the reigning world chess champion Garry Kasparov plays against Deep Blue. At the start of the sixth game of six, the match was a tie. Then, Deep Blue forced a mistake from Kasparov that cost him the match.

Checkmate

In 1996, Deep Blue faced Kasparov. Although it lost the six-game match, played under strict tournament time constraints, it did win one individual game – losing three and drawing the other two. After the match, Hsu and colleagues took what they had learned about Deep Blue's weaknesses and improved it. By 1997, Deep Blue had undergone numerous modifications. It was now capable of evaluating 200 million different board positions every second, twice as fast as the incarnation that played Kasparov in 1996, and it held the moves of some 700,000 games against grandmasters in its working memory.

In May 1997, Deep Blue was pitted against Kasparov again. This time, three of the six games were drawn, Deep Blue won two games and Kasparov won just a single game. Deep Blue had won the match. A machine had beaten one of the best human brains – it was a major milestone in AI research.

Since then, other chess computers have triumphed against grandmasters. Most notably, in 2005, a computer called Hydra won five out of six games against English chess player Michael Adams, only drawing the sixth. And scientists are now developing software that can trounce humans at other games too, such as poker and checkers.

Moore's law

The number-crunching power of computers continues to grow at an alarming rate, doubling roughly once every 18 months – a progression known as Moore's law, after Intel founder Gordon Moore who first spotted the trend. Some artificial intelligence researchers have argued that this will lead to a point at which computers will always be able to out-think the human brain. A computer that attains this ability is known as a 'superintelligence'. A superintelligence would be able to re-engineer its own architecture quicker and more effectively than any human designers. This, in turn, makes it more intelligent still, and able to improve itself at an even faster rate. The pace of advancement soon becomes infinite, leading to a state that American AI researcher and futurologist Ray Kurzweil has termed 'the singularity'. Kurzweil is convinced we will reach the singularity by the end of this century.

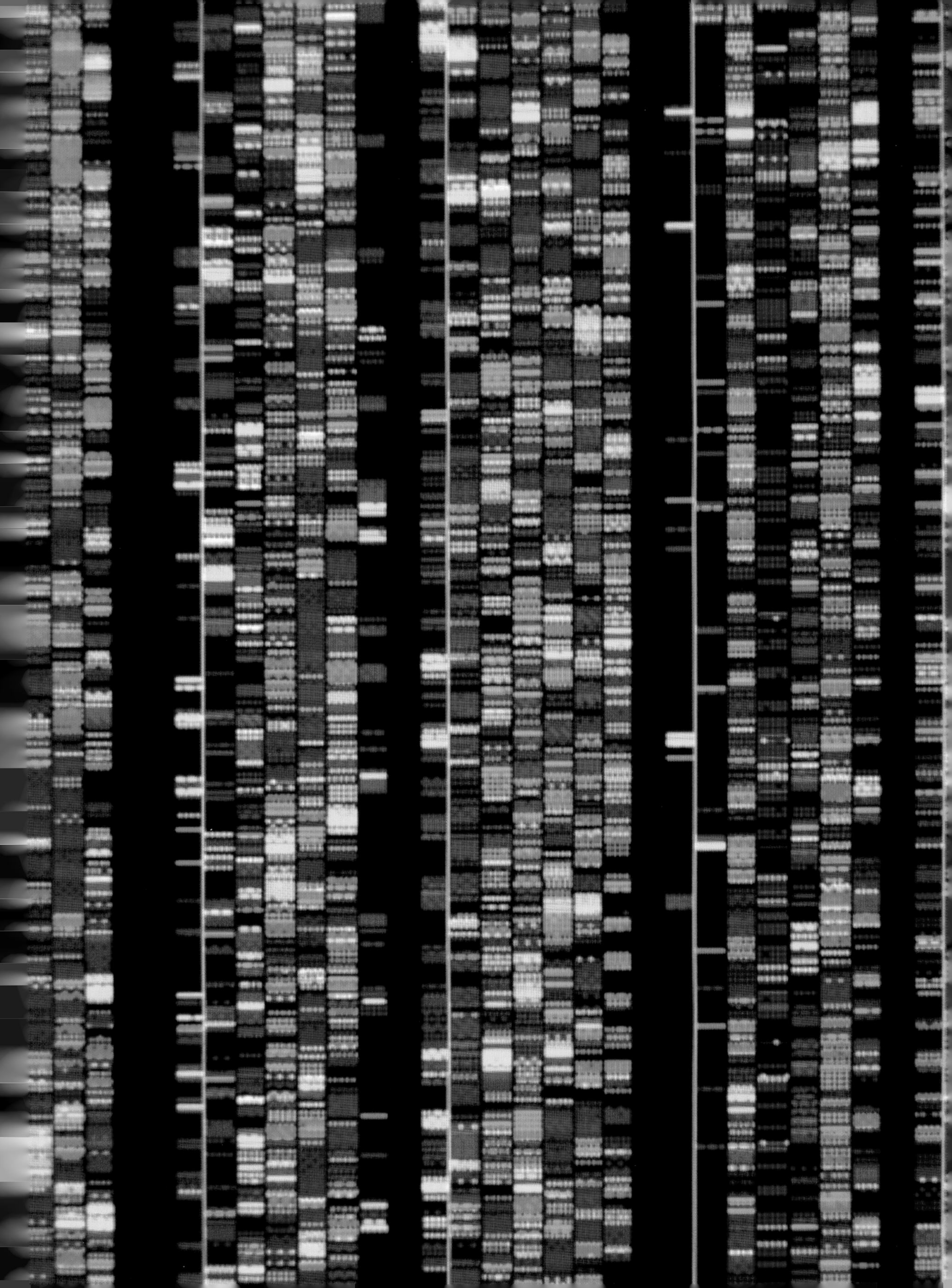

98 Human genome

DEFINITION THE SEQUENCE OF DNA CODE THAT LIES AT THE HEART OF EVERY CELL OF EVERY HUMAN BEING

DISCOVERY THE HUMAN GENOME WAS SEQUENCED IN 2003 BY THE INTERNATIONAL HUMAN GENOME PROJECT

KEY BREAKTHROUGH CREATION OF FASTER TECHNOLOGY TO READ AND RECORD THE THREE BILLION BASE PAIRS IN THE HUMAN CODE

IMPORTANCE SEQUENCING THE HUMAN GENOME SERVES MEDICINE, EVOLUTIONARY SCIENCE, FORENSICS AND OTHER FIELDS

The human genome is the genetic code that underpins the human species. It was first read off in 2003, the culmination of a 13-year project. The data reveals that 99.9 percent of every person's genetic code is the same and that individual characteristics are down to minuscule differences. Studies of these differences have ushered in a new field of genetic medicine.

Genes are the code of life – the language in which all the characteristics of an organism are written. When your body needs to make new cells, say to heal an injury, your genes form the blueprint from which those cells are made. We pass on our genes to our children, and that is why they look like us.

Genes are stored on a chemical molecule called DNA that sits at the centre of every cell in your body. The molecule is a long string of chemical 'bases' – labelled A, T, C and G – the precise sequence of which is what encodes the genetic information. The complete genetic code of any organism – that is, the complete sequence of all its As, Ts, Cs and Gs – is known as its 'genome'.

The Human Genome Project is an international effort to determine the sequence of chemical bases in the DNA of human beings, and place this information in readily accessible databases for other researchers around the world to use. Its goals include discovering the differences between the genetic make-up of humans and other species. Although we each have genetic differences that account for traits such as eye colour and intelligence, 99.9 percent of the human genome is common to all members of our species. The $3 billion project began in 1990 with the target of achieving a working sequence of the human genetic code in 15 years. However, technological

LEFT Section of the human genome. The colours denote specific chemical bases – the building blocks in DNA – the order of which determines everything from your susceptibility to certain illnesses to the colour of your eyes.

advances allowed the project to proceed faster than anticipated. A rough 'working draft' of the code was announced to the world in 2000 by the then US president Bill Clinton and British prime minister Tony Blair. The finished draft was published three years later, in spring 2003 – exactly 50 years on from the discovery of the double helix structure of DNA (see page 305).

'If the three billion base pairs from which it is composed were printed out on A4 sheets of paper, at the rate of 1000 pairs per sheet, then the list would take up three million sheets, or 900km (559 miles) of paper.'

Sequencing the three billion base pairs that make up the human genome was a long and complex process. Human DNA was cut into lengths, which were then sequenced using a method called 'chain termination'. This was first developed by English biochemist Frederick Sanger (who used it in 1977 to create the first-ever genome sequence – of a bacteriophage virus). It works by synthesizing a copy of the test DNA sample, base by base. After each stage in the construction, one of four chemicals is added to the sample. Each chemical reacts with one of the four nucleotide bases – A, T, C or G – to stop the synthesis of the new DNA copy. Observing which chemical does this reveals the next base in the sequence. And by repeating the process over and over again scientists can deduce the entire base sequence of the DNA strand.

Random slices

The Human Genome Project used a variant on the chain termination technique, called 'shotgun sequencing'. Chain termination is only really practical for short DNA sequences – much shorter than the lengths of human DNA being studied, which were typically 150,000 base pairs long. Shotgun sequencing involves randomly slicing each DNA strand into shorter fragments of between 100 and 1000 base pairs. Each fragment is then sequenced by chain termination. The process is repeated several times and the results from each run are fed into a computer, which matches up the overlaps between the fragments in different runs (remember, the original DNA strand is sliced up randomly each time) to reconstruct the sequence of the original strand.

The findings of the project make for interesting reading. The human genome itself is vast. If the three billion base pairs from which it is composed were printed out on A4 sheets of paper, at the rate of 1000 pairs per sheet, then the list would take up three million sheets, or 900km (559 miles) of paper – enough to stretch from Washington, DC to Atlanta, Georgia.

Junk genes

The human genome consists of around 25,000 individual genes. Each of these is made up of around 3000 base pairs of DNA, although that is an average – the longest is composed of an epic 2.4 million base pairs. The big surprise was how much of the DNA code is 'junk', unused sequences of bases that sit between the useful genes – 98 percent of the human genome appears

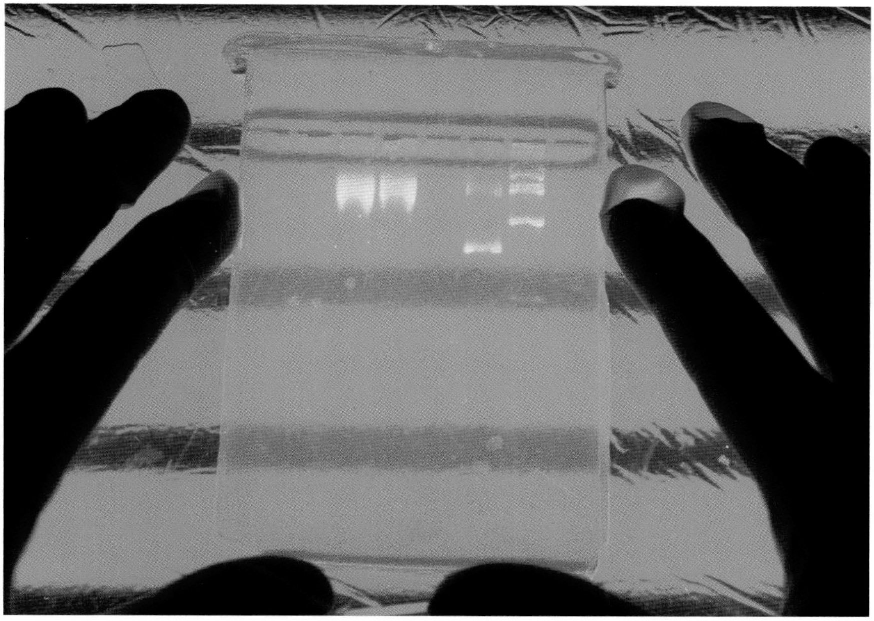

LEFT Electrophoresis gel is used to separate and study lengths of DNA according to their mass. DNA is embedded in the gel, then an electric field is applied. The charged DNA drifts through the gel in response to the field, leaving a trail (shown pink).

to serve no obvious purpose. The project has also identified over a million sites in the human genome where bases can differ between individuals. For example, you might have an 'A' base at the same location in the genome where I have 'C'. These sites are of particular interest because they have the potential to reveal why some people are susceptible to specific diseases. This information will ultimately be invaluable for developing gene therapies (see page 369). In particular, a subfield called 'pharmacogenomics' promises bespoke medicine, by reading the genomes of patients and then using the information locked away in the genetic code to tailor treatments that are as effective as possible while minimizing side-effects.

Secrets revealed

Already, some prominent figures in the development of genetics have had their genomes fully sequenced – including James Watson and Craig Venter, who invented the shotgun sequencing technique. The letters in Venter's own genome are associated with a physical tendency towards wet earwax and a predisposition to antisocial behaviour, amongst other things.

The Human Genome project is still ongoing. Scientists there are only just beginning the Herculean task of interpreting the vast pool of data they have created. In addition to the medical benefits, it is hoped that improved understanding of human genetics will bring new insights in molecular biology and to the theory of evolution – as well as better equipping scientists to address the pressing ethical and social issues that accompany a branch of science so close to the very core of humanity.

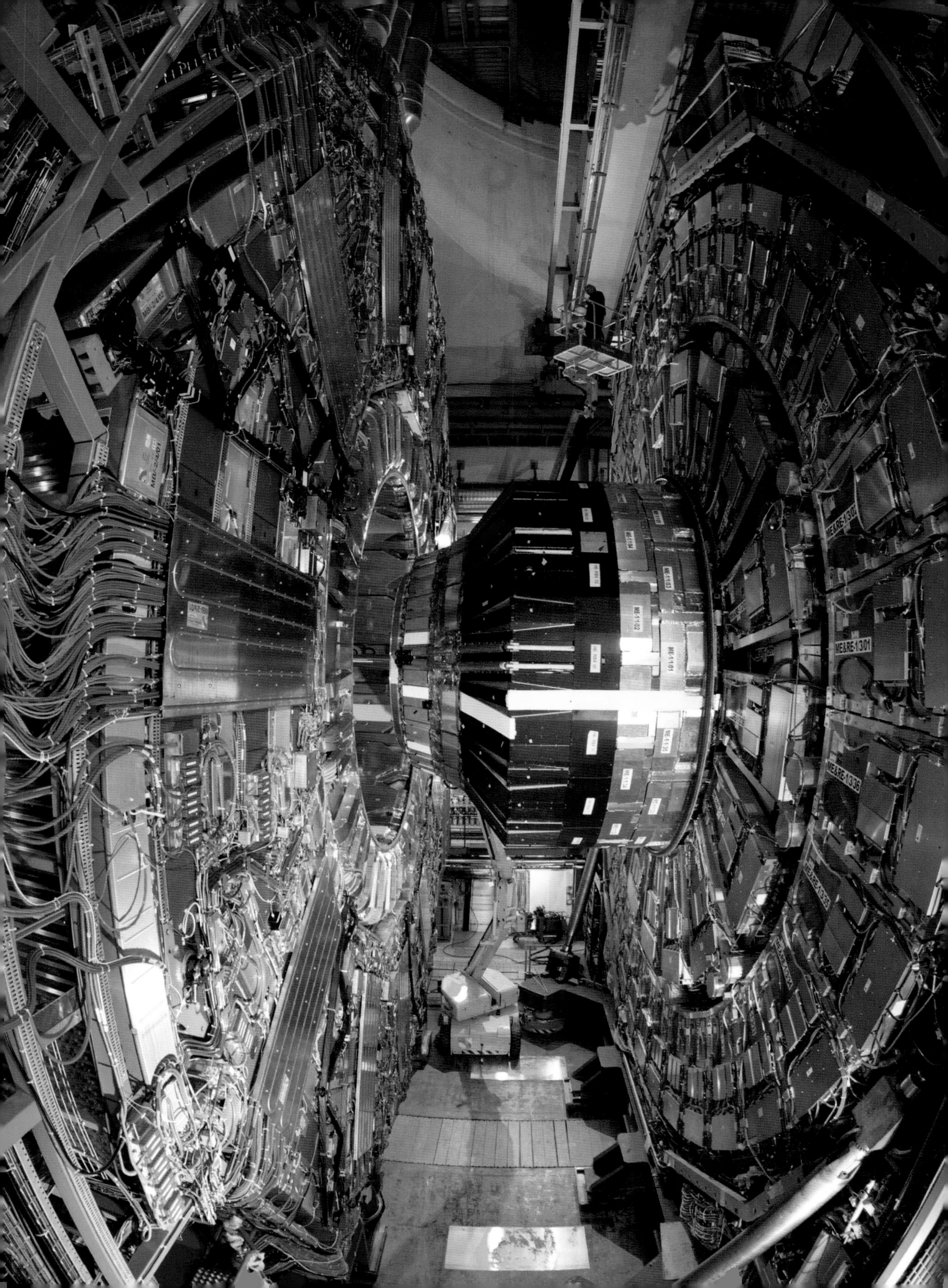
ME-1/1/04
ME-1/1/03
ME-1/1/02
ME-1/1/01
ME-1/1/35
ME&RE-1/3/01
ME&RE-1/3/36

99 Large Hadron Collider

DEFINITION THE WORLD'S MOST POWERFUL PARTICLE ACCELERATOR, LOCATED AT THE CERN LABORATORY, SWITZERLAND

DISCOVERY OFFICIALLY SWITCHED ON BY CERN SCIENTISTS IN 2008

KEY BREAKTHROUGH USING SUPERCONDUCTING MAGNETS TO CURVE A PROTON BEAM INTO A 27-KM (17-MILE) CIRCULAR PATH

IMPORTANCE THE LARGE HADRON COLLIDER IS TESTING THE NATURE OF MATTER AT HIGHER ENERGIES THAN EVER BEFORE

On the border between France and Switzerland lies one of the world's most expensive and powerful machines. Called the Large Hadron Collider, it is a particle accelerator designed to smash together subatomic particles at high speed in order to crack them open and reveal their workings. Switched on in 2008, physicists are now using its awesome power to investigate the fiery physics of the Big Bang, in which our universe was first created.

Particle accelerators work by using high-powered magnets to accelerate electrically charged particles. The magnets can either be arranged in a straight line as a so-called linear accelerator, or they can be configured into a ring-shape, taking advantage of the fact that electric charges in a magnetic field move in a circle, enabling circular accelerators to be constructed. These machines are typically the most powerful, because the particles can be circulated over and over again, gathering speed and energy with each lap.

Head-on collision

The idea behind the machines is that when two beams of particles collide, some of the collision energy is transmuted into other particles. This is in accordance with Einstein's famous $E = mc^2$ equation from special relativity, which says that mass and energy are equivalent. Studying rare particles produced in this way allows different particle physics theories to be tested.

One of the aims of particle-accelerator research is to recreate the conditions of the Big Bang in order to understand the physics governing how our universe was born. Tracing cosmic expansion back in time reveals that the universe was very small at the point of creation, and was thus governed by

LEFT The Compact Muon Solenoid particle-detector experiment inside the Large Hadron Collider. One aim of the experiment is to discover the elusive Higgs boson by examining the debris from high-speed particle collisions.

the quantum laws of particle physics. The energy pervading space at this time was enormous and, in order to recreate this, engineers are continually trying to build higher-powered accelerators. This means that the particles inside the accelerator must travel faster and faster but as they do so, bending their trajectories into a circle becomes harder and harder. Even with the most powerful magnets in the world to do the bending, making a particle accelerator more powerful generally means making it bigger.

Bigger and better

The first particle accelerators were built at the University of California in the 1930s. But the machines rapidly grew, reaching scales of metres and kilometres. Stanford University's linear accelerator, opened in 1966, measured 3km (1.9 miles) in length. And the powerful Tevatron circular accelerator at Fermilab, near Chicago – which was switched on in 1992 – measures 6.3km (3.9 miles) in circumference.

In 1989, the CERN particle physics laboratory switched on the largest particle accelerator ever built. Called the Large Electron-Positron Collider (LEP) it had a circumference of 27km (17 miles) and ran in a tunnel under the Swiss–French border. As the name suggests, it collided beams of electrons with their antimatter counterparts – positrons. Since electrons and positrons are extremely lightweight particles, when they run at high speed the amount of energy they pack is still relatively low.

Underground refurbishment

So in 2000, after 11 years of successful running, LEP was switched off. The accelerator machinery was stripped out and the underground tunnel was refitted with equipment for a new accelerator that could whirl much heavier particles around the 27-km (17-mile) tunnel. These included protons and whole nuclei of lead atoms. Such particles are known collectively as 'hadrons' – particles that experience the strong nuclear force that works to hold together atomic nuclei. The accelerator was thus renamed the Large Hadron Collider (LHC).

'When running at maximum power, the protons in the LHC will be moving at about 3 metres (3.2 yards) per second slower than light, making them circle the tunnel 11,000 times per second.'

The LHC requires extremely powerful magnets to bend such high-energy particles, and it achieves this by using superconducting magnets. These are electromagnets made from coils of wire through which a current is passed – setting up a magnetic field in accordance with Maxwell's equations of electromagnetism (see page 153). In a superconducting magnet, the wire is made of a superconducting material – one in which the resistance to electric current vanishes, enabling it to generate a much larger field. The LHC has 1600 superconducting magnets dotted around its circumference to keep the particle beam on track. And these in turn require a total of 96 tonnes of liquid helium, which serves as a cooling fluid to maintain the magnets in their superconducting state.

Power up

ABOVE Tunnel inside the Large Hadron Collider, showing the pipe containing the two opposing particle beams. Each pipe section contains a superconducting magnet that curves the beams into a circular path. The beams cross at four points around the ring, and these are where the particle detectors are located.

With a great international fanfare, the Large Hadron Collider was switched on in September 2008. But just a few days later an electrical fault in the circuitry connecting some of the magnets triggered a catastrophic leak of liquid helium, causing the accelerator to shut down. The mechanical damage was considerable and took over a year to repair, with the accelerator finally being restarted at low energy in November 2009. Engineers gradually ramped up the power, and by March 2010 the LHC had achieved a proton-to-proton collision energy about half the strength of the accelerator's maximum specification. This was still enough to make it the most powerful particle accelerator ever.

When running at maximum power, the protons inside the LHC will be moving at about 3 metres (3.3 yards) per second slower than light, making them circle the tunnel 11,000 times per second. At full power, the proton beams carry the same energy as is released by detonating over 170kg (374.8lb) of TNT – all packed into about a billionth of a gram of subatomic particles. And the big numbers do not stop with the physics. The total price tag of the LHC was about $4.4 billion. That makes it one of the most expensive scientific facilities ever built (although, to put the number in perspective, it is roughly equal to the cost of three weeks of the US war effort in Afghanistan).

What's it all for?

Physicists believe the Large Hadron Collider may answer a variety of unanswered questions: it could reveal extra dimensions of space and time, offer clues about how the forces of nature are unified and even reveal what the universe's dark matter is actually made of (see page 265). Perhaps most exciting of all is the search for the Higgs boson, which is the only particle in the generally accepted standard model of particle physics for which there is, as yet, no experimental evidence.

And these could be just the first of the discoveries made with the accelerator. Scientists at CERN are now working on plans to upgrade the LHC in 2019 to the even more powerful Super Large Hadron Collider. The proposed enhancements will pack more protons into the particle beam, increasing the number of collisions per second by a factor of ten, and improve the detector apparatus that is used to study the debris from the collisions – expanding our understanding of the particle world even further.

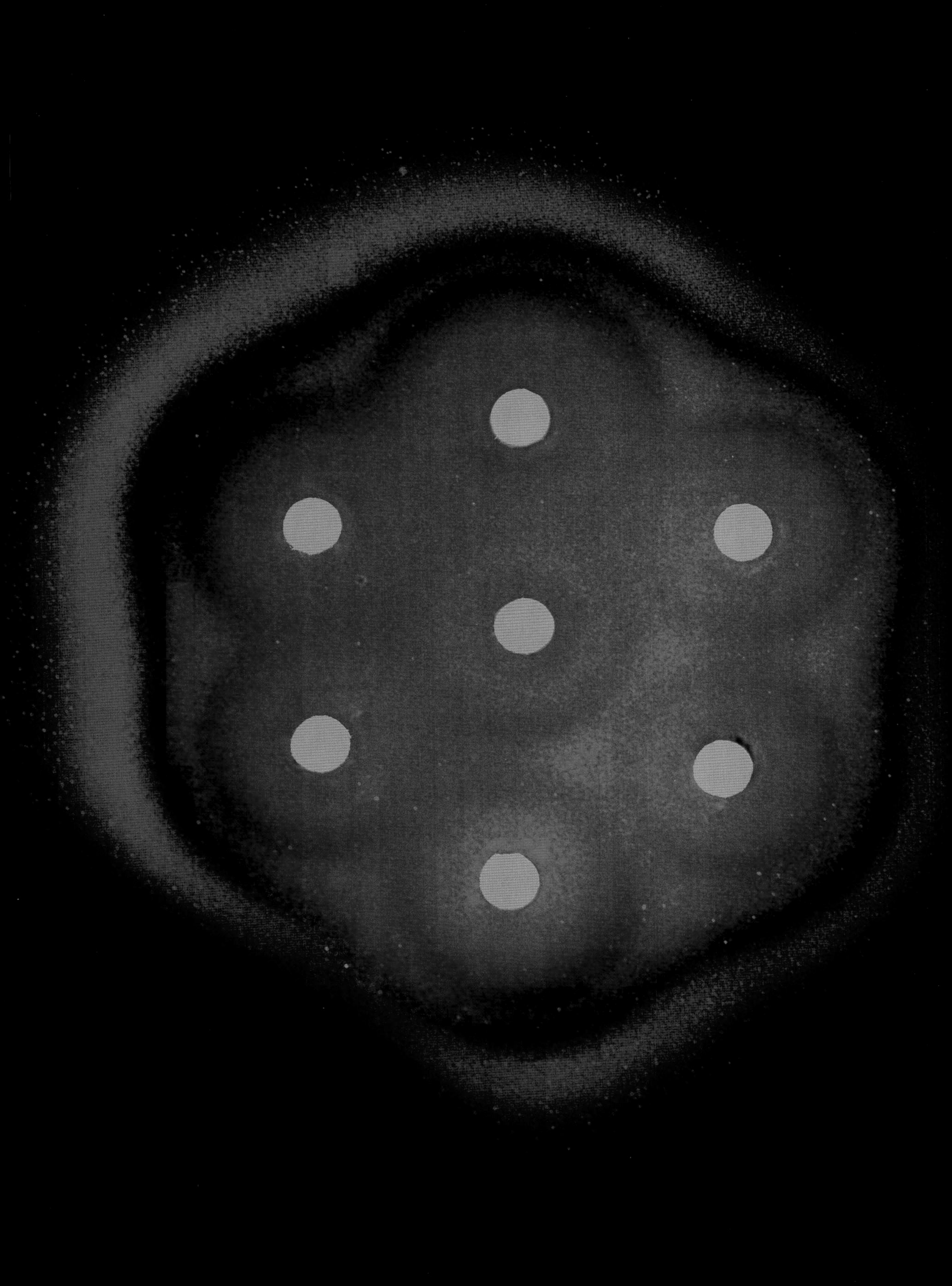

100 Synthetic life

DEFINITION ORGANISMS ARISING FROM ENTIRELY ARTIFICIAL DNA THAT HAS BEEN SYNTHESIZED FROM BASE CHEMICALS IN A LAB

DISCOVERY FIRST VIABLE SYNTHETIC CELL CREATED BY AMERICAN GENETICIST CRAIG VENTER AND COLLEAGUES IN 2010

KEY BREAKTHROUGH VENTER'S TEAM PERFECTED A TECHNIQUE FOR ASSEMBLING ARTIFICIAL GENOMES INSIDE YEAST CELLS

IMPORTANCE SYNTHETIC ORGANISMS CAN HELP MASS-PRODUCE CHEMICALS, FROM PHARMACEUTICAL DRUGS TO BIOFUEL

Biology is going to be the dominant technology of the 21st century. That was the message of Craig Venter and his colleagues when they created the first wholly artificial life form in 2010. The creation of synthetic life takes our ability to manipulate genetic material to an unprecedented level – and promises further enormous leaps forward over the years to come.

On 20 May 2010, scientists at the J. Craig Venter Institute – based in Maryland and California – announced that they had created life from scratch. The team had used a computer to design a genetic sequence, then in a laboratory they manufactured this sequence from DNA bases and implanted it into the nucleus of a bacterial cell, which then replicated and grew independently. The gene sequence of the new life form, called *Mycoplasma laboratorium*, was made entirely from laboratory chemicals. This was a departure from traditional genetic engineering, where pieces of DNA are snipped from existing organisms and pasted together. Perhaps that is why the science underpinning this procedure – a field known as 'synthetic biology' – has also been described as 'genetic engineering on steroids'.

The scientists were led by Craig Venter who, in 2007, had sequenced and published his own genetic code. That same year, his research team performed the first whole-genome transplant of DNA between two bacterial cells – implanting *Mycoplasma mycoides* DNA into the nucleus of *Mycoplasma capricolum*, successfully converting *M. capricolum* into fully replicating *M. mycoides*. They constructed the first synthetic genome in 2008, but it would take them a further two years to perfect the necessary skills to make the genome function within a living cell.

LEFT These *E. coli* bacteria cells have been reprogrammed to produce fluorescent proteins in response to certain chemicals, transforming them into chemical detectors. Building biological devices such as these are key aims of synthetic life technologists.

Engineering talk

The term 'synthetic biology' was first used in 1974 by Polish geneticist Waclaw Szybalski in reference to what has since become known as genetic engineering. In fact, modern synthetic biology probably is probably better deserving of the 'genetic engineering' title – it is all about innovation and creating solutions to problems using biological components. Indeed, many synthetic biologists are from engineering backgrounds – applying their skills at design and problem-solving first, and learning the biological knowledge they need later. They talk about 'parts' rather than 'genes' – while new cells are 'booted up' and 'debugged'. To them, genes are the software of life, nature's subroutines that can be clipped together to perform complex tasks.

Welcome to the real world

What sort of tasks can make use of synthetic biology? A simple example might be to make a temperature sensor to put on the packaging of perishable foods to tell you if a critical temperature has been exceeded, thus making the food unsafe. Let us say you have a bacteria species that is sensitive to the temperature range in which you are interested. If you engineer a stretch of DNA that manufactures a fluorescent protein and splice that into the bacterial genome at just the right place, then you get a new bacteria strain to put in your sensor that will glow whenever the storage temperature of the food has exceeded the threshold.

ynthetic ugs and ls, ifferent- there ne soil a that emicals he

Other potential applications for synthetic biology include drugs trained to sniff out and destroy cancer cells, plants that grow different-coloured foliage if there are land mines in the soil nearby and bacteria that can digest toxic chemicals and pollutants in the environment, such as polychlorinated biphenyls (PCBs). Venter's company, Synthetic Genomics, is collaborating with oil company Exxon to develop organisms to serve as a high-yield source of biofuel – providing petroleum and diesel fuel substitutes without the environmental impact of fossil fuels. Venter has been quoted as saying that synthetic biology will transform the chemical industry – with biologically engineered organisms becoming the standard method by which chemicals are mass-produced.

Designer DNA

Venter's *Mycoplasma laboratorium* genome is 1.08 million base pairs in length. The DNA was manufactured to a digital blueprint in 1078 batches of 1080 base pairs. Each batch had an 80-base-pair overlap with its neighbouring batch. These overlaps enabled the batches to be assembled into longer strands using the DNA manipulation machinery found naturally inside yeast cells – yeast was selected because of its ability to work with relatively short lengths of DNA. The complete genome was transplanted out of the yeast and into a vacant *Mycoplasma capricolum* nucleus, and the cells then multiplied to form a colony composed purely of the new bacteria.

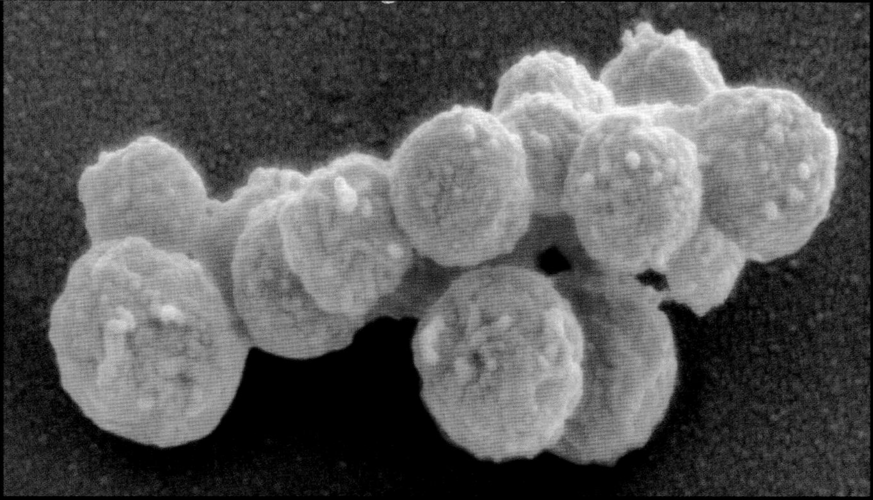

LEFT *Mycoplasma laboratorium*: the first entirely synthetic life form. The DNA of the cells was designed on a computer, synthesized in a lab and implanted into existing bacterial cells, which then grew according to the new genetic blueprint.

It was not all plain sailing. Initial attempts to create *Mycoplasma laboratorium* failed to produce viable cells, forcing the team to embark upon a lengthy debugging process, scouring the genetic code they had written, looking for biological errors. They eventually traced the problem down to a single missing base pair on one fragment of DNA. Once this was fixed, the cells functioned perfectly. Also spliced into the sequence, using the alphabet of DNA code (the letters A, C, G and T representing DNA's four chemical bases – see page 305), was the name of each scientist on the project, a secret web URL, and three famous quotations – including physicist Richard Feynman's poignant, 'What I cannot build, I cannot understand.' As well as being a scientific landmark, the final piece of DNA was also the largest custom-built chemical molecule ever created.

Scientists at Venter's institute are now taking synthetic biology to the next level. The Human Genome Project revealed in 2003 that much of the DNA code in human beings is redundant – 98 percent of base pairs in the genome do not seem to do anything useful at all. Venter's team are now investigating whether 'junk DNA' in *Mycoplasma laboratorium* is indeed useless, or whether it might have an undiscovered role. They are doing this by reducing the genome to its minimal form, stripping out redundant base pairs one by one to arrive at the shortest genome that preserves the functionality of the cells.

Genetic Lego

Researchers have set up a library of synthetic DNA fragments, each engineered to perform a specific function. Called BioBricks, these components are freely available to other scientists, rather like electronic components. In the spirit of open source, the scientists are encouraged to place their new creations back in the library for others to access. The Massachusetts Institute of Technology even hosts a yearly contest for the best BioBrick project by international undergraduate teams. If the high standard of entries is anything to go by, then it may well be that playing with these DNA building blocks will be how synthetic biologists make their biggest strides forward.

Glossary

Absolute zero The lowest temperature theoretically possible, equivalent to −273.15 degrees Centigrade (−459.67 degrees Fahrenheit).

Acceleration The rate of change of an object's velocity.

Acid A reactive chemical compound that, in solution with water, yields hydrogen ions (protons) that attract electrons from other atoms.

Active galaxy A galaxy that releases huge amounts of energy. The energy source of an active galaxy is probably a supermassive black hole.

Alkali A chemical compound, classed as a 'base', which is soluble in water.

Allele A variant of a gene that produces heritable traits, such as eye colour in animals or flower colour in plants.

Alloy A mixture of a metal and at least one other element, resulting in a new material. Steel is made from iron and carbon; bronze is an alloy of copper and tin.

Amplitude A measure of the energy in waves judged by the height of the wave peaks.

Antimatter For every particle of ordinary matter there is an identical particle of antimatter, but with an opposite electrical charge. When a particle meets its antiparticle, they annihilate.

Asteroid Rocky bodies in the solar system. They range from small boulders to flying mountains spanning hundreds of kilometres.

Astronomical unit (AU) A measure of distances in space, defined by the average distance between the Earth and the Sun.

Atom The smallest particle of an element that can take part in a chemical reaction without being permanently changed. Atoms consist of a positively charged nucleus that binds electrons around it.

Atomic mass The number of protons and neutrons in an atom. Protons and neutrons both weigh 1.67×10^{27} kg.

Atomic number The number of protons in an atomic nucleus. Protons are not dislodged except by extreme nuclear events. The atomic number is used to classify the charge of the nucleus.

Bacteria Single-celled organisms that are the most common form of life.

Carbon The element that defines the chemical properties of life. Carbon is the third most common element in cells.

Carbon cycle The circulation of carbon – from the ground, through plants and animals, to the atmosphere, then back down again. Ordinarily, the amount of carbon in the cycle is constant. But the burning of fossil fuels adds carbon to the cycle, causing climate change and the greenhouse effect.

Cell The smallest unit of life. Cells contain the compounds for life, such as proteins and DNA. Cells are either prokaryotic (simple) or eukaryotic (have a nucleus).

Chromosome Strands of genetic information within cells. In humans, chromosomes assemble in pairs in all cells except the reproductive cells.

Covalent bond A strong attraction between two or more atoms, formed by the sharing of electrons.

Electrical charge A property of electrons and protons that creates forces of attraction or repulsion. Electrons are repelled by other electrons, but attracted to protons.

Electrolyte A substance, usually liquid, that conducts electricity because of the presence of positive or negative ions.

Electromagnetic radiation Waves of electric and magnetic energy, vibrating at right angles to each other, and propagating through space at the speed of light. Gamma rays, X-rays, ultraviolet light, visible light and radio waves are types of electromagnetic radiation.

Electrons Negatively charged particles that exist around the nucleus of an atom. Electrons are negatively charged, and are 'fundamental' particles that cannot be divided.

Element A substance that cannot be split into other substances. There are 92 elements in nature, plus additional synthetic elements.

Energy The capacity to do work – work is done by transferring energy from one form to another. For example, chemical energy in vehicle fuel is converted to thermal energy as it burns. The thermal energy is then converted into mechanical energy to move the vehicle.

Enzyme A protein that can speed up a biochemical reaction without being consumed by it – essentially a biological catalyst.

Fission The splitting of an atomic nucleus, causing the nucleus to divide and release some of its mass as energy. Fission liberates more neutrons that then collide with other adjacent nuclei, causing a chain reaction.

Fluid Any substance that can flow. Both gases and liquids can be considered fluids.

Force Pushing and pulling influences that make objects move or change their shape.

Fusion The joining together of atomic nuclei to form a larger nucleus. Fusion powers the Sun and other stars. In the future, hydrogen-fusion technology may provide a new type of energy for use on Earth.

Galaxy A cluster of stars, typically containing hundreds of billions of them. Most galaxies contain supermassive black holes at their centres.

Gene A particular portion of base-pair sequencing within DNA that transmits specific heritable characteristics. A gene cannot be divided into smaller units without losing its informational content.

Genome The complete set of genetic information that defines an organism.

Globular cluster A spherical, localized star cluster, containing up to hundreds of thousands of stars held together by mutual gravity.

Inertia The resistance of a body to move, normally measured by its mass. The more inertial mass the body has, the harder it is to move it from a stationary position, slow it down or change the direction of its motion.

Infrared Electromagnetic radiation with wavelengths longer than visible light, but shorter than microwaves. All warm or hot objects emit infrared radiation.

Interstellar medium The gas and dust that float between stars. The interstellar medium is emptier than any vacuum on Earth but has a huge mass.

Ion An atom or molecule that has acquired an electric charge by either gaining or losing electrons. An atom with missing electrons has a positive charge and is called a cation. An atom with extra electrons is negative and is called an anion.

Isotope Different forms in which the same element can occur. Isotopes differ from one another according to the number of neutrons they contain. Neutrons have little or no effect on chemical reactions, except during nuclear processes, such as radioactivity and fission.

Kilogram An unit of mass, as defined by the mass of a platinum-iridium rod held in a special laboratory in Paris.

Law In science, a law is a theory based on its reliability as a predictive tool. Laws are revised when more detailed theories replace older ones. For instance, Newton's laws have been revised to take account of relativity theory.

Magnetism Magnets generate magnetic fields around them. They normally have a north and south pole, with a magnetic field between. Opposite poles attract and poles of the same polarity repel. Magnetic fields attract ferromagnetic materials.

Mass A measure of the total matter in an object. Weight, on the other hand, is a measure of gravity's effect on an object, which varies according to gravitational force. Mass stays the same, even in outer space.

Milky Way The name of the galaxy in which our Sun and solar system lie.

Molecule A aggregate of at least two atoms linked by covalent bonds. Almost all matter on Earth is made up of molecules. A molecule is the smallest amount of a compound that can exist and still retain the characteristics of that compound.

Momentum The tendency of a moving object to keep moving in the same direction. Momentum is given by multiplying speed by mass.

Nanotechnology The manufacture of materials by manipulating single atoms. 'Nano' comes from the size of molecules, measured in nanometres, or one-billionth of a metre.

Neutron One of the particles in the nucleus of an atom. Neutrons have substantial mass but no electrical charge.

Newton The unit of force. 1 Newton (N) is defined as being the force required to give a mass of 1 kilogram an acceleration of 1 metre per second squared ($1 m/s^2$).

Nucleus In biology, the nucleus is the organelle inside a cell that contains its DNA. In physics and chemistry, the word refers to the core of an atom, where neutrons and protons are bound and most of the atom's mass resides. Nuclei are thousands of times smaller than the atom – indeed, atoms are comprised almost entirely of empty space.

Particles The smallest components of matter are the so-called subatomic particles. The most common types of particles are protons, neutrons and electrons. Protons and neutrons cluster in different numbers to form the atomic nuclei of elements. Electrons orbit the nuclei to form atoms, and these bond together to form molecules.

Plasma In biology, it is the fluid in which blood cells are suspended. In physics, plasma is a gas-like state of matter consisting of positively charged ions, free electrons and other subatomic particles. Plasma occurs naturally in stars and the solar wind, as well as in lightning and fire.

Plastic Synthetic material made from hydrocarbon macromolecules. 'Polymer' is an another term for macromolecules, but not all polymers are plastics. 'Thermoplastics' soften when heated repeatedly, whereas 'thermosetting' plastics stay solid after being heated once.

Prokaryote A single-celled life form lacking a nucleus. Prokaryotes are thought to be the oldest forms of life.

Proteins Found in all life, proteins are amino-acid polymers that contribute to the operation of cells.

Proton One of the particles that make up the nucleus of an atom. They are formed from triplets of quarks and are positively charged.

Quantum mechanics The theory that describes the behaviour of atoms and subatomic particles. It is based on the idea that energy comes in discrete, indivisible bundles called 'quanta'.

Radioactivity Spontaneous and unpredictable emission of particles or electromagnetic radiation from the nuclei of unstable atoms.

Resistance The extent to which an electrical conductor hinders the flow of an electric current. Its is measured in ohms, with the symbol R.

Scientific notation A system for writing very small or very large numbers as a decimal number between 1 and 10, multiplied by a power of 10. For example, the number 30,000 can be written as 3×10^4.

Sound waves Hit a solid object and, assuming it has enough elasticity to vibrate, the disturbance will travel through it as a wave. If the frequency of the wave is in the range we can hear (20–20,000 Hz), then this is a sound wave. Sound is measured in decibels (dB).

Standard model The overarching theory of subatomic particle physics. It is assembled from quantum-field theory descriptions of the four forces of nature and specific particle families.

Substance The name given to matter so that it can be identified by its chemical make-up. A chemical remains the same substance regardless of what state it is in – solid, liquid or gas.

Theory An explanation for one or more natural phenomena, confirmed by experimental tests.

Thermodynamics The study of how heat transmutes into other forms of energy.

Uncertainty principle In the subatomic world, the uncertainty principle states that it is impossible to know the position and momentum of a particle simultaneously. The more accurate one measurement is, the less accurate the other becomes.

Universe The sum of all that exists physically, including matter, energy, physical laws, space and time. Multiple universes may exist.

Virus The smallest of all organisms. The status of viruses as living things is debatable because they do not eat or excrete and they depend on host organisms in order to replicate themselves.

Voltage The measure of 'electromotive force' available to drive an electric current around a circuit. Voltage is measured in volts and is represented by the symbol V.

Index

Picture credits

SPL – Science Photo Library

2 SPL/Martin Krzywinski; **8** The Bridgeman Art Library/© Christian Larrieu – La Licorne; **11** SPL/Science Source; **12** SPL/Mark Garlick; **15** SPL/Frederica Georgia; **15** Frederica Georgia**;** **16** SPL/NASA; **19** SPL; **20** NASA, ESA, and the Hubble SM4 ERO Team; **22** SPL/Adam Hart-Davis; **24** SPL/Dr Fred Espenak; **27** SPL/J-L Charmet; **28** SPL/Dr Gary Settles; **31** SPL/US Air Force; **32** SPL/R. Masionneuve, Publiphoto Diffusion; **35** SPL/Saturn Stills; **36** SPL/Dr J.Durst; **39** SPL/Pekka Parviainen; **40** SPL/Simon Fraser; **43** SPL/Peter Menzel; **44** SPL/NASA/JPL/Space Science Institute; **47** SPL/Detlev Van Ravenswaay; **48** SPL/David Parker; **51** SPL/Adam Hart-Davis; **53** SPL/D.Roberts; **55** SPL/Dr Gary Settles; **56** SPL/Mark Clarke; **59** SPL/Dr Gary Settles; **60** SPL/Gustoimages; **63** SPL/Anthony Mercieca; **64** SPL/Dr. A Yazdani & Dr. D.J.Hornbaker; **67** SPL/National Institute of Standards and Technology; **68** SPL/Simon Fraser; **70** SPL/Martyn F.Chillmaid; **72** SPL/Stephen & Donna O'Meara; **75** SPL/European Southern Observatory; **76** SPL/David Parker; **79** SPL/Tony Craddock; **80** Roy Williams; **83** SPL/Roger Harris; **84** SPL/Eye of Science; **87** SPL/Charles D.Winter; **88** SPL/Martin Bond; **90** SPL/Jeremy Walker; **92** SPL/Loren Winters, Visuals Unlimited; **95** SPL/CERN; **96** SPL/Dr Klaus Boller; **99** SPL/CDC; **100** SPL/Phil Jude; **103** SPL; **104** SPL/Berenice Abbott; **107** SPL/Giphotostock; **108** SPL/Mike Agliolo; **111** SPL/US Department of Energy; **112** SPL/Charles D.Winters; **115** SPL/Andrew Lambert Photography; **116** SPL/NASA; **118-119** SPL; **120** Tony Craddock; **122-123** SPL/NASA; **124** SPL/Bernard Edmaier; **127** SPL/Ralph Lee Hopkins; **128** SPL/Dr Gopal Murti; **130** SPL/Dr David Furness, Keele University; **132** SPL/Kent Wood; **135** SPL/Jerry Mason; **136** SPL/Dr Arthur Tucker; **139** SPL/Farrell Grehan; **140** SPL/Larry Landolfi; **143** SPL/NOAA; **144** SPL/Claude Nuridsany & Marie Perennou; **147** SPL/John Beatty; **148** SPL/Martin Dohrn; **151** SPL/E.R.Degginger; **152** SPL/Pasieka; **155** SPL/Steve Gschmeissner; **156** SPL/Eye of Science; **159** SPL/CNRI; **160** SPL/Science Pictures Ltd; **162** SPL/Bluestone; **164** SPL/Paul Gunning; **167** SPL/Munoz-Yague/Silkeborg Museum, Denmark; **168** SPL/Dr Erwin Mueller; **171** SPL; **172** SPL/D.Roberts; **175** SPL/Javier Trueba/MSF; **176** SPL/Steve Gschmeissner; **179** SPL/3D4Medical.com; **180** SPL; **182** SPL/George Steinmetz; **184** SPL/Dr Mitsuo Ohtsuki; **186** SPL/Patrick Landmann; **188** SPL/Pasieka; **191** SPL/Eye of Science; **192** SPL/Russell Kightley; **195** SPL/Eye of Science; **196** SPL/RIA Novosti; **199** SPL/Mark Williamson; **200** SPL/Sandia National Laboratories; **203** SPL/MIT/J.Joannopoulos; **204** Dreamstime.com/Mommamoon; **207** NASA/Carla Cioffi; **209** SPL/Bernhard Edmaier; **211** SPL/NASA; **212** SPL/Gianni Tortoli; **215** SPL/Bernhard Edmaier; **216** SPL/Alan Sirulnikoff; **219** SPL/Alan Sirulnikoff; **220** SPL/Eye of Science; **223** SPL/IMI/Univ.of Birmingham High TC Consortium; **224** SPL/Dr Ken MacDonald; **227** SPL/Daniel Sambraus; **228** W. Couch & R. Ellis; **231** SPL/NASA/ESA/STScI; **232** SPL; **235** SPL/Steve Horrell; **236** Dr. David Wexler, coloured by Dr. Jeremy Burgess; **239** SPL/Andrew Lambert Photography; **241** SPL/Dr Fred Espenak; **243** SPL/Victor De Schwanberg; **244** SPL/Andrew McClenaghan; **247** SPL/Pasieka; **248** Corbis/Radius Images; **250-251** SPL/George Steinmetz; **252** SPL/NASA/ESA/STSCI/S.Beckwith, HUDF Team; **255** SPL/David A.Hardy; **256** SPL/Roger Harris; **258-259** SPL/Take 27 Ltd; **260** SPL/Lawrence Berkeley Laboratory; **263** SPL/CERN, P.Loiez; **264** SPL/Mehau Kulyk; **266** SciencePhotoLibrary/NASA/ESA/STSCI/W. Colley & E.Turner, Princeton; **268** SPL/Planetobserver; **271** SPL/Dr Gene Feldman, NASA GSFC; **272** SPL/NASA/JPL/ESA; **275** SPL/NASA; **276** SPL/James King-Holmes/Bletchley Park Trust; **278** Patrick Landmann; **280** SPL/US Department of Energy; **283** SPL/Robert Brook; **284** SPL/Royal Greenwich Observatory; **287** SPL/Mark Garlick; **288** SPL/Lawrence Berkeley National Laboratory; **291** SPL/Carolyn A.McKeone; **292** SPL/Pasieka; **295** SPL/Eye of Science; **296** SPL/Mehau Kulyk; **299** SPL/Gregory Macnicol; **301** SPL/Jean-Francois Podevin; **303** SPL/ArSiMed; **304** SPL/Pasieka; **306** SPL; **308** SPL/Olivier Vandeginste; **311** SPL/Science Source; **312** SPL/Massimo Brega, The Lighthouse; **315** SPL/James King-Holmes; **316** SPL/NASA; **319** SPL/Alexander Tsiaras; **320** SPL/Lionel Bret/Eurelois; **323** SPL/Isaac Chuang/IBM Almaden Research Center; **324** SPL/Jim Amos; **327** SPL/Matt Mazloff; **328** SPL/Science Source/NASA; **331** SPL/2Mass Project/NASA; **332** SPL/David Parker; **335** SPL/Science Source; **336** SPL/Roger Harris; **339** SPL/Peter Menzel; **340** SPL/Mehau Kulyk; **343** SPL/Laguna Design; **344** SPL/Martin Oeggerli; **347** SPL/Chris Knapton; **348** SPL/NASA; **351** SPL/NASA/DOE/International LAT Team; **352** SPL/Geological Survey of Canada; **355** SPL/Martin Bond; **356** SPL/David Scharf; **358** SPL/Eye of Science; **360** Volker Steger; **363** SPL/Christian Darkin; **364** Michael W. Davidson; **367** SPL/Victor Habbick Visions; **368** SPL/Dr Gopal Murti; **370** SPL/Jackie Lewin, Royal Free Hospital; **372** SPL/Sovereign, ISM; **375** SPL/BSIP, Laurent; **377** SPL/UCSF Chimera; **379** SPL/BSIP, Cavallini James; **380** European Southern Observatory; **383** SPL/NASA/Troy Cryder; **385** SPL/Bahman Kalantari; **387** SPL; **388** SPL; **391** SPL/PH.Plailly/Eurelios; **392** SPL/Peter Menzel; **394-395** Corbis Saba/Najlah Feanny; **396** SPL/James King-Holmes; **399** SPL/Peter Menzel; **400** Massimo Bregal/The Lighthouse; **403** SPL/David Parker; **405** SPL/Ron Weiss; **407** SPL/Thomas Deerinck, NCMIR;

Page 2 picture caption

Circular genome map showing shared genetic material between humans (outer ring) and (from inner ring outwards) chimpanzees, mice, rats, dogs, chickens and zebrafish. Each ring is based on the colour-coded representation of one chromosome. Similar colour patterns within each ring reveal 'hot spots' of shared genetic material. The more fragmented patterns indicate greater evolutionary divergence from humans.

A FIREFLY BOOK

Published by Firefly Books Ltd. 2011

First printing

Publisher Cataloging-in-Publication Data (U.S.)
Parsons, Paul 1971–
Science in 100 key breakthroughs / Paul Parsons.
416 p. : 200 col. photos. ; 246 × 185 mm.
Includes index.
Summary: Brief accounts of the key concepts and discoveries of science through world history.
ISBN-13: 978-1-55407-808-0 (pbk.)
1. Science – Popular works. I. Title.
500 dc22 Q162.P378 2011

Library and Archives Canada Cataloguing in Publication
Parsons, Paul, 1971–
Science in 100 key breakthroughs / Paul Parsons.
Includes index.
ISBN 978-1-55407-808-0
1. Discoveries in science--History--Popular works.
2. Science--History--Popular works. I. Title. II. Title: Science in one hundred key breakthroughs.

Q180.55.D57P37 2011 509 C2011-900025-3

Published in the United States by
Firefly Books (U.S.) Inc.
P.O. Box 1338, Ellicott Station
Buffalo, New York 14205

Published in Canada by
Firefly Books Ltd.
66 Leek Crescent
Richmond Hill, Ontario L4B 1H1

Printed in China